Strategic Survey 2020

The Annual Assessment of Geopolitics

published by

Routledge
Taylor & Francis Group

for

The International Institute for Strategic Studies

The International Institute for Strategic Studies
Arundel House | 6 Temple Place | London | WC2R 2PG | UK

Strategic Survey 2020
The Annual Assessment of Geopolitics

First published November 2020 by **Routledge**
4 Park Square, Milton Park, Abingdon, Oxon, OX14 4RN

for **The International Institute for Strategic Studies**
Arundel House, 6 Temple Place, London, WC2R 2PG, UK

Simultaneously published in the USA and Canada by **Routledge**
52 Vanderbilt Avenue, New York, NY 10017

Routledge is an imprint of Taylor & Francis, an Informa business

DIRECTOR-GENERAL AND CHIEF EXECUTIVE Dr John Chipman

EDITOR Dr Nigel Gould-Davies

ASSOCIATE EDITOR Alice Aveson
ASSISTANT EDITOR Gabriel Everington
EDITORIAL Nick Fargher, Jill Lally, Michael Marsden, Jack May
GRAPHICS RESEARCH Catherine O'Connor, Jack May
KEY EVENTS RESEARCH Clara Etchegaray, Catherine O'Connor
COVER/PRODUCTION/CARTOGRAPHY John Buck, Carolina Vargas, Kelly Verity
COVER IMAGES Getty

British Library Cataloguing in Publication Data
A catalogue record for this book is available from the British Library

Library of Congress Cataloguing in Publication Data

ISBN 978-0-367-70118-5
ISSN 0459-7230

Contents

Chapter 1

Introduction

Editor's Introduction

Science fiction has long speculated that if Earth faced an alien invasion, humanity would put aside divisions and unite against it. But the deadly common threat of COVID-19 that emerged in late 2019 did not bring the world together. On the contrary: even as the virus spread to almost every country, divisions among them deepened.

By mid-2020, relations between the United States, Europe and China had all fallen to their lowest point for decades. Russian–Western relations remained locked in suspicion. Sino-Indian tensions flared in deadly border clashes. Institutions, laws and norms of cooperation suffered multiple setbacks. The US denounced or withdrew from several organisations and treaties, including the World Health Organization. The United Kingdom left the European Union. China altered the special status of Hong Kong.

COVID-19 exacerbated such tensions. By inviting the question of responsibility, the pandemic became a contentious source of blame and denial. By testing the political leadership, government capacity and social cohesion of many countries, it enhanced or diminished the soft power of national reputation. By dominating attention and straining state budgets, it risked undermining the preparedness and resources that underlie hard power.

But there were also more hopeful developments. The EU achieved unprecedented cooperation in tackling the economic challenges caused by the pandemic. In Afghanistan, the US reached a deal with the Taliban to end the longest war in American history. Nor was COVID-19 the only common challenge: the climate crisis increasingly preoccupied international opinion. It dominated the World Economic Forum's annual meeting at Davos in January 2020 – just as the virus was becoming a pandemic.

In every region, issues of power, conflict, interests and rules – the essence of geopolitics – continued to play out, refracted but not displaced by the pandemic. They included:

- the expansion of influence by Turkey in the Middle East and beyond; by China in the South Pacific; and by Brazilian organised-crime groups in Latin America
- the prospects for connectivity: in Japan's Free and Open Indo-Pacific concept, and among Central Asian states
- the significance of the Red Sea region for local and external actors; and of Ethiopia's giant dam for Nile-dependent states.

Strategic Survey 2020: The Annual Assessment of Geopolitics explores all these themes and many more. It analyses how each has developed, the significance it holds, and how it may evolve over the coming year.

To complement our essays, we have rethought and refreshed the 'Drivers of Strategic Change' that open each regional section. This carefully selected data provides a rich and vivid guide to forces underlying geopolitical change.

COVID-19 will pass and will leave its mark on history. Geopolitics endures. We hope you will find the insights and arguments of *Strategic Survey 2020* a valuable guide to an exceptionally turbulent year.

Drivers of Strategic Change

Geopolitics is driven by changes in the ability of states to use and resist power. The first depends on power resources, and the second on domestic resilience. Our Drivers of Strategic Change measure and compare key trends in both areas. They illuminate recent shifts in geopolitics and sources of potential future change.

Geopolitics is a craft, not a science: judgement, skill, chance and other immeasurable factors also shape international relations. But they do so within a range of possibilities set by the underlying domestic and external capacities of states. We encourage you to explore the rich data in our Drivers and the insights they yield.

The Drivers begin each geographical chapter. Unless otherwise stated, they chart change over 20 years by plotting data from 1999, 2009 and 2019.

Regional Share of Global Population, GDP and Defence Budget

(Sources: World Bank; Worldometers.info)

The first Driver depicts the region's share of global population, GDP and defence budget. These are key power resources: the more of each that a country or region possesses, the greater its potential power, especially in combination. This Driver thus shows how the relative power of each region has changed over the past two decades.

The next six Drivers depict data for key selected countries in each region.

Population

(Sources: United Nations Department of Economic and Social Affairs, Population Division; World Bank)

The second Driver shows population, age structure and median age. These are important for several reasons. Population is a power resource. A high proportion of young people – a 'youth bulge' – is a strong predictor of civil violence. It also presages a 'demographic dividend' of higher economic growth through future workforce growth (especially if fertility rates subsequently fall). Conversely, an ageing population means a high dependency ratio of economically inactive to active citizens, creating fiscal and productivity challenges that can limit resources needed to sustain power. The most recent age structure given is for 2020.

GDP

(Source: World Bank)

The third Driver shows GDP (constant 2010 US dollars) and global ranking. The larger a country's economy, the more of other forms of strength, including military hardware, it can procure.

GDP per Capita

(Source: World Bank)

The fourth Driver shows GDP per capita (constant 2010 US dollars), which has been shown to have a significant impact on the development of social values. Rising affluence leads to robust and predictable changes in political orientation – in particular, a decline in deference towards authority and a rise in demands for inclusion and participation.

Defence Budget and Active Military Personnel

(Sources: IISS, Military Balance+; IISS, *Military Balance* 1999, 2000, 2010)

The fifth Driver shows defence budget (constant 2015 US dollars) and active military personnel, which are indicators of hard power.

Human Development Index (HDI)

(Source: UN Development Programme)

The sixth Driver shows Human Development Index scores, a composite measure of human well-being. This indicates a country's ability to provide well-being and life chances for its population, with positive implications for governmental legitimacy and stability.

Political System

(Source: Freedom House, 'Freedom in the World')

The seventh Driver shows how democratic a political system is. Democratic legitimacy tends to produce stable and responsive government that is more resilient in a crisis. Conversely, the recent decline of democracy in some high-income countries, where the underlying demand for accountability remains high, may presage declining stability.

The final Driver for each chapter uses a range of data to illuminate region-specific trends.

For Asia, Europe, Latin America and North America:

Trust in Institutions

(Source: Edelman Trust Barometer)

This Driver shows the general public's average percentage of trust in non-governmental organisations, business, government and media. Falling trust in institutions – a recent feature of many countries – implies a decline in stability and cohesion.

For Russia and Eurasia:

Domestic Approval Rating for President Vladimir Putin, and Russia's GDP Growth Rate

(Source: Levada Center)

This Driver shows approval ratings for Russian President Vladimir Putin and Russia's GDP growth rate, 2012–20. It highlights Putin's declining popularity and the poor systemic performance that has contributed to this.

For the Middle East and North Africa:

Breakeven Oil Prices

(Sources: BP, Statistical Review of World Energy 2020; IMF)

This Driver shows the oil price per barrel needed to ensure that planned government spending will not incur a budget deficit, together with the average annual oil price, for 2015–20. It highlights the impact of post-2014 oil-price decline on the fiscal sustainability of oil-export-dependent states.

For sub-Saharan Africa:

Percentage of Children in Education

(Source: World Bank)

This Driver shows the gross percentage of children in education for 1999, 2009 and 2015. The strong recent growth of education is a major investment in human capital that should lead to significant development and growing prosperity.

Chapter 2

Key Events

July 2019–June 2020

July 2019

9 The United Arab Emirates (UAE) announces a 'strategic redeployment' of its forces away from the city of Hudaydah in Yemen, as well as a partial tactical retreat in other parts of the country.

15 Beijing hosts the first China–Africa Peace and Security Forum, attended by delegations from the African Union and 50 African states.

20 22,000 people protest in Moscow after Russian authorities ban several independent and opposition candidates from taking part in September's municipal elections.

24 Boris Johnson replaces Theresa May as prime minister of the United Kingdom.

25 United States President Donald Trump asks Ukrainian President Volodymyr Zelensky to announce that Ukraine will investigate allegations that Joe Biden had engaged in corrupt practices relating to his son's business activities in Ukraine.

August 2019

2 The US withdraws from the 1987 Intermediate-Range Nuclear Forces Treaty.

5 The Indian government announces that it is changing the status of Jammu and Kashmir from a state to a union territory, removing its 'semi-autonomous' status.

5 Amid ongoing protests, Hong Kong is hit by its first general strikes since 1967.

26 Indonesia announces plans to relocate its capital city from Jakarta to East Kalimantan province.

September 2019

4 Hong Kong leader Carrie Lam announces that she will withdraw the controversial extradition bill that incited months of protests.

14 Drone attacks target two Saudi Aramco sites in Saudi Arabia. Yemen's Houthi movement (Ansarullah)

claims responsibility. The US and Saudi Arabia blame Iran.

24 Singapore Prime Minister Lee Hsien Loong and US President Trump sign an amendment that extends to 2035 the Memorandum of Understanding Regarding US Use of Facilities in Singapore.

26 Tokyo's latest Defense White Paper explicitly identifies China as posing the most serious military threat to Japan.

October 2019

6 US President Trump orders a partial withdrawal of US troops from northern Syria, allowing Turkey to launch a ground offensive in the region against Kurdish forces that had been a US ally in the fight against the Islamic State, also known as ISIS or ISIL.

11 The US announces that it is sending 3,000 additional troops and several missile-defence systems to Saudi Arabia to protect it from Iranian aggression.

14 Ecuadorian President Lenín Moreno signs an agreement to restore fuel subsidies after nearly two weeks of widespread violent protests.

18 Chilean President Sebastián Piñera declares a state of emergency in a televised address to the nation following days of violent protests over inequality.

22 Canada's Liberal Party retains power in a narrow election win. Justin Trudeau remains prime minister as head of a minority government.

22 Russian President Vladimir Putin and Turkish President Recep Tayyip Erdogan agree a 32-kilometre-deep 'safe zone' for the Kurds along the Syria–Turkey border and affirm the territorial integrity of both countries.

23 Russia hosts the first-ever Russia–Africa Summit. Forty-three African heads of state attend.

25 Incumbent Bolivian President Evo Morales is announced the winner of the country's recent presidential election. Protests take place amid claims that the election results are fraudulent.

27 ISIS leader Abu Bakr al-Baghdadi is killed in Syria during a raid by US special forces.

27 Alberto Fernández is announced the winner of Argentina's presidential election.

November 2019

10 Bolivian President Morales resigns after nearly a month of protests, amid accusations of electoral fraud.

18 US Secretary of State Mike Pompeo announces the reversal of a long-standing US policy, declaring that Israeli settlements in the West Bank are not illegal.

December 2019

1 Ursula von der Leyen succeeds Jean-Claude Juncker as president of the European Commission.

9 Russian President Putin and Ukrainian President Zelensky meet for the first time to discuss the conflict in eastern Ukraine.

11 India's parliament passes the Citizenship (Amendment) Act (CAA), which would grant citizenship for six religious minorities (excluding Muslims) from Afghanistan, Bangladesh and Pakistan who entered India illegally prior to 2015.

12 Following the ousting of long-time ruler Abdelaziz Bouteflika, Algerians elect former prime minister Abdelmadjid Tebboune as president. The election ignites widespread protests amid claims of electoral fraud.

13 The Conservative Party wins the UK general election with a clear majority, paving the way for Brexit.

18 The US House of Representatives votes to advance two articles of impeachment against President Trump.

20 US President Trump signs into law the Caesar Syria Civilian Protection Act of 2019, a comprehensive bill that imposes heavy sanctions on economic dealings with the Assad regime.

January 2020

1 North Korean leader Kim Jong-un announces that Pyongyang will no longer be 'unilaterally bound' to any prior commitments, including the moratorium on nuclear and long-range-missile tests.

3 Gen. Qasem Soleimani, commander of Iran's Quds Force, is killed in Baghdad by a targeted US drone strike.

11 Tsai Ing-wen is re-elected as president of Taiwan on a platform of preserving the country's sovereignty amid pressure from Beijing.

15 The US and China sign an initial deal that seeks to alleviate their two-year trade dispute.

15 Russian President Putin proposes major constitutional changes. Russia's government resigns. Mikhail Mishustin replaces Dmitry Medvedev as prime minister.

29 The USMCA trade deal between the US, Canada and Mexico takes effect, replacing NAFTA.

30 The World Health Organization (WHO) declares coronavirus a public-health emergency.

31 The UK leaves the European Union.

February 2020

5 The US Senate votes to acquit President Trump of two impeachment charges relating to abuse of power and obstruction of Congress.

10 Chinese President Xi Jinping declares a 'People's War' aimed at containing COVID-19.

11 The WHO names the novel coronavirus strain 'COVID-19'.

18 Incumbent Afghan President Ashraf Ghani is announced the winner of the 2019 election with 50.6% of the vote.

23 During protests against India's CAA in northeast Delhi, 53 people are killed in Hindu–Muslim communal riots.

28 An attack by Syrian government forces in opposition-held northwestern Syria kills at least 33 Turkish soldiers, in a significant escalation of the conflict.

29 The US and the Taliban sign a peace agreement to end the war in Afghanistan.

March 2020

1 Turkey launches a major counter-offensive in northern Syria against the Assad regime, for the first time openly declaring war on the government.

6 Russia rejects an OPEC plan to slash oil production in response to falling global demand caused by the COVID-19 pandemic, triggering an oil-price war with Saudi Arabia.

10 Russia's parliament approves major constitutional changes which, among other measures, allow President Putin to seek re-election in 2024. A nationwide vote on the amendments is scheduled for 22 April.

11 The WHO declares COVID-19 a pandemic.

26 The US Department of Justice indicts Venezuela's President Nicolás Maduro, charging him with 'narco-terrorism' and offering rewards up to US$15 million for information leading to his arrest.

April 2020

9 OPEC and other major oil-producing countries agree to reduce oil production by almost 10m barrels a day to stabilise oil markets.

15 The G20 nations agree to suspend debt service payments for the world's poorest states until 2021.

20 Israeli Prime Minister Benjamin Netanyahu and Blue and White party leader Benny Gantz agree to form a national emergency government.

26 The Southern Transitional Council declares a state of emergency in southern Yemen and announces that it will implement 'self administration' in Aden and other southern provinces.

27 US President Trump suggests that the US should seek 'substantial' compensation from Beijing for its supposed COVID-19 policy failures.

28 Libyan National Army leader General Khalifa Haftar is accused of attempting a *coup d'état* after he announces his intention to place eastern Libya under direct military rule.

May 2020

12 The Afghan government resumes offensive operations against the Taliban and other armed groups, following two separate attacks that killed dozens of people.

15 The US Bureau of Industry and Security tightens export controls against Huawei, announcing plans to limit the company's ability to use US technology and software to make semiconductors abroad.

21 The US announces that it will withdraw from the Open Skies Treaty, an agreement that authorises signatory countries to conduct unarmed aerial surveillance flights over each other's territory.

22 Argentina defaults on sovereign debt for the ninth time in its history after missing a deadline to pay US$503m in interest on dollar bonds.

25 The killing of George Floyd by a Minneapolis police officer sparks global protests.

25 The European Commission calls for a €750 billion (approximately US$852bn) recovery fund to tackle the impact of COVID-19.

29 US President Trump announces that the US will leave the WHO.

June 2020

15 A violent clash between China's People's Liberation Army and the Indian Army takes place along the border in Galwan Valley, eastern Ladakh region, killing 20 Indian soldiers. They are the first fatalities along the Line of Actual Control in 45 years. The extent of Chinese casualties is unclear.

16 North Korea destroys the inter-Korean liaison office in the border town of Kaesong.

18 Viktor Babariko, widely seen as the leading opposition candidate in the Belarusian presidential election scheduled for August, is arrested and charged with money-laundering and tax evasion.

22 European Commission President von der Leyen and European Council President Charles Michel meet with Chinese Premier Li Keqiang and President Xi in a virtual summit.

25 Russia holds a national vote on constitutional amendments, including a provision that would make it possible for President Putin to serve two further terms in office. According to official results, nearly 78% of respondents support the measures.

29 China passes a controversial security law for Hong Kong, apparently undermining the 'One Country, Two Systems' framework. The law criminalises 'treason, secession, sedition (and) subversion' and targets protesters.

Chapter 3

Strategic Prospects

The COVID-19 pandemic of 2020 accelerated the atomisation of international society of the last few years. Global institutions continued to wither and international rules-based governance was further diluted. Multilateralism melted into competitive unilateralism. Alliance politics consumed goodwill more than they generated effective cooperation. Many armed conflicts continued even as various regions went into different forms of lockdown.

The pandemic will have enduring effects on geopolitics. International suspicion is rising and trust is in limited supply. Strategic adventurism by some states further unsettles the status quo. The world will enter a period of even greater strategic flux. Diplomatic and security crises will erupt regularly with only intermittent and unsatisfactory efforts to resolve them. As 2020 draws to a close, several trends are evident that will shape the international environment in 2021 and place greater burdens on both states and institutions.

The great postmodern competition is unfolding in the technological space. The long-term commanding lead of the United States in this area is being rapidly diminished by Chinese innovation and capacity to penetrate other markets effectively. China's Digital Silk Road remains the most intriguing element of its Belt and Road Initiative, and the one that could bring it the greatest long-term benefits as it aspires to data-superpower status. The security and data-privacy implications of China's technological gains trouble the US. It has gone to great lengths both to prevent Chinese access to US technologies and to persuade other states, especially in Europe, that adoption of Chinese technology would endanger their relationship with Washington. For its part, China is keen to accelerate its capabilities and deepen the reliance of other regions on its products. In essence, a Digital Great Game is being played, and the competitors are playing rough.

The US is trying to maintain its leading position in information and communications technology (ICT) by controlling the microchip and semiconductor industry. Its policy of sanctions, however, carries risks. China has an internal market of one billion internet users – greater than the US and European Union combined – and will in time be able to create its own products. Chinese technology is attractive to states in the developing world that appreciate both its low cost and high surveillance characteristics. Eventual US exclusion from the Chinese internal market will also carry costs for American firms. Qualcomm is, after Huawei, the second-largest customer of China's Semiconductor Manufacturing International Corporation, which is targeted by the US Department of Commerce. The inability of any Chinese company to operate in US markets will force those same companies to embrace all the more firmly their Chinese government sponsors. China's progress will be slowed, but US ICT dominance will suffer from self-inflicted wounds, and most other parts of the world will become 'battleground states' in this technological rivalry. Digital spheres of influence are in prospect. The course and outcome of this Digital Great Game will be hugely consequential and escalate competition across a wide range of issues.

Washington's use of economic sanctions to pursue its interests in this sphere intensifies an emerging trend of 'economic statecraft'. This has dissolved the barriers between security and business. Trade has become enveloped in national security and has sometimes been suffocated by it. Companies have had to extend their political antennae and become alert to foreign-policy issues. States can adopt new attitudes that suddenly compromise well-prepared commercial plans. The extraterritorial reach of US sanctions reached new frontiers in 2020. US members of Congress, for example, threatened crushing legal and economic sanctions on the German port of Mukran in Sassnitz, and on the board members and employees of the company that owns it, because of its role in completing the Nord Stream 2 pipeline to bring Russian gas to Germany via the Baltic Sea. Even German politicians who were opposed to the pipeline were taken aback by these pressure tactics.

China, too, has embraced the extraterritorial approach. The national-security law in Hong Kong effectively applies criminal penalties to anyone from outside the Special Administrative Region (SAR), including to a person who is not a permanent resident of the SAR, who may breach its provisions. Such assertion of extraterritorial jurisdiction applying to non-citizens is highly unusual and has led many countries to question the viability of their extradition treaties with Hong Kong. The claims of extraterritorial jurisdiction by the US and China add an additional element to their own competition. It will persist as a technique used against third countries that will likely be both unable and unwilling to retaliate in kind.

But economic power, sanctions and regulatory innovations will evolve to be part of the strategic arsenal of other states. The EU has also looked at more ways to flex such muscles. European efforts to tax global digital firms demonstrated a determination to take action at a time when the US had decided not to continue multilateral talks on the matter. The EU as an institution has repeatedly tried to develop mechanisms to blunt the effect of US secondary sanctions. The unique role of the dollar as the global reserve currency means that the EU cannot compete with the United States' capacity to impose financial sanctions. But the EU can act in other increasingly sensitive areas, such as the digital domain, to try to set standards for quality and security that would pre-empt, or at least moderate, geopolitical tensions. The absence of proper standard-setting in the 5G area led to such troubles. But there is still time for states and corporations to start thinking about how to set standards for the 6G era.

Better protocols are needed to manage extraterritorial jurisdiction, and more organised standard-setting is required to cope with digital international politics and security. There is a need, as this *Survey* argues, for more wholesale consideration of the gaps in international law that have arisen over the past few decades, and of how best to fill them. For 'like-minded' states concerned about the decline of the rules-based order, there is a need to refocus efforts on legal diplomacy to update the international legal order. Ritual calls to revive, reinvent or create new multinational institutions are not enough on their own. The structures of the state-centric Westphalian system that generated modern public international law are under threat. Non-state groups and private corporations are now strategically important actors. States need to shape the international legal environment while they can. The practice of carefully designed and well-explained statecraft can still serve to modernise international law and make it more attuned to the nature of contemporary challenges.

This can only take place when there is an open public debate about how best to address the new requirements that have emerged in public international law because of the changing nature of international relations. Legal diplomacy should be conducted to embed carefully considered principles of international law. Such diplomacy is needed across several fronts. Just as domestic law can get out of step with an emerging social consensus, so international law (which is reformed at a slower pace) can find itself out of sync with the evolving needs of legitimate statecraft. The champions of a liberal international order should understand that good statecraft, democratically rooted, creates sound law. Yet outdated law, left untouched, imprisons ethical statecraft. Actions to deal with emerging threats need to be conducted within a strong legal context. But an ossified international law cannot be allowed to render illegal ethically sound and legitimate actions carried out to defend national interests against emerging threats.

Violent transnational actors in particular are increasingly acting outside the scope of modern international law. They ignore the state system, and the state system passes them by. They have sought to exploit the spaces between the law, and the state system has not responded by rendering their activities illegal. The laws of armed conflict should be refreshed to engage more effectively with transnational non-state armed groups and to endow responsible states with the legal powers to contain and defeat them. In the past, such groups became familiar enough in their structures and composition to be penetrated and dismantled. Now, and in the near future, terrorist ideas and methods will travel through emulation and inspiration, and non-state groups may not display obvious 'command-and-control' features. Such threats are more difficult to counter and disrupt. Opportunities to do so will be fleeting and may require remote intervention to prevent attack. Modernising legal appreciation of when, in the digital age, threats may be considered imminent, and how authority to deal with them can be established, should become a priority.

Other elements of international law may also need review, including cyberspace, space, agreements governing asylum and refugees, and the regulation of private military companies and their activities. The treaty-based 'regulatory infrastructure' of international security would also benefit from more attention. The dismantling of much of the arms-control architecture of the Cold War period leaves too much to chance. Balances of power are harder to maintain if no accounting or audit is done, and no effective limits set. Perhaps, too, the principles of global public health should be subject to a wider range of norms, if not actual law.

The experience of the COVID-19 pandemic has invited calls for more transparency about sources and transmission. The existence of a multinational institution in the form of the World Health Organization is an asset. But as trust in its governance has been depleted in some quarters, its management must be strengthened, and the perception and reality of its independence re-established. Given the current fragility of multilateral institutions, especially those within the United Nations structure, it is important both to preserve what is good and to update what has been overtaken by events and by globalisation. In January 2020, before the pandemic took hold, climate change had become the world's number one issue. It was quickly relegated by the global health crisis but revived in the public imagination in the second half of the year as more climate-induced catastrophes took place from California to Australia and the

Arctic. Climate-change politics, and efforts to regulate carbon emissions, will regain centre stage when the politics of the pandemic recede.

The year to June 2020 has shown the power of the national approach and impulse in geopolitics. Strategic self-determination is fuelling more efforts by states to develop their own strategic identity rather than have this shaped for them by regional or institutional affiliations. The decision by the United Arab Emirates (UAE) to establish diplomatic relations with Israel in September 2020 was an iconic act of strategic self-determination. The state's leaders had decided that they no longer wished to subsume their regional strategic interests within or under the Palestinian cause. Nor would they accept being tied to an Arab League consensus against any foreign-policy move inspired by national considerations. The previous 'operating presumption' – that the Palestinians negotiate peace with the Israelis, and then the rest of the Arab world follows – was reversed with this decision. The UAE, and some others, have decided that recognising Israel does not necessarily preclude pressing for peace with the Palestinians. The decision upended some of the unwritten rules of the region, creating new facts on the ground.

This search for national strategic identity and departure from regional assumptions are strikingly evident in Asia. China has announced its great-power identity and wrapped it in exuberant language. More quietly, and partially in response, Japan is exerting itself strategically. It is more publicly defining its threat perceptions and more openly defending its territory from what it perceives to be provocative incursions. Most intriguingly, it is campaigning for further multilateralisation of its security arrangements. The nascent Quad security consultations were reinforced in 2019. Japan, the US and Australia are more readily being joined by an India further shedding its non-alignment garb. The Japanese defense minister asserted Tokyo's interest in becoming more closely affiliated to the Five Eyes intelligence network comprising the US, UK, Canada, Australia and New Zealand. While starting from a low and modest base, Tokyo's efforts to improve cyber capabilities are increasing. So long as there is doubt in Asia, and elsewhere, about US strategic willpower, 'mini-lateral' arrangements of this kind will proliferate.

Cutting through all these trends is another: the shapelessness of conflict. Conflicts appear to have no clear beginnings, edge, character, contours or endings. In the cyber domain, conflict is constant and ever-present – cyber tools are forward-deployed, are placed in the systems of others, occasionally pop up and are displayed to opponents for deterrence purposes, or are used in ways intended to mask attribution and confuse responses. Miscalculation is inherent in the conduct of 'persistent engagement' by cyber powers. Yet, without extensive situational awareness, cyber defence also becomes more difficult. This formlessness is more maliciously true of the activities of terror groups who are able to withdraw from theatres of battle, reconstitute themselves elsewhere, work remotely, play jurisdictional arbitrage to immunise themselves against state attack, and otherwise game the international system to their advantage. States that prioritise influence operations and use proxy partners to camouflage their strategic intent also benefit from the strategic ambiguity inherent in these techniques.

Increasingly, we are entering an age of 'tolerance warfare', which the IISS defines as the 'constant effort to test the tolerances for different forms of intervention against settled states'. Sometimes tolerance warfare is conducted overtly and is in effect 'declared'. But often it is executed through foreign networks or private partners, especially in theatres of operation that are adjacent to the power executing this technique. A favoured tool of actors who wish to change the status quo, tolerance warfare is difficult to counter because it generates conflict below the threshold of traditional war, outside the confines of established laws, yet above the acceptable limits of stability. Tensions along the Line of Actual Control in the Himalayas between China and India were heightened by tolerance-warfare tactics on each side. Many countries probe weaknesses and seek to provoke responses to test what the tolerable limits might be, and this strategic method of warfare, as it becomes more prevalent, will ensure that tensions in fragile geopolitical relationships remain high.

The COVID-19 pandemic has barely allowed for a strategic pause in these trends. National resilience and self-sufficiency are being prized as key goals. Reputation is being resurrected as an important element of national power. Competence in governance is demanded. The resurgence of public protest, whether or not it is permitted, is an indicator of impatience with 'whole-of-government' strategies that do not deliver the goods promised. Governance has become more complex domestically, and nearly absent internationally. Anarchy seems so often only a few mistakes away.

In this environment, international actors – governments, businesses and others – need to cope with profound changes in the nature of war, power and rules. The characteristics of warfare, the shape of conflicts, the strategies employed, the actors engaged and the weapons used are all changing rapidly. The nature of power that states, companies and transnational actors deploy digitally and economically is altering at speed. Regional orders are mutating. The interplay of norms, guidelines, standards, regulations and laws that govern international society is in flux. International society remains insufficiently regulated and is almost becoming an 'ungoverned' space.

The future of war, power and rules is contested. An organised understanding of how they interact with each other is what the IISS intends to provide, as we are confronted by a new anarchical international society.

Chapter 4

Strategic Policy

2020: The Politics of Pandemic

How has the most severe disease in a century shaped geopolitics?

2020 will go down in history as the 'year of coronavirus'. COVID-19 is the greatest global health crisis since the Spanish flu pandemic at the end of the First World War. The economic, social and political consequences, both domestic and international, have been profound and will endure. Major economic and financial powers shut down vast sectors of their economies to tackle the virus, creating the deepest and fastest global economic contraction in modern times.

The crisis hit at a time when relations between China and the US had dramatically deteriorated, erasing the level of trust that would have been needed to underpin effective cooperation of the kind achieved to tackle the 2008 global financial crisis. Controversy around the origins of COVID-19, and China's mishandling of the first phases of the crisis, sharpened tensions between Washington and Beijing, adding a major geopolitical dimension to the pandemic.

Phase one: China and East Asia

While it is impossible to predict where and when infectious-disease outbreaks will occur, the prevalence of open-air wet markets selling a wide range of live poultry, fish, reptiles and mammals in China and Southeast Asia has long been identified as a principal vector for such diseases to 'jump' from animals to humans. COVID-19 is the latest example. International experts agree that the key to effective containment of such disease outbreaks is early identification and action. But China's political system failed to achieve this. As early as December 2019, doctors in Wuhan were aware of the outbreak of a mysterious, apparently new, disease. But this information was not quickly acted upon or disseminated.

This lack of immediate action was due to Communist Party officials' fear of getting on the wrong side of the top leadership in Beijing. As a result, nearly two key weeks in January were lost in addressing the Wuhan outbreak. To make matters worse, the outbreak coincided with the onset of the Lunar New Year season, when middle-class Chinese citizens travel for holidays or to visit relatives. Wuhan's mayor stated that five million people (nearly half its population) left the city in the week before the imposition of the quarantine designed to limit the spread of COVID-19. For this reason, many Western epidemiologists believe that China's official count of roughly 90,000 cases of COVID-19 as of June 2020 may have significantly underestimated its incidence.

Following their initial missteps, Chinese authorities moved rapidly to quarantine Wuhan and surrounding Hubei province, isolating some 60m people. They also implemented a nationwide lockdown to stop the uncontrolled spread of the disease across the country. These efforts largely succeeded, and China began the process of opening up again in April 2020. China's leaders viewed their handling of COVID-19 as a great success. Beijing's propaganda machine worked hard to ensure that this positive story replaced the negative one of China's initial delay and mishandling of the outbreak in Wuhan.

In mid-February 2020, American media close to the administration of President Donald Trump, and some Republican members of Congress, began to propagate the theory that COVID-19 originated in the laboratory of the Wuhan Institute of Virology, which is administered by the Chinese Academy of Sciences. This explosive claim pitted scientists, broadly sceptical of the claim, against Trump administration officials who rallied around the theory as the White House intensified its criticism of China and called for an international investigation of its handling of the outbreak. In response, China initiated a major disinformation campaign, targeted principally towards the Middle East and Africa, which picked up a Russian-originated claim that the US was the source of COVID-19, and that visiting soldiers had brought it to China.

Phase two: Europe and the United States

Until well into February, COVID-19 was considered largely a Chinese, or at most an East Asian, problem. Global concerns about it were limited to travel restrictions to the region and market worries about the impact that disruptions to the Chinese economy would have on the global economy. But in late February and early March, COVID-19 spread with extraordinary speed beyond East Asia to become a global phenomenon.

Initially, the epicentre of the virus shifted to Iran and southern Europe, with Italy and Spain particularly badly affected. But the rest of Europe and the US were not far behind, while China and other north-east Asian countries saw a dramatic drop in new cases. In March, new daily cases in Europe rose from near zero to close to 30,000 per day. On 11 March, the World Health Organization (WHO) declared a pandemic. From mid-March, an almost exactly parallel surge took place in the US. Growing financial alarm broke the bull market that had been previously invulnerable to both economic weaknesses and geopolitical tensions. In the last week of February, global stock markets recorded their biggest fall since the 2008 global financial crisis. On 24 March, Japan postponed the 2020 Summer Olympics.

Few expected that Western democracies would be able to impose control measures on the scale of an authoritarian regime like China. More surprising was the increasingly divergent paths of the US and Europe. With a similar GDP, healthcare capacity and climate, the two regions might have been expected to handle COVID-19 with comparable effectiveness. Indeed, it would have been reasonable to predict that the US would do better since it had spent more on infectious-disease detection and response. In October 2019, the Johns Hopkins Center for Health Security had ranked the US as the country most prepared to face a pandemic.

But while the new case load in Europe began a steep decline from a peak of 30,000 per day in early April to less than 5,000 per day in mid-May, the downward slope of the infection curve was much flatter in the US. Infections only dropped to 20,000 per day over a similar time frame and then began to climb again, reaching levels similar to those in April by the end of June. The reasons for the strikingly different trajectories lie in policy and politics.

In Europe, government guidelines for easing lockdown varied between countries, but were cautious and generally observed, leading to gradually reduced transmission rates. The US Centers for Disease Control and Prevention (CDC) compiled similar guidelines. But these were often sped up or ignored by state-level governments under political pressure from a public that had grown weary of quarantine and confinement. But Trump, despite initially dismissing COVID-19 and then downplaying the crisis, did support the general policy of lockdown that his health advisers proposed. As a result, many areas of the country hardly impacted by the virus still faced restrictions not dissimilar to those in the Boston–Washington corridor, where the bulk of cases were found in the initial wave in the US.

America's already highly polarised political environment exacerbated the challenge of tackling COVID-19. Imposition of, and adherence to, restrictive public-health measures became intensely politicised. Nearly all the US states and cities that suffered the worst initial outbreaks in March and April were solidly Democrat, and Democrat governors and mayors were most supportive of lockdowns. Republicans were far more likely to resist or oppose lockdown, social distancing and other measures. Mask-wearing became a political issue, with Trump and senior advisers insisting it was a matter of personal choice and rarely wearing them in public.

As the US infection rate began to diminish in its initial epicentre in the northeast, and states there began to open up, political pressure for parallel opening became overwhelming in less-affected parts of the country.

But this proved premature. By the end of June, the US was in the midst of a second wave of cases and hospitalisations, including in many states, notably in the south, that had not previously suffered a high incidence of COVID-19.

Phase three: developing countries

By early summer 2020, COVID-19 had spread to the global south: South Asia, parts of the Middle East and North Africa, sub-Saharan Africa and Latin America. Data reliability in many of these countries made it harder to assess the course of the pandemic. Nonetheless, by the end of June, the scale of outbreaks in the developing world led most epidemiological forecasters to predict that, before the end of 2020, the US would be the only developed economy in the top-ten countries for COVID-19 cases. This third phase did not have an epicentre. But high population density and relative lack of testing facilities meant that South Asia produced the largest numbers of new cases. India was most affected, though its youth-skewed demographics may enable it to avoid a high death rate.

Indian Prime Minister Narendra Modi was initially one of the most adept world leaders in handling the domestic politics of COVID-19. Due to early action in ensuring a strict national lockdown, his approval ratings soared to over 80% in May 2020. India's border clashes with China in mid-June, and a weakened and divided political opposition, ensured overwhelming nationalist support. Yet by the end of June, India had become the fourth-worst-hit country after the US, Brazil and Russia, and the pandemic began to inflict severe economic damage on the country. More than 120m people reportedly became unemployed. The World Bank expected India's economy to contract by 3.2% in the 2020–21 financial year.

In Latin America, Brazil handled the crisis least effectively and became the most severely affected country. President Jair Bolsonaro consistently downplayed the seriousness of COVID-19, organising rallies during the pandemic and touting the 'miracle cure' of chloroquine, which led to the resignation of two capable health ministers in one month. By June 2020, Brazil had the highest casualty rate in the world (before being overtaken by the US), with over 1,000 deaths per day.

Like Bolsonaro, Russian President Vladimir Putin did not make COVID-19 a priority in the early weeks of the pandemic. His decision in March 2020 to reject an OPEC+ agreement, triggering an oil-price war with Saudi Arabia and US shale producers in the face of collapsing global demand, reflected a gross underestimation of the economic impact of the pandemic. But as Russian case numbers rapidly increased in April, Putin reassessed and imposed limited forms of lockdown while agreeing new OPEC+ production cuts that helped stabilise global prices despite the sharp contraction in the global economy. By late June, new COVID-19 case numbers were declining, though they remained among the highest in the world.

Russia remained active on the anti-Western misinformation front throughout the pandemic, with signs of growing collaboration with China. However, China attributed its largest outbreak since reopening began to the return of infected Chinese citizens from Russia.

Nationalism and the quest for a vaccine

Treatment regimes can be important disease mitigators, but even the most promising may merely reduce time spent in hospital. A vaccine will be essential to hasten a return to social and economic normality. Soon after COVID-19 was identified, research on developing one began.

As of late June 2020, some 100 vaccines were being researched, with more than 20 having entered human clinical trials. Despite expectations of a high failure rate, and the fact that new vaccines have always taken at least three years to develop, epidemiologists were optimistic that one or more effective vaccines would be developed over the coming year for emergency use and for front-line healthcare workers, if not the wider public. The key questions are who will develop these vaccines and who will have access to them. This will be driven as much by economic and geopolitical factors as by medical and public-health ones.

The risks around developing a COVID-19 vaccine are manifold. These include high financial risks to scaling up production under unprecedented time constraints; the dangers that governments cannot or will not pay for them in a constrained economic environment; and competition from other producers and (perhaps even more) from governments potentially undercutting financial returns with an alternative viable vaccine weeks or months later.

Vaccine research is taking place in an international environment that lacks significant cooperation or an organisation with the credibility to create and legitimise guidelines for this. As the pandemic crisis deepened, Trump escalated criticism of the WHO, and in May announced his intention to withdraw the US from it. The US and others claimed China and Iran were conducting cyber espionage on vaccine-research efforts, further undermining willingness to cooperate.

Several competing efforts got under way. The US Biomedical Advanced Research and Development Authority (BARDA) invested in several potential vaccines to reduce financial risks for pharmaceutical companies and ensure first access to a successful vaccine. BARDA's investments include foreign firms like France's Sanofi and the UK's GlaxoSmithKline. French President Emmanuel Macron publicly attacked the idea of 'preferencing', irrespective of the source of investment funds.

Oxford University's effort to adapt a vaccine originally designed for Middle East respiratory syndrome coronavirus (MERS-CoV) showed significant progress and gained widespread public- and private-sector backing. In June, France, Germany, Italy and the Netherlands agreed to purchase 400m doses of the Oxford vaccine which, if approved by regulators, would be produced by the Anglo-Swedish drug company AstraZeneca. The company reached similar deals with the UK and the US, where the Trump administration supplied US$1.2 billion in funding. Several Chinese companies, as well as China's Academy of Military Medical Sciences, made progress in their human trials. The WHO pooled resources with the EU to support several companies worldwide, with the aim of supporting more equitable distribution of any successful vaccine. But the success of any one of them will likely lead to efforts to lock in access by countries with a direct financial stake in the development and production of the vaccine.

Commercial and geopolitical tensions seemed likely to grow around not only the race to discover a vaccine but also its manufacturing chain, given the billions of doses needed and the challenges of a global vaccination campaign. Whether or not China develops a successful vaccine, it will be critical to supplying related medical devices for administering the vaccines, unless other governments with potential capacity – Australia, Canada, France, Germany, Israel, the UK and the US – can radically ramp up production.

The world economy: is a new depression looming?

Economists were slow to recognise the implications of COVID-19. This was partly due to China's efforts in early 2020 to downplay the impact on its growth, lest this create a self-fulfilling prophecy. But Beijing's announcement of a 6.8% contraction in GDP year-on-year in the first quarter of 2020, and the simultaneous actions of all other major countries in substantially shutting down their economies, generated a faster and deeper global economic contraction than almost anyone expected. The US Bureau of Labor Statistics reported that 20m Americans lost their jobs in April 2020, with unemployment reaching nearly 15%. In the second quarter of 2020, US GDP fell by 9.5% despite the sharp rise in consumer spending in late May and June. In Europe, over 30m people in France, Germany, Italy, Spain and the UK applied for state support, while the eurozone economy contracted by 12.1% year-on-year in the second quarter of 2020, the largest quarterly decline since collection of this data began in 1995. Industrial production in the eurozone fell by more than 18% in March and 11% in April before turning sharply upwards in the following months.

The scale of the contraction led to increasingly pessimistic assessments of the longer-term implications by most economic forecasters, who had initially hoped that governments would be able to mitigate and then reverse what they had put in train, resulting in a 'V-shaped' recovery. When the IMF hosted its virtual spring meeting in April, it predicted a 3% contraction of global GDP in 2020. In June, it revised this to 4.9%. Nonetheless, financial markets diverged sharply from these forecasts. In the US, where the Dow Jones had lost more than one-third of its value between mid-February and early April, by early June some three-quarters of those losses had been recouped. Most other global markets followed a similar pattern, buoyed by the ending of lockdowns and signs of increased spending in both the US and Europe. This contrasted sharply with China, where successful efforts to put supply chains back to work in March and April were not sustained by growing consumer spending or by large-scale stimulus, thereby 'flattening' the economic-recovery curve.

The main likely source of this difference was the unprecedented level of fiscal and monetary support Western central banks and governments provided to businesses. While the financial sector had benefited from government bailouts during the 2008 global financial crisis, it was commerce and consumers that benefited from such support during the COVID-19 crisis. A University of Chicago study in May 2020 estimated that 68% of Americans who had lost their jobs would receive more from government support than they previously had from their wages. The pattern in Europe was similar. In China, by contrast, the scale of the urban informal economy, with scores of millions of unregistered workers, made such an outcome impossible. Additionally, the authorities in Beijing did not want to repeat the experience of 2008, when massive economic stimulus – worth 13% of GDP in 2008 – led to a surge in total debt. The management of this debt has been a priority under President Xi Jinping.

Expansionary Western fiscal and monetary policies have so far made developing-country debt issues less challenging than expected. The IMF spring 2020 meeting focused on creating a framework for dealing with debt issues in both the poorest countries, where 'official' debt dominates, and emerging markets, where commercial debt dominates. G20 finance ministers agreed to declare a one-year debt moratorium for the poorest countries (with the largest proportion in Africa), which were due to pay an estimated US$12bn during the remainder of 2020. No agreement was reached on a debt-restructuring framework for the emerging-market countries and wealthier developing countries. But very low or even negative interest rates due to enormous volumes of new liquidity allowed these countries to continue to refinance their debts without exorbitant financing costs.

This ability to refinance forestalled what many had expected to be a confrontation between China and other creditors (official and commercial) about potential debt restructuring in countries that had accumulated large volumes of debt to China through their participation in the Belt and Road Initiative (BRI). But the risk remained of a broader debt crisis that would include larger, less poor countries with far greater debts. This would test Beijing's ability to work with other major creditors, both multilateral development banks and traditional bilateral donors.

By mid-2020, most global economic forecasters expected a relatively slow and uneven recovery, despite growing confidence of a fairly sharp first stage of recovery in the second half of 2020 based on the rebound in consumer demand in the US and Europe. This caution reflected greater pessimism about several factors: the economic impact of COVID-19 in developing countries; the likely continued restrictions on travel and tourism, which before the pandemic made up roughly 10% of the world economy; and fears around protectionism and nationalism in the US–China relationship.

COVID-19 and the future of globalisation

COVID-19 has changed key dynamics of globalisation. The most dramatic change is the shift in thinking about supply chains. The disruption of supply chains as a result of the crisis, and increased sensitivity around over-dependence on China-based supply chains, has already begun to shift business and government calculations away from efficiency towards resilience. When supply chains seize up, components cannot move along the production and assembly process. Managing this resilience requires optimising stocks of components. Storing more than necessary imposes costs, but not storing enough exposes production to supply-chain risks. This risks losing sales opportunities, angering customers and jeopardising long-term market share. Supply-chain resilience must balance cost and efficiency.

The pandemic may also widen the range of industries that are considered strategic because they are critical to national security. Against the background of the recent sharp decline in US–China relations, high-end technology products have been classified as strategic. COVID-19 is feeding bilateral tensions, making it more likely that the world will develop two sets of standards for technology, surveillance, and data filters and collection: one used by the most advanced industrial economies, and the other by China and many developing countries. Beyond high-tech products, COVID-19 is likely to expand the definition of 'strategic' to include pharmaceuticals and medical supplies. This could lead to higher tariffs, legal incentives for repatriation of capital and tax incentives for local production. All these will impair globalisation.

COVID-19, global politics and great-power relations

Although COVID-19 is an epic common global challenge, there has been little international cooperation among the biggest powers. The pandemic struck at a moment when the US was already stepping away from global leadership; when international organisations were structurally weakened; and when China was both coming under intense pressure from the US and sensing an opportunity to play a much larger role in world affairs.

As COVID-19 spread across Europe and the US in March and April, so did blame of China for the 'original sin' of the virus, along with calls for an international investigation of its origins. The US escalated its criticism of the WHO and accused it of enabling Beijing's misdeeds. China rejected the charges against it and accused the US of violating Beijing's sovereignty. It initiated a multifaceted public-diplomacy campaign highlighting its success in curbing the virus in China and its contributions to curtailing the global spread of the disease, especially in the provision of protective equipment and other health-related goods and services.

But China's efforts at promoting humanitarian support were compromised by problems with the quality of some of the exported personal protective equipment (PPE) products and the comparatively limited extent of Beijing's financial support. Some foreigners also faced mistreatment. In particular, Africans working and studying in Guangzhou became the targets of widespread racial abuse and shunning from the public and the authorities. This was strongly condemned by several African and Western countries.

In response to the pandemic, Xi expanded his personal diplomacy, initiating dozens of telephone calls to counterparts around the world. His message was simple: in this time of crisis, China was stepping up to a leadership role. Beijing announced a further addition to its BRI in the form of the 'Health Silk Road'. At the World Health Assembly in May, Xi gave a confident and effective speech in which he stated that China was in favour of transparency on the origins of the virus once the crisis had receded, and that a vaccine should be seen as a global public good available to all.

But in the months following Xi's speech, China did not prove adept at balancing its highly nationalistic 'wolf warrior' diplomacy (largely geared to a domestic audience) with its effort to lay claim to global multilateral leadership. Requests for Chinese financial leadership from international organisations to help alleviate the economic distress were largely ignored. China ratcheted up regional tensions over Taiwan and Hong Kong. It chose the day of Taiwan President Tsai Ing-wen's inauguration following her re-election to announce its intention to assert more direct control over Hong Kong by transferring key legislative powers on national-security issues away from Hong Kong's assembly to the mainland's National People's Congress.

China's assertiveness on the global scene was matched in Washington, where the Trump administration increasingly focused on painting Beijing as the main source of global instability. The rhetoric from both the US and China took on an increasingly ideological tenor, raising the prospect of a new cold war. These tensions tipped the balance in the US in favour of those in the Trump administration pushing for aggressive measures against China's technology sector.

Meanwhile, the EU succeeded in putting together a major financial package to support southern European states most heavily impacted by COVID-19. At the same time, Europe's top leaders – Macron, German Chancellor Angela Merkel and EU Commission President Ursula von der Leyen – all spoke about creating a 'geopolitical' Europe as a major force in world affairs. In all these ways, COVID-19 both highlighted and accelerated the increasing fluidity of world politics and great-power relations.

Looking ahead

By mid-2020 there was a growing consensus that the disease would be present in a fairly substantial way for at least two to three more years. Nonetheless, there were reasons for cautious optimism. It looked likely that several viable vaccines would be publicly available by early 2021, if not earlier. A spate of new epidemiological research suggested that the number of people with natural or acquired immunity to COVID-19 might be much greater than those already diagnosed with it.

But even if that is the case, the economic recovery promises to be extended and inconsistent. The global travel and tourism industry is unlikely to recover rapidly. Supply-chain uncertainty will take years to resolve. And the world's two largest countries by population – China and India – both face enormous challenges in restoring the health of their informal sectors, which provide most jobs in both countries. In view of these challenges, China is adopting what it calls the 'dual circulation model' in which overseas markets are subsidiary to the domestic one. By turning inward, Beijing hopes to make long-term development less reliant on factors beyond its control. This is a response to growing geopolitical tensions and the erosion of confidence among the Chinese leadership about external vulnerabilities.

The experience of COVID-19 has damaged the reputations of both the US and China. But China's losses may be greater. It has suffered substantial deterioration of relations with five critical middle powers with which it had been improving them: Australia, India, Japan, the UK and Germany. The EU, in contrast, has emerged on a stronger footing as member states have come together in unprecedented ways to address the crisis. COVID-19 will continue to overlay global politics with uncertainties and challenges. But it will also provide opportunities for countries to put aside differences and cooperate more closely in the quest for medical and economic solutions – if only they can take them.

Shifting Attitudes Towards Multilateralism

Cooperation is needed more than ever: why are its institutions under strain?

It is a sad irony that the great common global challenge of COVID-19 struck just as multilateral cooperation was facing its own serious challenges. As a result, the most serious pandemic in a century provoked discord and rivalry rather than the collective response it demanded. This has shed light on the larger malaise of multilateralism. In recent years this key principle of post-war order has been undermined by the most powerful states. What is the nature of this challenge, how is it manifesting itself and what can be done?

States have always communicated, cooperated and engaged with one another through diplomacy, treaty and alliance. Only from the nineteenth century have they have done so through international institutions. And only since 1945 has such multilateralism come into its own, driven by the disastrous experience of the breakdown of international order in the 1930s and the rise of interdependence. Its growth was bumpy – notably when the US left the gold standard, a key part of the Bretton Woods international economic order, in 1971 – but appeared inexorable. The number of international organisations, of their members and of the issues they covered all grew steadily.

In the 1980s the collapse of the Soviet bloc, and China's economic reforms, accelerated the multilateralist trend. In the decade that followed, globalisation drove states to manage more of their economic relations through organisations – exemplified by the creation of the World Trade Organization (WTO) in 1995. On security issues, US dominance of a unipolar world without sharp ideological division enabled broader cooperation – exemplified by the ouster of Iraq from Kuwait by a 34-country coalition, authorised by the United Nations, in the First Gulf War in 1991. Multilateralism, it seemed, was the wave of the future.

But since 2001 this trend towards ever-greater multilateralism has come under strain. Resistance to it accelerated after 2016, the year of Donald Trump's election as US president and the Brexit referendum.

Attacks on international institutions intensified further in 2019–20. This was not just a series of setbacks to particular organisations, but a challenge to the idea of multilateralism itself.

Attacks on, and failures to defend, multilateralism

Three sources of challenge to multilateralism stand out. The first is the growing assertiveness of non-Western states, above all of a rising China and resurging Russia, that do not accept the international status quo. While they do not reject all institutions and practices that regulate the current global order, neither do they assume these are legitimate and beyond challenge. As they become more confident, such revisionist states continue to undermine or violate norms of behaviour – for example, China's mass incarceration of Uighurs in Xinjiang and Russia's military intervention in Ukraine.

In some cases, authoritarian states seek to influence or control international institutions rather than resist or defy them. A recent example is China's attempt to fill leadership roles in international organisations. In the strategically critical area of telecommunications, for example, China has continued to take advantage of light Western representation at the International Telecommunication Union (ITU) to shape ITU standards on surveillance technology to enable Chinese exports to the developing world.

The second challenge to multilateralism comes from powerful states in the West. This challenge is newer and less expected, and therefore more disconcerting. It was the West that led the way in building the postwar multilateral order. But rather than defending this order against new challenges, powerful Western voices have added their own criticism and pressures. These are heard on both sides of the Atlantic, though their specific complaints vary.

The most significant such revolt comes from the most powerful country, the US. Its break with the past, though significant, should not be overstated. America's active involvement in world affairs after 1945 was itself a striking departure from the isolationism of the 1930s. As Graham Allison has noted, the view that the US even in this period has been the guarantor of a 'liberal order' is, if not quite a 'myth', then an exaggeration.

Even before the Trump presidency, the US on occasion excused itself from organisations, treaties and norms to which other Western states adhered. President George W. Bush's administration was notably sceptical of some constraints on American unilateralism. The US-led invasion of Iraq in the Second Gulf War in 2003 without a UN mandate exemplified this. More recently, its calls on China to observe a 'rules-based order' in the South China Sea sits at odds with its own failure to ratify the UN Convention on the Law of the Sea (UNCLOS).

But even if US ambivalence towards multilateralism is less surprising than it seems, its escalation into more overt hostility under Trump is striking. From the start, his 'America First' presidency has been marked by a strong antipathy towards multilateralism. Trump has withdrawn the US from the 2015 Paris Agreement on climate change, the Trans-Pacific Partnership, UNESCO, the Joint Comprehensive Plan of Action (JCPOA, or the 'Iran nuclear deal') and the North American Free Trade Agreement – though he agreed a successor to the last of these, the United States–Mexico–Canada agreement.

Rejection of multilateralism intensified in the year to June 2020. At the November 2019 NATO summit, Trump threatened to use US trade sanctions against NATO members that did not meet their defence-spending obligations. His use of a threat to harm fellow members of an alliance was a vivid disavowal of the spirit of multilateralism. By December 2019, the US had effectively paralysed the WTO by persistently blocking the appointment of members of its Appellate Body on the grounds that the organisation had overreached member states and discriminated against the US. In May 2020, the US announced that it would withdraw from the 1992 Open Skies Treaty. In June, it announced it would impose sanctions on officials of the International Criminal Court involved in investigating alleged war crimes, including by US forces, in Afghanistan. Most striking of all, as the COVID-19 pandemic escalated in April 2020, Trump cut US funding to the World Health Organization for what he alleged was its 'role in severely mismanaging and covering up the spread of the coronavirus', and its too China-friendly view. In May, he announced that the US would leave the organisation.

Some of the motives driving Trump's attack on multilateralism seem personal, particularly his determination to undo Barack Obama's legacy. But it is also an intrinsic part of his America First platform, with its belief that international organisations are 'unfair' to the US, and its preference for unilateral action complemented with transactional 'deals' on more favourable terms.

In addition, for the Trump administration, many multilateral institutions dilute, rather than enhance, US power. America's formidable strength – especially in two areas of coercive statecraft: financial sanctions and remote power projection – means not only that the US can act effectively on its own, but that, without the US, the multilateral efforts of others will have limited effect. Policy towards Iran is a case in point. In 2019–20 the US layered successive sanctions onto the Iranian regime, made doubly effective by secondary sanctions on third parties. In January 2020, the US killed the Islamic Revolutionary Guard Corps (IRGC) Quds Force Commander Qasem Soleimani through a bold projection of military force in Iraq. Multilateral efforts to contain the IRGC were symbolic by comparison.

For many, Britain's populist attack on multilateralism parallels that in the US, with the Brexit referendum in June 2016 prefiguring the election of Trump five months later. The eventual departure of the United Kingdom – a leading multilateralist state – from the European Union – the most ambitious multilateral organisation – on 31 January 2020 was both symbolically and substantively significant. Both the UK and the US have broken the near taboo among major states against walking out of major organisations.

But the comparison should not be overdrawn. There is no sign yet of a more general British retreat from active involvement in, and support for, international institutions. No major political force in Britain has targeted any other organisations from which to 'take back control'. The UK was also reminded of the benefits of multilateralism in 2018 when it worked through the Organisation for the Prohibition of Chemical Weapons, the EU and NATO to mount an effective international response to the Russian nerve-agent attack in Salisbury.

Officially, the post-Brexit vision for foreign policy is a 'global Britain'. It is not yet clear what this will mean in practice. Part of the Brexit electorate, especially in less-prosperous parts of the country, might in due course prefer a role more akin to a 'Britain first' that protects the UK from foreign competition and the costs of wider engagement. But for now, the country is committed to an active international role, including through multilateral organisations. The pursuit of stronger relations with more distant countries is manifested especially in a determination to develop the British Commonwealth, most of whose members share a common past as former British colonies.

Second thoughts in new Europe

Just as the United States' isolationist past should make its current retreat from multilateralism less surprising, so Britain's long scepticism of the European project – both before its entry into the European Community in 1973 and thereafter – makes its departure less anomalous than that of a founding member, or a more consistently pro-EU country, would be. But some newer, initially enthusiastic, EU members are also creating tensions, notably populist–nationalist governments in Hungary and Poland. This is a third source of challenge to multilateralism.

Since 2010, Hungary's Prime Minister Viktor Orbán has moved to concentrate power in the hands of the ruling Fidesz party. His 'illiberal democracy' has challenged EU norms and laws, notably on immigration. In March 2020, he used the COVID-19 pandemic to justify sweeping new powers to rule by decree. But there are limits to Hungary's rejection of multilateralism: Orbán's goal is not to leave the EU, from which Hungary has greatly benefited, but to carve out space within it for national governments whose political values diverge from those of Brussels, and to limit the power of Brussels to enforce all membership obligations uniformly. Poland's democratic backsliding under the ruling Law and Justice Party (PiS) also continued. In April 2020, the government came into open conflict with the EU after adopting a new law that further undermined judicial independence.

Against this background of Brexit and other populist tensions, the EU sought to marshal an effective collective response to COVID-19. Initially, this seemed to falter. In April, President of the EU Commission Ursula von der Leyen offered a 'heartfelt apology' to Italy for failing to provide the medical help it had requested at the start of the crisis. Negotiations on an EU-wide financial response to COVID-19 then exposed familiar tensions between northern states – especially the 'frugal four' of Austria, Denmark, the Netherlands and Sweden – and southern states.

Nonetheless, EU member states looked set to agree unprecedented new measures, notably the 'New Generation EU' plan. Proposed in May 2020, this would allow the EU to borrow up to €750 billion (approximately US$883bn) to distribute as grants and loans to member states. Ironically, Hungary and Poland looked set to be big winners, while threatening to veto any attempt to tie disbursements to EU 'rule of law' criteria that their actions otherwise challenged. But even as German Chancellor Angela Merkel championed an EU-wide approach to the crisis, the country's own Federal Constitutional Court issued a ruling requiring the government to obtain a justification from the European Central Bank of its strategy of purchasing public-sector bonds of member states. Hungary and Poland welcomed this implicit sovereign challenge, from an unexpected quarter, to the European law on which EU multilateralism is founded.

The Arab world: multilateralism outcompeted

There is less evidence of declining support for multilateralism among smaller powers than among the largest ones. But this does not necessarily mean that they are able to contribute to effective cooperation through international organisations, even on a regional scale.

A case in point – one not only important to the region, but a continuous concern of major powers – is the Arab world. The main multilateral organisations of the Middle East – the Gulf Cooperation Council (GCC) and the Arab League (AL) – have long suffered from schism and operational under-delivery. The AL has failed to play an effective role in resolving any of the disputes that afflict its members, from Libya and Syria to Yemen.

The GCC failed to contain a dispute between Qatar and other members that has inflicted huge unnecessary cost on the regional economy and further polarised the region between competing power blocs (Saudi Arabia and the GCC plus Egypt on one side; Turkey, Iran and Qatar on the other). It was unable to build on the truce established in the early stages of the COVID-19 pandemic between GCC states to foster a more lasting reconciliation. Its limited role as convener and facilitator has suited the uneven nature of its membership but it has also denied the region, at a time of grave crisis, a model of effective multilateralism. In fact, the GCC's most powerful members, Saudi Arabia and the United Arab Emirates (UAE), appear to rely on it less and dismiss its role, preferring to partner together to pursue their objectives. In contrast, the GCC's smaller powers, such as Kuwait and Oman, appear more keen to ensure its relevance and survival. Its failure to play an effective role in the most urgent security issue in the Arabian Peninsula – Yemen – exposes its limitations and in turn those of Gulf multilateralism.

Similarly, when it has come to meeting defence or security needs, states in the region continue either to reach for unilateral solutions, rely on bilateral relations with external security partners, or take the 'posse' approach and act with or through partners but in an operation led clearly by one dominant player. This allows for greater flexibility but also creates more uncertainty, as demonstrated by the UAE's unilateral withdrawal from its involvement in the war in Yemen.

The limitations of the Arab world's multilateralism have been demonstrated most recently in the AL's performance over Libya. It has been unable to support the UN-mandated Government of National Accord (GNA) in Tripoli, as powerful AL members Egypt, the UAE and Saudi Arabia continue to support the opposition Libyan National Army (LNA) led by General Khalifa Haftar. Although the AL renewed its support at foreign-minister level in June 2020 for a UN-led process, it has been criticised for continuing to weight its efforts and statements in favour of Egypt and the LNA.

Other multilateral organisations have also struggled to deliver results in the Middle East. The UN's mediation role also suffered: in each of these theatres, multilateral processes fell victim to ambitious international powers (Russia in Syria), regional ones (the UAE in Libya or Turkey in Syria) and local ones (Haftar in Libya) willing to flex muscles and derail UN-led diplomacy.

The UN enjoyed success in Yemen in avoiding an all-out battle over Hudaydah but there was no carry-over effect: the UN remains unable to impose a nationwide ceasefire, let alone drive a political process. In Libya, the tension between multilateralism and its alternative has taken military form as the GNA struggles to fight off the rebel forces under Haftar. That the former is backed militarily by Turkey and the latter by Russia and the UAE, with uneasy French diplomatic endorsement, indicates the degree to which decisive action has passed into the hands of unilateral actors.

International and regional powers have begun to assert control over zones of influence outside or in contradiction to multilateral processes. Turkey and Russia have done so in Syria and are in the process of doing so in Libya. The US also maintains a small presence in Syria, where Iran has developed a large footprint through partner militias. In Syria, Russia, Turkey and Iran coordinate through the Sochi process to shape the politics of the war-ridden country, while the UN orchestrates the drafting of a new constitution without the full commitment of the relevant local and international parties.

The failure of multilateralism in the Middle East risks reinforcing in a generation afflicted by the current conflicts a preference for the alternatives: authoritarianism, unilateralism or the intervention of regional actors such as Iran or Turkey, all of which have become features of regional conflicts. In June 2019, a survey of 25,000 people across the Middle East and North Africa by the Arab Barometer research network showed support for Turkish President Recep Tayyip Erdogan across the region 'dwarfing' support for either Russian President Vladimir Putin or Trump. In addition to Turkey's intervention in Syria, it also intervened in Libya and asserted itself in the Eastern Mediterranean in 2019–20.

Multilateralism: more demand than supply?

COVID-19 is the most urgent collective challenge inviting global cooperation. But it is not the only one and, in the longer term, not the most serious. While the pandemic will abate, the threat of climate change will endure. It dominated the global agenda at the annual meeting of the World Economic Forum at Davos in January 2020. In the months leading up to the event, international awareness had intensified, led by the advocacy of Greta Thunberg and the radical activism of the Extinction Rebellion movement. Thunberg did not spare either the UN or the EU in her criticism, both of which she blamed directly for the failure to act. Thunberg's demands are setting a new standard for delivery by existing structures. If multilateral organisations lose credibility as a way of delivering on a key issue for the rising generation, the latter may turn to more direct action against commercial and government targets.

The management of the international order through systematic cooperation now faces a series of converging challenges, from different sources and to different degrees. This complex picture defies easy generalisation about a 'retreat from multilateralism'.

The biggest difficulties arise from a small number of large countries. For the first time since the 1930s, status quo and revisionist military powers alike are retreating from multilateral commitments. But the decline of the idea or practice of multilateralism should not be overstated. The EU – the first 'post-modern power', albeit one with 'strategic' ambitions under von der Leyen's presidency – has deepened cooperation in ways unthinkable before COVID-19, despite facing significant internal challenges. Elsewhere, few smaller states share the growing scepticism or hostility towards international institutions evident in the US and China. But this may be of limited reassurance. For it is precisely the largest powers that have traditionally been, if not indispensable to multilateralism, then effective leaders and sustainers of the cooperation that underpins it.

Multilateral institutions and practices are still very strong by historical standards. Dealing with COVID-19, managing economic recovery and combatting climate change are all best tackled by more, not less, systematic cooperation. Demand for multilateralism from most states is likely to remain strong. Less clear is whether it will be met.

Key International Legal Gaps
What are they, and how can they be closed?

International law develops slowly. Treaties, especially broad multiparty treaties, are generally negotiated to address past international problems, not emerging ones. Customary international law – unwritten rules that all countries observe out of a sense of obligation – develops at an even more glacial pace. As a result, gaps can develop in international law where no agreed rules exist to govern state actions that may affect other states and their populations.

Some foundational international agreements, such as the United Nations Charter, the Geneva Conventions and the Refugee Convention, were negotiated immediately after the Second World War and have remained unchanged for 70 years. Although many of their core principles – such as prohibitions on use of force against other states, mistreatment of non-combatants in armed conflicts and forced return of persons who face persecution – remain relevant to the global security system, these agreements are not well equipped to address some twenty-first-century challenges such as transnational terrorism, nuclear proliferation and mass atrocities committed by states against their own populations. Meanwhile, technological developments, including advances in digital communications and space capabilities, have created new security vulnerabilities and means of waging conflict. Gaps between states' security imperatives and existing legal agreements have led some states to ignore or adopt strained interpretations of their terms, or to urge recognition of additional non-treaty principles to fill grey areas.

It would not be feasible to revise or replace major global-security and humanitarian treaties such as the UN Charter or Geneva Conventions in the foreseeable future. The 193 UN members have such disparate interests that it would be impossible to negotiate and reach a meaningful consensus – particularly in areas where technology is changing rapidly. However, rather than leave it to each state to act unilaterally where there are gaps or grey areas in international law, it can still be valuable for states to work diplomatically to narrow their differences in interpretation and for smaller groups of states to agree on non-binding principles to guide their actions, with the potential for these principles to attract broader support.

Despite gaps or outdated provisions in many security treaties, legal rules could be usefully expanded or updated in at least five areas. These are: the use of force by states, operations against non-state actors, refugee protection, cyber security, and competition and conflict in outer space.

Adapting rules on force to contemporary security threats

One of the bedrock provisions of the UN Charter is Article 2(4), which prohibits member states from using force against other states. There are three exceptions to this prohibition. Firstly, a state may consent to the use of force in its territory by another state (for example, to help the state repel an invasion or fight an insurrection). Secondly, the UN Security Council (UNSC) may authorise member states to use force against another state (as the Council did in 1990 when it authorised the use of force against Iraq after Iraq invaded Kuwait). And thirdly, Article 51 of the Charter provides that member states continue to enjoy an 'inherent' right to individual or collective self-defence in response to an 'armed attack'. Most states and international lawyers interpret Article 51 to include a right to use military force in or against another state to defend against an 'imminent' attack.

This framework has helped to limit and regulate the use of military force by states over the last 75 years. However, it has come under considerable strain from several types of security crisis, most notably humanitarian catastrophes, nuclear proliferation and threats from non-state groups operating in the territory of another state.

The ongoing civil war in Syria remains a stark example of a massive humanitarian tragedy for which outside military intervention might seem justified, while the UNSC has been paralysed. To some observers, Security Council paralysis points to a need for international law to accommodate forceful humanitarian intervention. Others, meanwhile, argue that outside intervention might only make an intractable situation worse. Such sceptics see the ongoing civil war in Libya as a cautionary tale. In a rare move, in 2011 the UNSC authorised the use of force to prevent the Libyan government's slaughter of civilians. The resulting NATO-led bombing campaign led to the fall of the Libyan government, resulting in a decade of chaos and violence in Libya and accusations by China and Russia that the Security Council's authorisation had been abused.

The Syrian government's use of chemical weapons against civilians has highlighted apparently different legal positions even among countries supporting humanitarian intervention. In 2018, the United States, United Kingdom and France jointly launched military strikes against Syrian targets following Syrian government gas attacks. Neither the US nor France explained publicly how their operations complied with international law, while the UK justified them legally as necessary to address an urgent humanitarian emergency. Many other states offered political backing – but not explicit legal backing – for the strikes, while many others criticised them as illegal.

That divide resembles past humanitarian interventions. In NATO's Kosovo intervention in 1999, only the UK claimed that airstrikes were permissible under international law as a humanitarian intervention. Many countries, including China and Russia, condemned the strikes as illegal. The US and many other like-minded states justified them as appropriate in the circumstances of this case, without stating that they were legal.

Although the US and some other countries want the flexibility to intervene in such circumstances, they are reluctant to claim that international law allows a humanitarian-intervention exception to the prohibition on force, partly out of concern that China, Russia or other non-democratic countries would exploit such a legal doctrine as a pretext to justify aggression. To the former states, operating in a legal grey area or claiming that moral imperatives justify some violations of law is preferable to opening a door to revision of international law.

Nuclear proliferation raises related legal questions about the use of force in certain extreme circumstances. North Korea's continued development of nuclear weapons and ballistic missiles, and international disputes with Iran over its nuclear programmes, are two contexts in which this issue might arise. Those cases also highlight some limitations of the nuclear non-proliferation regime.

In 2002, the US provoked international legal debate by declaring a 'pre-emption' doctrine: namely, that when dealing with 'rogue states' armed with weapons of mass destruction (WMD), it might be too dangerous to wait until an armed attack is imminent before striking first in self-defence. Although the UK, US and most coalition members ultimately cited previously adopted UNSC resolutions as the legal basis for their 2003 military intervention in Iraq, the US continued to argue that a state's inherent right to self-defence includes the right to use force to preempt the development of nuclear weapons by a hostile state. Intelligence failures in assessing Iraq's WMD and the terrible consequences of the Iraq War hardened the position of many states against widening the scope of legal self-defence in this way. Nevertheless, even as they declined to embrace the pre-emption doctrine, many Western allies have acknowledged that nuclear weapons – on account of their speed and destructiveness – might require more flexibility in interpreting 'imminence' for self-defence purposes.

It will not be possible to reach a global consensus on the circumstances in which humanitarian intervention or the use of force against nuclear threats is permissible. Even among like-minded Western allies, sharp differences will remain. Rather than try to resolve these legal debates or ambiguities, democratic allies could try to agree on minimum conditions that should guide any resort to force in dealing with humanitarian catastrophes or catastrophic threats, while also acknowledging that some states might interpret international law to impose additional requirements. Without committing states to legal doctrines they are not prepared to accept, reaching a consensus on minimum conditions could help reduce disagreements (at least among like-minded states) over future military interventions that are not clearly permissible under the UN Charter.

Another contested question is whether and when the UN Charter allows a state to use military force against terrorist groups (or other non-state actors, such as pirates) that launch attacks or train from the host state. After 9/11, the US administration of George W. Bush embarked on a controversial 'Global War on Terrorism', in which it asserted a right to use force against members of al-Qaeda and associated groups not only in Afghanistan but in other countries that were 'unwilling or unable' to prevent terrorist groups from using their territory. US president Barack Obama dramatically increased the use of drone strikes against terrorists in areas outside active hostilities, launching more than 500 strikes against terror suspects in Pakistan, Somalia and Yemen. Beginning in 2014, the Obama administration, in concert with several other states, commenced bombings and drone strikes in both Iraq and Syria against members of the Islamic State, also known as ISIS or ISIL. To justify such use of force under international law, the US relied in some cases on the consent of the governments involved (including Iraq, Pakistan and Yemen), while in others it cited a right to use force in self-defence where the government was unwilling or unable to stop terrorists operating within its borders (such as Syria).

A growing number of other Western governments, including Australia, Canada, Germany, the Netherlands, Turkey and the UK, have explicitly cited the right to use force against terrorist groups in other states, including when conducting or supporting strikes against ISIS members in Syria. In 2012, the UK Foreign Office's former senior legal adviser published a list of 16 'Principles Relevant to the Scope of a State's Right of Self-Defense Against an Imminent or Actual Armed Attack by Nonstate Actors', which he said had emerged from discussions among governments with experience operating against terrorist groups. Although these principles appear to have been accepted and relied on by half a dozen Western governments, they have not gained much broader explicit approval.

Given the lack of international consensus on whether and when the UN Charter permits states to use military force in self-defence against non-state actors in other states, countries that participated in the development of the self-defence principles could work to build additional support for them from other states. Such legal diplomacy might include efforts to narrow differences over when a non-state group poses an 'imminent' threat triggering a right of self-defence and when a state is 'unwilling or unable' to prevent the threat posed by the group operating in its territory.

New rules for waging war against non-state actors

The Geneva Conventions – a group of treaties negotiated after the Second World War to define how combatants and non-combatants should be treated in armed conflicts – are among the best known of all treaties. They include four agreements of 1949 covering treatment of prisoners of war, wounded and sick soldiers and sailors, and civilians in conflict zones, as well as three additional protocols negotiated in 1977 and 2005. The original four conventions have been ratified by 196 countries, including all UN member states.

During the last 20 years, some states and commentators have questioned whether the Geneva Conventions are outdated or have gaps as they apply to contemporary conflicts between states and non-state groups. Soon after the 9/11 attacks, Bush administration officials attracted withering public criticism for calling some of the 1949 Geneva Convention provisions 'obsolete'. In 2006, then UK defence minister John Reid called for the Conventions to be re-examined to determine whether they were appropriate for modern conflicts between states and non-state groups, noting 'we risk continuing to fight a 21st century conflict with 20th century rules'.

Although some of the more than 400 provisions of the 1949 Geneva Conventions are indeed outdated (such as the requirement to provide tobacco to detained persons), most of the core provisions, including the requirement of humane treatment of detained persons and the prohibition on torture or cruel, inhuman or degrading treatment, remain sound. States should continue to endorse them to promote more universal compliance. That said, the Geneva Conventions do have significant gaps and ambiguities in coverage, especially with respect to the detention of persons in so-called non-international armed conflicts. These gaps have made it difficult for states engaged in detention operations – such as the US, UK, Canada, the Netherlands and other countries operating in Iraq and Afghanistan – to determine how international law applies. One question, for example, is whether ISIS members detained in Iraq and Syria should be treated as prisoners of war (with a special set of privileges), as civilians who pose security risks (with another set of privileges), or neither. Other questions include whether detained terror suspects should be allowed access to lawyers to challenge their detention, and whether, when released, detainees should be sent to their countries of nationality, to the country where they were detained, or elsewhere.

In recent years, the International Committee of the Red Cross (ICRC) has acknowledged that there are gaps in the Geneva Conventions when applied to such contexts. In 2016 it convened a group of states to agree on principles to fill the gaps. The ICRC effort foundered over disagreements among the participants and officially ended in December 2019.

The existing Geneva Conventions should not, and realistically could not, be opened up for amendment. Instead, a smaller group of like-minded states could work towards agreement on principles to govern detention in conflict situations with non-state groups for which the Geneva Conventions do not provide detailed rules. As the more formal ICRC-sponsored process could not reach consensus, like-minded states could strive to announce and promote jointly a set of principles of humanitarian treatment that they would be prepared to support as a matter of customary international law.

International law and refugee flows

The principal international agreement governing refugees and asylum – the 1951 Refugee Convention – was developed immediately after the Second World War to help the millions of persons displaced by the Nazis and the war in Europe. Although it continues to provide vital protections for persons who qualify under its terms, many refugee advocates view the Convention as an anachronism that fails to address the needs of modern migrants and the security and economic concerns of many refugee-receiving states.

The Convention was originally designed to protect individuals displaced 'because of events before 1 January 1951', and was focused on persons 'in Europe'. It defines a refugee as a person who is unable or unwilling to return to his or her country of origin owing to a 'well-founded fear of being persecuted for reasons of race, religion, nationality, membership of a particular social group, or political opinion'. In 1967, a protocol was negotiated to remove both the temporal and geographic limitations. Today, 146 states are party to the Convention. Several large countries, including India and Pakistan, are not.

The Convention continues to provide important protections for persons fleeing persecution by authoritarian regimes. But it does not address contemporary causes of large-scale migrations, such as civil wars, gang violence, climate change and natural disasters. Nor does it explicitly apply to persecution based on gender. The Convention also provides no rights to refugees until they reach a state that is party to the treaty, which both limits its value to displaced persons and contributes to migrant trafficking.

The Convention also does not adequately address the concerns of refugee-receiving states. It places no obligations on states to prevent mass migrations and it does not consider the impact on or provide for burden-sharing among states disproportionately affected by refugees. States have adopted different interpretations of the definition of 'refugee', which has led to disparate treatment of refugees in different areas.

Although the Refugee Convention ideally should be updated to address these problems, it would be impossible to reach consensus on a replacement. Many states would undoubtedly want to limit their obligations to refugees rather than expand them. As a result, the Office of the UN High Commissioner for Refugees has opposed any effort to reform the existing Convention. In recent years, states have instead attempted to reach agreement on non-binding agreements, including a Global Compact for Migration and a Global Compact on Refugees, which were adopted by the UN General Assembly in 2018. Although these agreements have important provisions that attempt to balance humanitarian concerns with state sovereignty, some key refugee-receiving states, including Australia and the US, have declined to sign one or both of them.

The inability to reach consensus even on non-binding principles augurs badly in the near term for formal international-law solutions to the problem of refugees. As with the Geneva Conventions, rather than attempt to reach an agreement among all governments (as the Global Compacts attempted to do), it may be more effective for a group of especially affected states (those that accept a large number of refugees or provide significant financial support for refugee operations, for example) to attempt to agree on principles and policies they are willing to apply that go beyond the requirements of the Refugee Convention, including policies on burden-sharing.

Cyber security and cyber conflict

Most governments and legal experts agree that existing international legal frameworks, including the UN Charter and the law of armed conflict, apply in cyberspace. There is much less agreement, however, about what that means in practice. Salient questions include how to characterise cyber attacks under international law and attribute responsibility; what actions states may take in response to cyber attacks; how rules to protect civilians and civilian property in armed conflicts apply; what rules govern cyber operations that fall well below the threshold of armed conflict; and how international law regulates cyber activities that transit the networks of states that have not consented to such use of their networks.

Although most states share an interest in preventing hostile cyber activities, negotiating new treaties is difficult because the major international players have different capabilities, vulnerabilities, political systems, and particular interests in certain offensive and defensive cyber activities. Whereas democracies generally favour a free and open flow of information, politically closed systems define cyber threats quite differently and seek to control the flow of information. Divergent visions for the internet, uncertainty about how technology will evolve and the challenges of verifying agreements in this area all stymie development of international rules.

The only major cyber-security-specific treaty is the 2004 Council of Europe Convention on Cybercrime. This agreement is limited in scope and participation. It seeks only to harmonise parties' criminal law and procedure to facilitate prosecutorial cooperation against individuals who commit crimes in cyberspace. More than 60 states are now party, and many others have adopted substantive and procedural cyber-crime laws that are consistent with its provisions. However, several major international players – including China, India and Russia – are not members, and Russia is now championing an initiative to create an alternative cyber-crime treaty. The Convention does not tackle issues – like intelligence activities – at the core of national security; indeed, it contains exceptions for them.

Instead of negotiating treaties, most of the international legal work regarding cyber conflict and states' core cyber-security interests has emphasised interpreting established bodies of international law (such as rules regarding the use of force) to fit activities in cyberspace. Those interpretation efforts include multilateral forums, non-governmental initiatives and unilateral declarations.

In 2018, the UN General Assembly created two separate forums for states to discuss these issues: a Group of Governmental Experts from select states and a more inclusive Open-Ended Working Group, both of which started convening in late 2019. A previous Group of Governmental Experts had made some important progress in establishing that international law applies in cyberspace and the imperative for negotiating additional non-binding norms. That process ended acrimoniously in 2017 when state members failed to reach agreement on basic principles governing states' rights to respond to cyber attacks. These forums are important for airing and debating views, but they are unlikely to result in consensus on many key issues.

In parallel with these multilateral initiatives, the NATO-accredited cyber-defence centre based in Estonia has also sponsored an effort by international-law experts to recommend interpretations of existing international law in cyber contexts. The resulting 'Tallinn Manuals' are useful for identifying issues and a range of possible interpretations of existing international law. However, they do not necessarily reflect the views of the most influential states, and it remains to be seen how the Manuals' interpretations will match actual practice.

In recent years, several states – including Australia, France, the Netherlands, the UK and the US – have publicly outlined their views of how existing international law applies to cyber operations. These statements are very general because states want to preserve flexibility and are reluctant to resolve tough hypothetical questions in advance. Yet these statements – and similar pronouncements by other states that may follow – are important because they help reinforce interpretations of existing legal frameworks for cyber activities, promote transparency of cyber operations and likely responses to them, and serve as the basis for further diplomatic dialogue.

In addition to translating existing legal frameworks to cyberspace, states should also continue working towards the agreement of non-binding norms. For example, the US has pushed for pledges against attacking critical infrastructure or impeding emergency response to cyber instances. It has also worked with close allies to jointly and publicly attribute cyber attacks to states, in order to hold them accountable. Such efforts can help bolster deterrence and may lead to binding agreements.

Space conflict and activities

The international-law challenges regulating the use of outer space are similar to those of cyberspace. States are increasingly reliant on space systems for military, economic and other societal activities. The most technologically advanced states view space as a key war-fighting domain in future conflicts. And expanding activities in space, including commercial uses and resource exploitation, may result in international friction.

In December 2019, the US created US Space Force as a new military branch, and other states have in recent years reorganised their armed forces to better address future conflict in space. India conducted a successful anti-satellite-missile test in early 2019, while China and Russia have pressed ahead in developing capabilities to target space-based assets. Several states – including France and Japan – have published defence-strategy documents in the past year placing greater emphasis on space operations, and they have also enhanced allied cooperation in this area. Amid these developments, there are many (sometimes competing) efforts to develop new international rules, or at least non-binding guidelines, for space activities, while space technology is moving and spreading quickly.

It is widely agreed that general international law, including the UN Charter and the international law of armed conflict, applies to space activities. But what this means in practice remains uncertain. There is little space-specific treaty law, and new binding agreements are unlikely in the foreseeable future. The most important treaty, the Outer Space Treaty of 1967, dates back to the Cold War space race. It outlines some broad principles, including a call that space be used for peaceful purposes, but it contains few hard prohibitions on military activities. For example, the treaty prohibits placing nuclear weapons or other WMD in space, but neither it nor other space treaties bar other weapons that may be used in space, from space or against targets in space.

In recent decades states have proposed various types of agreement to regulate the responsible use of space from either a security or a safety and sustainability perspective. But none has gained broad support among major powers. China and Russia have proposed a binding treaty that would restrict some types of space weaponisation. The US and others have objected to this proposal as unverifiable and a diplomatic ploy as its sponsors press ahead with their own space-weapons programmes. The European Union proposed a non-binding code of conduct to limit certain types of space weapons. That effort did not garner wide consensus either.

Like the Tallinn Manuals for cyberspace, alongside these state efforts there are ongoing efforts led by academic institutions and composed of non-governmental experts to lay down how extant international law applies to the military uses of space. Again, although these processes are useful for identifying issues and possible legal interpretations, and may be influential in shaping some states' views, they do not necessarily reflect the views of powerful states, and the proposed manuals that result may not match actual practice.

With space capabilities proliferating worldwide, one immediate priority should be to develop voluntary international guidelines to reduce accidents and promote stability in space. As with cyberspace, states could also work towards a set of non-binding norms, such as prohibitions against targeting certain types of assets in peacetime and understandings of what constitutes provocative or dangerous behaviour in space.

These efforts would help mitigate friction, build mutual confidence and perhaps lead to formal agreements.

Paths forward

When gaps in international law become apparent over time, states and other stakeholders must decide whether and how to address them. As noted above, it is often impossible or impractical for states to try to revise and update major multilateral treaties. It is also rarely possible even to forge non-binding agreements among all states. Instead, it can be more effective for states to address international legal gaps or ambiguities in other ways, such as through agreements among smaller groups of states on non-binding principles, best practices or interpretations of vague treaty terms.

No single formula or process is appropriate in all cases. Some key questions and trade-offs include participation (widely inclusive, perhaps including international organisations or civil-society representatives in addition to states, versus tightly exclusive and limited to like-minded or similarly affected states), proposed output (general frameworks versus detailed rules or guidelines) and obligation (legally binding commitments versus political pledges).

For example, in 2012, a group of 24 governments and five international organisations that had participated in the multi-year, Danish-led 'Copenhagen Process on the Handling of Detainees in International Military Operations' endorsed 16 non-binding principles and good practices for states and international organisations that detain persons in the course of military or peacekeeping operations. Similarly, the ICRC and the Swiss government sponsored an initiative that led in 2008 to the 'Montreux Document' governing the use of private military and security firms in war zones. The document sets forth 70 best practices that have now been endorsed by 56 governments and all major private security firms. Both initiatives were criticised by advocacy organisations that were not included in the discussions. Nevertheless, these kinds of non-binding 'soft law' agreements among groups of like-minded or specially affected states can help align practices among participating states and strengthen political pressure on others to conform. Over time such agreements may attract more participating states or develop into binding 'hard law' agreements.

Progress towards filling gaps or addressing ambiguities can also be achieved in other ways, such as through public statements of legal interpretations and best practices by states, and even unilateral statements of principles that a state is prepared to support as customary international law.

No amount of diplomacy can address all the gaps in international law. Gaps and disagreements of interpretation are inevitable because international agreements cannot be designed to anticipate all future circumstances, including changes in technology. Often, state negotiators must accept some gaps and uncertainties in order to reach consensus on other questions. Moreover, even if states can agree on how to adapt or supplement international law, the usual challenges of ensuring compliance are exacerbated when a state's security or other core interests are threatened.

Despite these limitations, international law remains integral to promoting international stability and balancing security imperatives with other values. At a time when the liberal international order is fraying, redoubled diplomacy and leadership in addressing gaps and responding to emergent threats is needed to update the international legal system and ensure its continued vitality.

Prospects for Arms Control
Will the future resemble the past?

The year to June 2020 was particularly grim for arms control. On 2 August 2019, the United States officially withdrew from the 1987 Intermediate-Range Nuclear Forces (INF) Treaty. And on 21 May 2020, US President Donald Trump announced plans to withdraw from the 1992 Open Skies Treaty. The US and Russia continued to accuse each other of violating old and newer agreements. Unless the 2010 New Strategic Arms Reduction Treaty (New START) is extended before it expires on 5 February 2021, the world is approaching an era without arms control.

But even as existing agreements break down, states are also attempting to build new arms-control mechanisms. The US is pursuing trilateral arms control with China and Russia, and in April 2020 appointed a special envoy, Marshall Billingslea, to lead this work. Beijing has declined to engage thus far. For China and Russia, missile defences will be an essential component of future arms-control efforts, which appears unlikely given US strategic postures. Experts are calling for arms control to incorporate emerging technologies such as hypersonics and cyber. Amid these bleak prospects for arms control between the biggest nuclear powers, Europe – quite literally caught in the middle – is exploring opportunities to play a greater leadership role in reviving arms control, though perhaps in a new form.

These recent trends raise deeper questions about the future of arms control. Does it have a future and what might that look like, particularly with regard to increasing geopolitical competition? What challenges do emerging technologies present for arms control, particularly around definitions and verification? Finally, can states make progress on arms control, or should preparations be made for a post-arms-control world?

Objectives of arms control

Arms control can be defined as the management of weapons to reduce the risks associated with their use. This does not always entail the reduction or elimination of weapons – arms control is not disarmament. Rather, it involves reciprocal agreements to limit the development, possession or use of weapons in order to increase the transparency and predictability of a relationship. Arms control is a product of its geopolitical context, and typically accompanies a broader easing in tensions or a rapprochement. It is most often associated with US–Soviet and US–Russia bilateral nuclear/strategic agreements but can also include multilateral treaties such as the 1968 Nuclear Non-Proliferation Treaty (NPT), which now has 188 signatories. But arms control does not always take the form of treaties: it can also include informal agreements and confidence-building measures (CBMs), such as the 1991 Presidential Nuclear Initiatives (PNIs).

Why do states engage in arms control? During the Cold War, nuclear experts developed the concept of 'strategic stability', which incorporated concepts of crisis stability and arms-race stability. Strategic stability could mean that, despite their rivalry, neither the US nor the Soviet Union had an incentive to use nuclear weapons first, escalate a crisis to the point where these might be used or increase the numbers they deployed.

Changes in the context of strategic stability, such as new technologies, produce corresponding changes in the requirements for arms control. Hypersonic missiles, for example, reduce warning time and might inhibit a state's ability to retaliate against a nuclear first strike. Alternatively, non-kinetic technologies such as cyber or artificial intelligence (AI) could impact nuclear command and control, undermining the credibility of a nuclear deterrent. These technologies will increase pressure on both crisis escalation and incentives for arms racing.

Arms control also has an economic and cost-saving role. During the Cold War, some experts, such as Bernard Brodie, argued that the economic benefits were more significant. Since the US and Soviet Union continued to engage in geopolitical competition, the security benefits of arms control were by contrast fleeting. The US and Soviet Union (and then Russia) have reduced their arsenals by nearly 80% since the height of the Cold War in the late 1980s. This has reduced nuclear-related defence spending as a proportion of national income, thus easing its burden on the economy.

A final objective of arms control is ethical. Arms-control efforts not only reduce the likelihood of war by encouraging trust and cooperation between states, but also have the potential to limit the scale of destruction in the event of war. Arms control might include, for example, restrictions on weapons designs that mitigate the humanitarian consequences of their use. This has motivated recent arms-control and disarmament efforts, such as a ban on blinding lasers and efforts to ban landmines.

Waves of arms-control successes and failures

The first major achievements in arms-control efforts were driven by the experience of the 1962 Cuban Missile Crisis, which instilled in both superpowers a deeper realisation of the very real risks of a nuclear crisis escalating. Concerns were also growing about the economic costs of the ongoing arms race, and the environmental and humanitarian damage caused by extensive nuclear testing. These trends led to a series of agreements that culminated in unprecedented arms-control activity for a decade. The most notable agreements included the 1963 Partial Nuclear Test-Ban Treaty, the 1968 NPT, the 1972 Biological Weapons Convention and the 1972 Strategic Arms Limitation Talks (SALT) Interim Agreement and the Anti-Ballistic Missile (ABM) Treaty.

Following the decline of superpower detente in the late 1970s, no real progress was made in arms control until Mikhail Gorbachev became Soviet leader and began to transform Soviet foreign policy. The 1987 INF Treaty eliminated ground-launched missiles with a range of 500–5,500 kilometres. In 1990, 22 countries from NATO and the former Soviet bloc signed the Conventional Armed Forces in Europe (CFE) Treaty, which set limits on conventional weapons and set the 'sufficiency' rule requiring any state to limit their armaments in relation to all other European states. In 1991, the US and Soviet Union concluded the Strategic Arms Reduction Treaty (START), which dramatically cut strategic delivery vehicles, nuclear warheads and missile throw weight. All three treaties included unprecedented verification measures, including on-site inspections.

Since the end of the Cold War, arms control has had a mixed record, with some treaties failing, new ones emerging and others remaining in limbo. In 2002, the US withdrew from the ABM Treaty but also concluded a new strategic arms-limitation agreement with Russia, the Moscow Treaty, which neither explicitly included verification nor required the elimination of weapons. Rather, it committed Washington and Moscow to reduce their warhead count to between 1,700 and 2,200 by the end of 2012. During the 2000s, Russia's record of compliance with arms-control agreements worsened. It stopped observing the CFE Treaty in 2007 and military leaders began to openly discuss withdrawal from the INF Treaty as early as 2008. These trends suggested a possible end to traditional treaty-based strategic arms control.

The US attempt to 'reset' relations with Russia early in the Obama presidency gave hope of the renewal of arms control. This resulted in the 2010 New START Treaty. The US and Russia agreed to continue START-style verification, albeit with fewer and less-intrusive inspections, and to maintain strategic arsenals of 1,500 warheads and 800 delivery vehicles. At the time, however, experts and many US senators acknowledged that it was likely the 'last of its kind'. The US wanted future arms-control agreements to incorporate Russian non-strategic nuclear weapons with shorter ranges because of Russia's numerical superiority in these categories. Conversely, Russia insisted that future agreements must include missile defence, multiple actors and/or advanced conventional weapons. Indeed, bilateral strategic arms control was already at risk before the collapse of the INF Treaty.

At the end of the Cold War there was also a rise in *multilateral* efforts to control nuclear weapons and materials. Amid, and immediately following, the collapse of the Soviet Union, the US and other states were concerned about the security of Soviet nuclear weapons. The US launched the Nunn–Lugar Cooperative Threat Reduction Program in 1991 to secure and dismantle the nuclear weapons and materials of Russia and other former Soviet states. Nearly all states joined the Comprehensive Nuclear-Test-Ban Treaty (CTBT) and Chemical Weapons Convention in 1996 and 1997 respectively. Other multilateral agreements included the 2011 reissuing of the Vienna Document between members of the Organization for Security and Co-operation in Europe to promote transparency and confidence, and the 2015 Joint Comprehensive Plan of Action (JCPOA) between the five permanent members of the UN Security Council, Germany and Iran, which set limits on Iran's nuclear production under international verification.

While the first two decades following the end of the Cold War included numerous bilateral and international successes in arms control (with the exception of US withdrawal from the ABM Treaty), in recent years numerous Cold War and post-Cold War agreements have come under threat or dissolved. These include US withdrawal from the JCPOA in 2018 and from the INF Treaty in 2019 following alleged Russian violations.

Arms control state of play, 2020

At least four major trends present challenges for a future arms-control agenda: competing interests among the major powers; the breakdown of existing agreements; new technologies; and international pressure for nuclear reductions.

Firstly, if arms control is a product of the geopolitical environment and great-power relations, the current competition between China, Russia and the US does not bode well for arms control based on the traditional conceptualisation of strategic stability. The US and its European allies now face two major military powers that are hostile to them and are more interested in pursuing strategic gains than in cooperation. Eventually, this competition may come to a head, either through a dangerous crisis such as the Cuban Missile Crisis or an economically unsustainable arms race, prompting Washington, Moscow and/or Beijing to seek a new detente, which could include arms control. For the time being, however, geopolitical competition presents a significant challenge for the future of arms control.

Secondly, the breakdown of existing agreements not only diminishes trust between states, exacerbating geopolitical tensions, but also undermines the *practice* of arms control. If one state can cheat without repercussions, or another may withdraw from a treaty at minimal cost, how reliable is arms control as a tool for strategic stability? The demise of the INF and Open Skies treaties leaves a limited number of remaining agreements. The New START Treaty is due to expire on 5 February 2021. The end of this agreement with no follow-on would mean the end of on-site inspections, regular data exchanges and consultations between Russia and the US. New START has the option of a five-year extension, but this too faces challenges. The US has said that it hopes to expand the treaty to include China. Beijing has repeatedly stated, as recently as mid-2020, that it will join arms-control agreements if the US first reduces its own arsenal, which seems unlikely by the time New START expires. Another treaty at risk is the CTBT. The Trump administration has claimed that China and Russia have violated the agreement, and reports have suggested that the administration is considering a return to nuclear testing. The demise of remaining arms-control agreements could undermine international support for future arms-control efforts.

A third trend to watch is the impact of emerging technology on strategic stability and arms control. China, Russia, the US and potentially other states are developing high-speed hypersonic missiles that could potentially carry conventional and nuclear warheads. These weapons could evade missile defences and shorten decision-making time, upsetting crisis stability. Other technologies, such as cyber and AI, present an entirely new set of challenges. Predicting the impact of many of these technologies on strategic stability is difficult. For example, the term 'artificial intelligence' can cover a wide variety of applications from drone targeting to pattern recognition in intelligence. Understanding the impact of these technologies on strategic stability requires further research and a greater degree of sophistication than simple calls to 'ban killer robots'. Are they offensive or defensive? Stabilising or destabilising? These definitional challenges are crucial to the future of arms control. Furthermore, many cyber activities are difficult to attribute and would therefore present a challenge to future arms-control regimes and verification efforts.

A final challenge will be meeting the rising expectations of many non-nuclear states and civil society to deliver an ambitious arms-control agenda, particularly one that demonstrates tangible progress towards nuclear disarmament. As arms control becomes more difficult, the wider international community beyond the nine nuclear possessor states is increasing the pressure to do more. Much of this pressure is due to frustration with a lack of progress towards nuclear disarmament. This was manifested in the 2017 Treaty on the Prohibition of Nuclear Weapons (TPNW). As of mid-2020 this had 40 members. It requires 50 to enter into force and may achieve this in 2021. The COVID-19 crisis and increasing focus on human security is likely to put pressure on defence spending, including on nuclear weapons.

The future of arms control

With these challenges in mind, what might the future of arms control look like? Are there realistic opportunities for further agreements, or should governments prepare for a post-arms-control world? There are three major scenarios. Some are long-term endeavours, while others could be achieved in a matter of weeks. They are neither mutually exclusive nor exhaustive, but they do set out new approaches that will be needed if arms control is to have a future.

In Scenario One, states would commit to preserving existing agreements. Specifically, Russia and the US would immediately extend New START either for the full five-year option or on a year-by-year basis with an annual review process. The US would commit to remain in the Open Skies Treaty. America's allies, particularly in Europe, would regularly engage with Washington to express their national interests in preserving arms control. Remaining agreements would not be taken for granted. There would be concerted international efforts to recommit Russia and the US to the practice of arms control.

In Scenario Two, existing strategic arms control would expand to include China in a New START follow-on. This would be done separately from attempts to extend New START, as Beijing is unlikely to agree to join any extension for the reasons mentioned above. One option for trilateral arms control is a multistage approach to arms control. In the first stage, Russia and the US would agree to an overall ceiling for the number of all delivery vehicles, both conventional and nuclear, to include hypersonic missiles. In the second stage, once the two largest nuclear powers reached a certain level, such as 600 delivery vehicles, China would join and agree to limit its own delivery vehicles and participate in a verification regime. Another option for trilateral arms control would be ratios on ceilings of nuclear-capable delivery vehicles. For example, Washington, Moscow and Beijing might agree that only 10% of their nuclear delivery vehicles would be hypersonic missiles.

This scenario would help manage the destabilising effects of hypersonic missiles and strengthen crisis stability by improving transparency and predictability. However, it would be difficult to develop verification mechanisms capable of distinguishing nuclear- from conventional-armed delivery vehicles, including hypersonic ones. Furthermore, China might still be reluctant to be join such an agreement, particularly one involving verification, if Russia and the US did not also reduce their warhead numbers.

While Scenarios One and Two would represent progress against the status quo, neither of them addresses the major priorities of the three biggest nuclear powers. The US wants a future agreement to include shorter-range Russian systems, including those that fall within the ranges of the INF Treaty. Moscow and Beijing would likely insist that missile-defence limitations be included in any future agreements. As ambitious as these two scenarios might seem, they would not resolve what all sides perceive to be the primary challenges to strategic stability.

Scenario Three would offer a broader approach to arms control with a focus on multilateral initiatives and CBMs, rather than verifiable treaties between a relatively limited number of parties. To some extent, this scenario is already playing out, with efforts by the United Nations and state groupings to develop 'rules of the road' to strengthen strategic stability, particularly with regard to emerging technologies. For example, in 2016 the UN established a Group of Governmental Experts on Lethal Autonomous Weapons Systems that is exploring and developing recommendations for various emerging technologies. Europe is already emerging as a leader in proposing new approaches to arms control. In March 2019, Germany launched an initiative on exploring opportunities for arms control and emerging technologies. Similar efforts should be encouraged to generate new ideas and solutions to technical challenges in arms control, such as verification.

If arms control is to have a future, it must not only sustain existing agreements and practices but also address the asymmetries created by emerging technology and multipolar geopolitical competition. While arms control in the past was typically the business mainly of the US and Soviet Union (and then Russia), the rest of the world is increasingly involved in its future.

Future agreements will have to involve China and other nuclear actors to promote strategic stability, although this will be a long-term project. But non-nuclear states such as Germany and Sweden are playing an increasing role in arms control. For these countries, however, arms control is not necessarily about strategic stability but rather is part of the pathway to nuclear disarmament. This aligns with the fourth trend mentioned above, whereby international pressure is increasing on major powers to reduce their nuclear arsenals and demonstrate progress towards disarmament. In the absence of leadership by the major powers towards arms control, this new approach could gain traction.

There is, of course, one more scenario: a post-arms-control world. In this world, the major powers would reject cooperative efforts to strengthen strategic stability and engage in an unbridled arms race. Some multilateral agreements, such as the NPT, might remain. But the core objectives of arms control would be at risk. Does arms control have a future? Yes, but it requires immediate action, and might look very different from arms control of the past.

Climate Change: Year of Reckoning, Postponed

COVID-19 deferred action – but did it create an opportunity?

The UN Climate Action Summit in New York in September 2019 capped what has been called 'the year the world woke up to climate change'. As more than six million people took part in a week of coordinated protests in 185 countries, 16-year-old Swedish campaigner Greta Thunberg chastised an audience of heads of state and high-level policymakers for their failure to act swiftly and decisively on climate change.

Yet the grim truth, which she said was too uncomfortable for national leaders to confront, was not new: 2020 was widely acknowledged as perhaps the last chance to close the gap between current policies and those needed to avoid disastrous global warming. To have even a possibility of success, national targets for cuts in greenhouse-gas (GHG) emissions would have to be increased by the time of the annual UN Conference of the Parties climate summit (COP26) in Glasgow in November 2020. But the COVID-19 pandemic changed the landscape. It postponed some urgent decisions (and the summit itself) but led to unplanned short-term emissions cuts. It also created the opportunity, if it could be seized, for a sustained green recovery that might improve prospects for averting dangerous climate change.

Elusive targets

The global response to the threat of climate change centres on the 2015 Paris Agreement, which aims to keep the global average temperature increase to well below 2°C, and ideally no more than 1.5°C, over pre-industrial levels. Each party (194 countries and the EU) submitted a formal emissions-reductions target ('Nationally Determined Contribution' or NDC) for 2025 or beyond; NDCs were to be replaced or updated by the end of 2020 (at COP26) and then every five years thereafter, reflecting increasing ambitions and extending the time frame first to 2030 and then 2035 and so on. As of July 2020, only eight countries (collectively accounting for less than 0.5% of global GHG emissions) had submitted new targets. A further seven countries, which together generate nearly a quarter of global GHG emissions, had indicated that they would not do so. All but one, however, were conforming to the letter, if not the spirit, of Paris, since their original NDCs ran until 2030. The exception was the United States, which had already announced that it would formally withdraw from the Paris Agreement in November 2020.

In November 2017 the United Nations Environment Programme (UNEP) concluded that the level of ambition reflected in the 2020 NDCs was likely to determine whether the 2°C target could be achieved, since the initial NDCs covered only about a third of the reductions necessary by 2030. In December 2019 the UNEP reported that an additional 24% reduction in emissions, beyond those agreed in the collective NDCs, would be needed by 2030 to meet the 2°C target, and a 54% reduction for the 1.5°C target. Moreover, seven G20 countries, accounting for nearly a quarter of global emissions, were not on target to achieve even their modest NDCs. The Climate Action Tracker consortium (CAT) estimated in December 2019 that the current NDCs would leave a 90% chance of exceeding 2°C warming. For current policies the chance was 97%, with a 50% chance that warming would be greater than 3°C.

In November 2015, just before the Paris Agreement was concluded, the UNEP had calculated that the gap between NDCs and greenhouse-gas reductions consistent with 2°C warming was 12 gigatonnes carbon dioxide equivalent ($GtCO_2e$). Global efforts have since failed to reduce this gap: the gap in emissions reduction consistent with 1.5°C warming leapt from 15 to 29 $GtCO_2e$ between 2015 and 2019. The UNEP reported that countries would have to increase their NDC ambitions for 2020–30 threefold to limit warming to well below 2°C and more than fivefold for the 1.5°C goal.

The fierce urgency of now

Hand in hand with the growing urgency of climate action, the year to June 2020 saw a convergence of trends in climate-change awareness and activism from the grassroots to corporate boardrooms. In September

2019 the third Global Climate Strike, timed to coincide with the UN summit in New York, was the second-largest protest event in history, surpassed only by the 2003 demonstrations against the war in Iraq. It grew out of the movement inspired by Thunberg, whose solo 'School Strike for Climate' in 2018 had gone viral. Citing the widening emissions gap, she told the summit: 'How dare you continue to look away and come here saying that you're doing enough, when the politics and solutions needed are still nowhere in sight … if you choose to fail us, I say: we will never forgive you.' In December 2019 she addressed the UN Climate Summit (COP25) in Madrid and in January 2020 she spoke to the World Economic Forum in Davos – the second year in a row she had been invited to both.

Critics dismissed Thunberg as an uninformed and naive teenager. But she was not wrong about the facts, and the protests she inspired reflected a shift in global public opinion on climate change. According to the 2018 Pew Global Attitudes Survey of 26 countries, 67% of respondents viewed climate change as a major threat, up from 63% in 2017 and 56% in 2013. In half of the countries, climate change was seen as the top international threat. New temperature records and extreme-weather events made the reality and severity of climate change increasingly difficult to ignore. The decade to 2019 saw eight of the ten hottest years on record, with 2019 ranking second only to 2016, and by the mid-year point 2020 threatened to be even warmer. Once-extreme weather was becoming the 'new normal', and new extremes appeared in turn. During the southern summer of 2019–20, the news was dominated by images of catastrophic fires in Australia that destroyed 186,000 square kilometres of bush and nearly 6,000 buildings. Record-breaking wildfires in Siberia and elsewhere in the Arctic in summer 2019 were followed in 2020 by the hottest spring on record, breaking the symbolic barrier of 100°F (38°C) for the first time.

The climate strikers, and organisations such as the global activist group Extinction Rebellion, were acting in part out of a sense of frustration and powerlessness in the face of what they saw as policymakers' foot-dragging. But others were voting with their wallets. The number of multinational corporations committing to 100% renewable energy by 2050 through membership of the 2014 RE100 initiative rose by a third to more than 200 in 2019, with 44% of that growth from the Asia-Pacific. Half the member companies reported they would be using their influence with policymakers and utilities to overcome barriers and push for a faster transition to a net-zero-carbon world, and three-quarters aimed to be 100% renewable by 2030. Such efforts – whether under the aegis of RE100, similar groupings, or made unilaterally – stimulate the market for green energy and green tech.

Some RE100 members launched their own clean-energy venture-capital funds, such as Amazon's US$2 billion Climate Pledge fund announced in June 2020. Overall global investment in green energy rose only fractionally in 2019, but the combined total of venture-capital and private-equity investment rose 6% to reach its highest level since 2010, after a 35% leap in 2018. The bulk of this investment is in expanded generating (and to a lesser degree manufacturing) capacity, with little devoted to start-ups, indicating that the green economy has achieved a degree of maturity after some growing pains in the first half of the decade. Around 40% of private-equity firms intend to increase investment in renewable energy and climate-friendly products or services over the next five years. At the New York Climate Action Summit, 14 investment firms and funds committed to setting science-based emissions targets consistent with 1.5°C warming and 38 (with some overlap) committed in principle to net-zero emissions by 2050. Much of the funding for Extinction Rebellion and the climate-strike movement has come from high-net-worth individuals and associated charities. The Climate Emergency Fund launched by three US philanthropists in July 2019 raised US$600,000 for activist groups.

Danger and opportunity

These trends were disrupted by the global spread of COVID-19 in the first quarter of 2020 and the resulting global recession. In March the World Health Organization declared COVID-19 a pandemic, and on 1 April COP26 was postponed indefinitely. On 28 May the summit was rescheduled to November 2021. The drop in economic activity caused by lockdowns around the world in response to the pandemic also brought a severe, if short-lived, drop in greenhouse-gas emissions, about half of which came from a decline in land transportation. In China, where the virus broke out and the first lockdowns were imposed, emissions fell by 25% in February – and global emissions fell by 6% as a result. By mid-March, as the economy recovered, China's emissions were down 18% year on year. But by mid-year, month-on-month emissions had not just recovered but were 4–5% higher than 2019, as 'brown' industries such as coal power and cement production recovered faster than the economy as a whole. China dropped its formal energy-intensity reduction target in May, but kept the carbon-intensity target that is central to its Paris NDC.

Other countries appeared to be following this pattern as COVID-19 hit them in turn. By early April, when countries accounting for 89% of global emissions were in some form of lockdown, total emissions over a 24-hour period were estimated to have fallen to 17% below the daily average during 2019, although by June they had increased again to only 5% below the 2019 daily average. With lockdown restrictions still in place in many countries at mid-year, it appeared that total global emissions during 2020 would be 5–8% down on the 2019 total, with the US and Europe showing above-average declines. In 2009, by comparison, during the global financial crisis, emissions in developed countries fell by 7–12% but were offset by large increases in China, India and South Korea, so emissions fell by only 1.3% globally and recovered within a year.

The COVID-19 pandemic also blunted the edge of climate activism. Coordinated global protests on a scale similar to those of 2019, planned for April 2020, were moved online, where they had little impact on their intended audience. When protesters returned to the streets in May and June, the issue was racism, not climate change. Yet in an Ipsos MORI poll in April, covering 14 countries, 71% of respondents said that climate change was as serious a crisis as COVID-19, and 65% said it was important that any post-pandemic economic stimulus should prioritise climate change. Even in the US, where climate scepticism remains strongest, there was majority support for both propositions.

The emissions reductions in the first half of 2020 came as part of a severe economic shock yet still fell short of the annual cuts that would have to be sustained for a decade to keep warming below 1.5°C. It was a stark demonstration that the inevitable disruption involved in achieving a net-zero-carbon world could be catastrophic if not properly paced or managed. As countries debated policies to revitalise their economies, international organisations such as the World Bank and International Energy Agency (IEA) were quick to call for plans that put climate-change goals at the centre of the recovery.

Green shoots

Of the US$3.3 trillion in fiscal and monetary stimulus packages disbursed globally in 2008–09, US$522bn (16%) went to green measures. Of the US$7.3trn in expenditures and guarantees allocated by G20 countries in response to COVID-19 by the end of April 2020, only around 4% was green, 4% was 'brown' (expected to increase net emissions) and the rest was 'colourless' or neutral. However, this initial tranche of support was meant to stop the bleeding more than start the recovery, and it was clear that significant further investment would be necessary. By mid-year the IMF estimated G20 recovery spending at US$9trn, of which 1–2% was green (and nine-tenths of the green component was from the EU, its member states and the United Kingdom). If all G20 states were to match the biggest stimulus packages as a percentage of GDP, the global total would easily double. If the green proportion of such a stimulus were roughly the same as in 2008–09, it would meet or exceed the US$2.1–4.7trn in annual energy-sector investment needed, according to the UNEP, to be consistent with climate-change goals. The sustainable recovery plan proposed by the IEA in June 2020, in contrast, called for US$1trn per year for the energy sector up to 2023, which would reduce annual emissions by 4.3GtCO_2e by 2025. The UNEP figure includes diversion of investment from brown to green energy as well as additional annual investment of US$220bn–1.67trn. Such investment, however, would have to continue each year for decades after the rest of the short-term recovery spending ends.

A climate-friendly recovery from the COVID-19 shock will require not just investment in green energy and green tech, but also avoidance of new subsidies for climate-unfriendly industries that would lock in the 'business as usual' high-carbon economy. Any stimulus measures should be colourless at least. But with the exceptions of the EU, France and the UK, at mid-2020 most countries' stimulus measures were skewed towards brown.

The economic and social disruption produced by the COVID-19 pandemic opens a window for fundamental changes to the global economy – and for shifts in values, norms, culture and world views that are recognised as preconditions for meeting the Paris goals and achieving a net-zero-carbon world by 2050. Some changes might be counterproductive, for example if there is a behavioural shift away from public to private transportation even after social distancing ends. Others, such as changing work habits and a decrease in air travel, could accelerate decarbonisation.

More concretely, most observers agree that the key to the necessary energy transformation is a robust system of carbon pricing, whether through carbon taxes (favoured by economists) or cap and trade (favoured by politicians). By the end of 2019, 46 countries had established some form of carbon pricing, covering about 20% of global emissions, but only 5% of emissions had been priced at levels consistent with the Paris goals. Coverage is set to expand (by mid-2020, countries accounting for 55% of emissions had included carbon pricing as an element of their NDCs), but only clear, near-universal and sufficiently high carbon prices can ensure that investment shifts quickly enough from high- to low-carbon energy.

Public opinion in most countries has opposed enacting carbon taxes, and where they have been introduced they have often been resisted, for example as part of the *gilets jaunes* (yellow vests) protests in France. But carbon taxes have proved more popular when they have included 'climate dividends' in the form of rebates to those least able to bear the financial burden. Low-carbon renewable energy has become increasingly competitive, and a robust carbon price as part of the COVID-19 recovery would both lock in this emerging advantage and prevent the demand-driven decrease in fossil-fuel prices from undermining recent gains. Some mechanism of this kind would be an essential part of a truly green stimulus package.

The path ahead

The breathing space offered by the drop in emissions in the first half of 2020 risked being offset by the pause in international climate negotiations as key meetings were postponed and governments focused on immediate pandemic responses and economic relief efforts, often at the expense of climate goals. With 2021 becoming the new 2020, a number of landmarks loomed.

G20 countries account for around 75% of greenhouse-gas emissions, and the group has served as a parallel track to the UN process for climate-change diplomacy almost from the start. Climate was high on the agenda for the G20 summit in Riyadh in November 2020, although at mid-year it was unclear whether or how the pandemic might affect the meeting, and Saudi Arabia's 'circular carbon economy' concept and proposals had yet to gain much traction. In 2021, however, Italy is set to assume the G20 presidency and looks likely to put climate change front and centre, while the UK – an international leader on climate issues – will be taking over the G7 presidency and hosting the postponed COP26. The two countries have been working together towards a seamless approach to climate leadership and negotiations.

The delay to COP26, moreover, means that the process might be revitalised by a change of administration and policy in the US. Washington's withdrawal from the Paris Agreement, effectively in 2018 and formally in November 2020, left a major gap in both policy and leadership. The Democratic candidate for president, former US vice president Joe Biden, signalled that, if he were elected, the withdrawal would be reversed on 'day one' and the US would come back with renewed vigour, enhanced ambition and dynamic leadership.

Besides being the deadline for new or revised NDCs with enhanced ambition for the next stage of the Paris process, COP26 was to be the last effective chance to agree the final part of the detailed Paris 'rulebook' covering Article 6, which provides mechanisms for creating an international carbon market. This had been a key goal of previous COPs in Poland and Spain in 2018 and 2019, but proved elusive. Agreement on Article 6 rules would open the door to an integrated system that would subsume existing national and transnational carbon taxes and cap-and-trade schemes, a vital ingredient for success.

Finally, postponement of COP26 meant that it would now coincide with the release of the sixth and final assessment report (6AR) from the International Panel on Climate Change, embodying the global scientific consensus on climate-change causes, impacts, adaptation, vulnerability and mitigation. Previous assessment reports, published at five- to seven-year intervals since 1990, had formed the basis for every major international climate agreement – the Kyoto Protocol, Copenhagen Accord and Paris Agreement – and underpinned the 2°C and 1.5°C warming targets.

It is not a cliché to say that 2020–21 will be a make-or-break period for climate-change policy. Against the backdrop of COVID-19, initiatives from civil society, businesses and local and regional governments will be critical to achieve even the modest current NDCs and to influence national governments to adopt green-recovery policies, ratchet up their ambitions to decarbonise and make the Paris Agreement work. Time, as Greta Thunberg told world leaders in New York, is running out.

CHAPTER 5

North America

NORTH AMERICA

Drivers of Strategic Change

REGIONAL SHARE OF GLOBAL POPULATION, GDP AND DEFENCE BUDGET

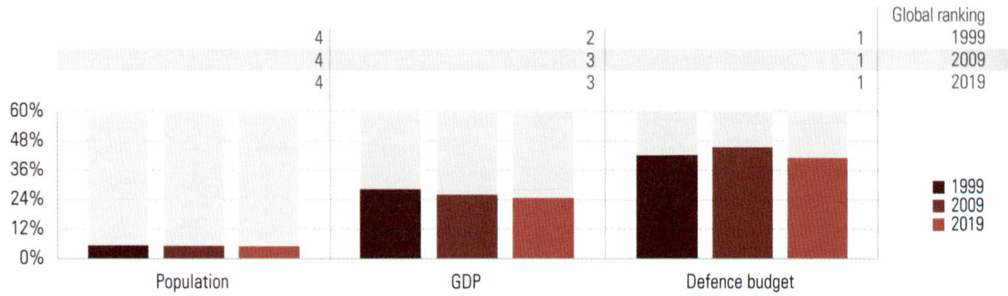

POPULATION

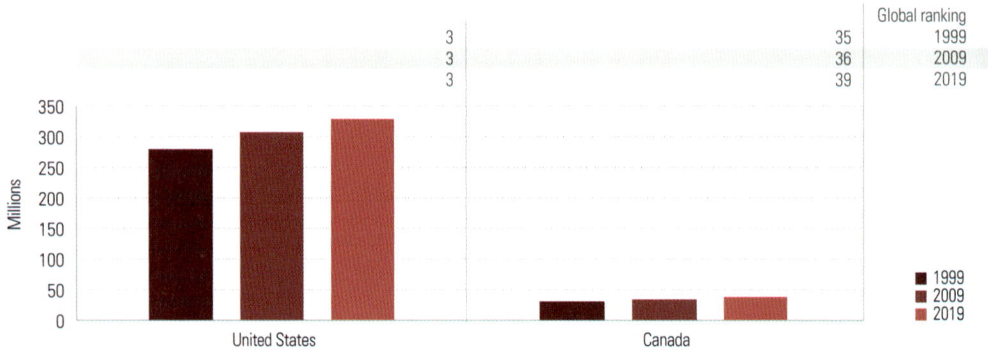

AGE STRUCTURE
(Percentage of national population)

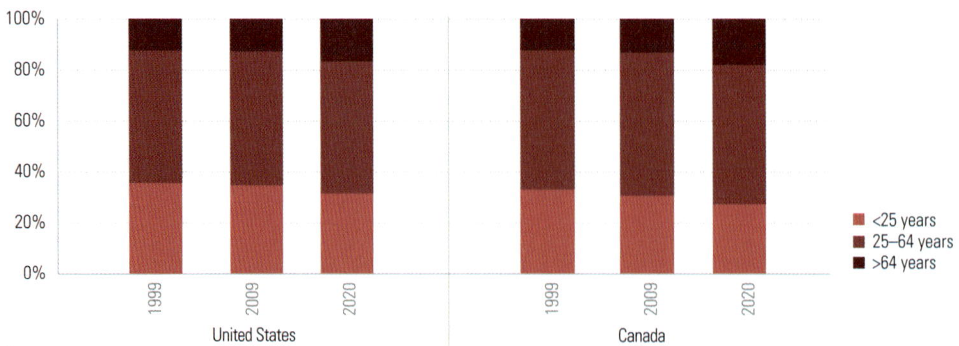

Median age	United States	Canada
2000	35.2	36.8
2010	36.9	39.6
2020	38.3	41.1

GDP

(Constant 2010 US dollars)

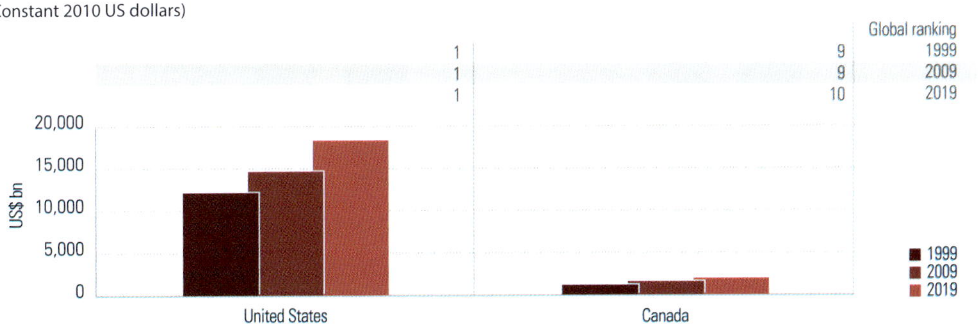

GDP PER CAPITA

(Constant 2010 US dollars)

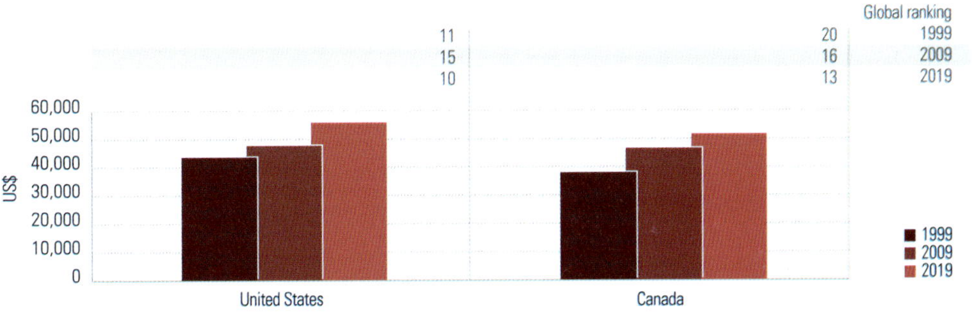

DEFENCE BUDGET

(Constant 2015 US dollars)

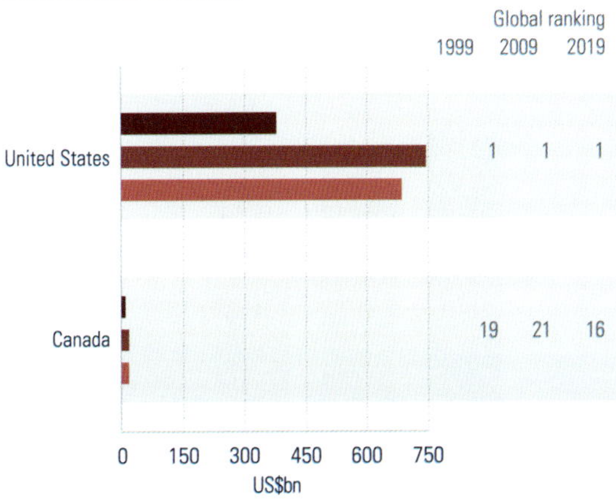

ACTIVE MILITARY PERSONNEL

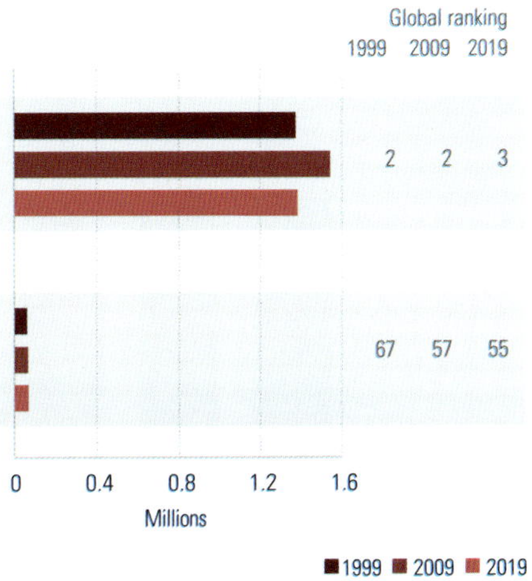

For explanation of drivers and sources, see page 7

HUMAN DEVELOPMENT INDEX (HDI)

(Score between 0 and 1, where 0 denotes a low level of development and 1 a high level of development)

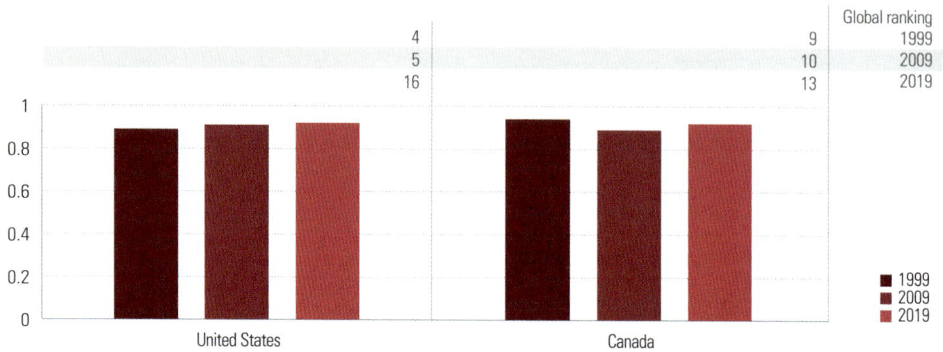

POLITICAL SYSTEM

(Score between 0 and 100, where 0 denotes no political freedom and 100 fully free)

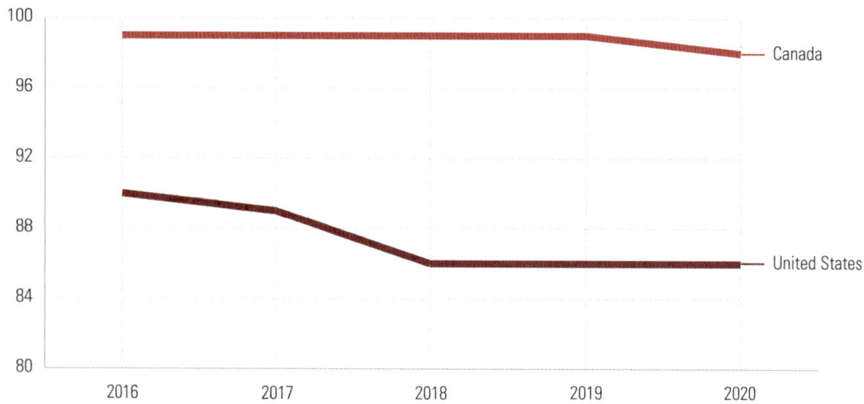

TRUST IN INSTITUTIONS

(Average level of trust)

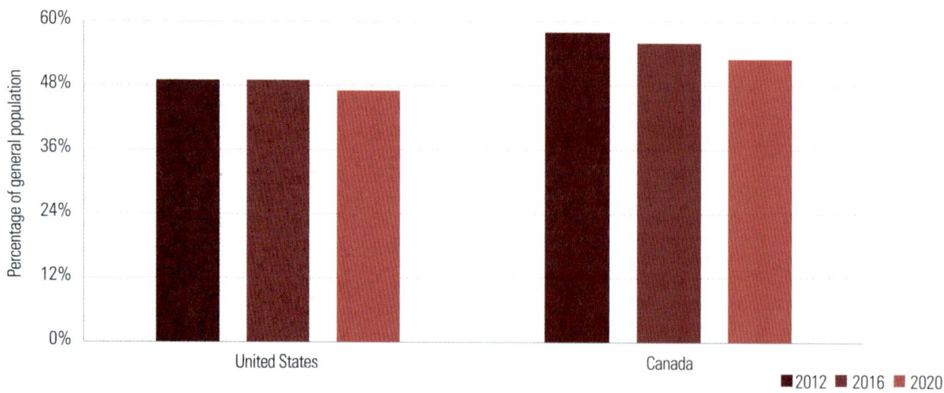

For explanation of drivers and sources, see page 7

2019–20 Review

The impeachment of United States President Donald Trump dominated domestic politics in the US during the latter half of 2019 and early 2020. Initially his acquittal seemed to illustrate his remarkable resilience to the sort of political setbacks that might have ended previous presidencies. But this triumph was short-lived. Soon after, the COVID-19 pandemic began to pummel the US. The pandemic's political, economic and social consequences dealt Trump a significant setback, as did his response to widespread protests against racial injustice that erupted in May 2020.

In international affairs, Trump's desire to confront Iran while avoiding all-out war contributed to a serious but inconclusive military skirmish in early 2020. The COVID-19 crisis aggravated US–Chinese relations and highlighted the degree to which transatlantic relations had deteriorated over the previous three years. While Trump's behaviour in both domestic and international spheres could appear erratic and impulsive, certain key coordinates in his worldview remained fixed: hostility towards immigration, an electoral strategy based on polarisation and culture war, and suspicion or even animosity towards traditional US allies.

US domestic politics

The Democrats' capture of the House of Representatives in the 2018 midterm elections had made Trump's impeachment a practical possibility. Yet Speaker of the House Nancy Pelosi had resisted such demands, maintaining that impeaching Trump would serve only to mobilise his supporters in the 2020 presidential election. Trump's actions in the summer of 2019, however, forced her hand. On 18 December, the House of Representatives approved two articles of impeachment against Trump, making him only the third president in US history to be impeached.

The impeachment case rested on a phone call Trump made on 25 July to Ukrainian President Volodymyr Zelensky. Trump wanted Zelensky to announce that Ukraine would investigate baseless allegations that the Democrats had been involved in malfeasance during the 2016 presidential election and that former vice president (and likely Democratic candidate in the next presidential election) Joe Biden had engaged in corrupt practices relating to his son Hunter's business activities in Ukraine. If he did so, Zelensky would be invited to the White House. By making these demands, Trump sought to tar his political opponents with similar allegations to those made against him: of receiving foreign assistance during the 2016 election and corruptly enriching his family.

Soon after the Trump–Zelensky phone call, a White House official ordered that US$250 million in military aid to Ukraine be withheld, in line with a command issued by Trump earlier that summer to place a hold on a total of US$391m in such aid. On 12 August, a CIA official assigned to the White House filed a 'whistle-blower' report claiming that he had 'received information from multiple US Government officials that the President of the United States is using the power of his office to solicit interference from a foreign country in the 2020 US election'. The House launched a formal impeachment investigation into Trump's actions on 24 September.

While there was little dispute over the essential facts of the case, the interpretation and portrayal of these facts differed starkly along partisan lines. While the White House transcript of Trump's call with Zelensky plainly illustrated the president's desire for Ukraine to announce an investigation into his political rivals, Trump and his allies denied that there had been a 'quid pro quo'. On 17 October, his acting chief of staff, Mick Mulvaney, admitted that the reason that US$391m in military aid had been held back was, at least in part, to encourage Ukraine to investigate the false claim that it, rather than Russia, had hacked the Democratic National Committee in 2016, and that the imposition of such conditions was a regular feature in US foreign policy. Although Mulvaney later retracted his remarks, they were typical of a line of argument often advanced by Trump's supporters: that furthering presidential electoral interests was a legitimate goal of statecraft. Indeed, days after the House investigation began, Trump openly called upon both Ukraine and China to investigate Biden and his son.

Since Trump's conviction would require a two-thirds majority in the Republican-controlled Senate, the outcome of the trial that began on 16 January 2020 was never in real doubt. On 5 February, the Senate voted to acquit Trump of abuse of power by 52–48, and of obstruction of Congress by 53–47. Only one Republican senator, Mitt Romney of Utah, voted to convict Trump on the first charge. Trump and his allies presented the impeachment process as an attempt to overturn his electoral victory and his acquittal as vindication. The following months saw the administration replace perceived enemies or obstacles within the so-called 'deep state': firing the inspectors-general of five government departments; dismissing officials who had testified about their concerns over Trump's Ukraine policy; and installing a vocal Trump loyalist, Richard Grenell (previously ambassador to Berlin), as acting director of national intelligence (DNI).

As an incumbent presiding over a strong economy, Trump had some reason to feel confident about his re-election prospects. The consequences of the COVID-19 pandemic, however, would soon shake his administration. Trump reportedly ignored warnings from the intelligence community of COVID-19's impending threat and publicly downplayed its risks to the US, in late February claiming that 'like a miracle, it will disappear'. Despite advance warning of the disease's arrival on US territory, there were crucial missteps in national preparation, including a faulty Centers for Disease Control and Prevention-designed COVID-19 test and shortages of personal protective equipment. By the end of June 2020, the US had suffered more than 2.5m confirmed cases of COVID-19 and almost 120,000 deaths, the highest absolute numbers of any country.

The pandemic brought economic cataclysm. By late March, the S&P 500 index had lost almost a third of its value. While it recovered much of these losses by mid-June, this did not reflect a recovery of the COVID-19-ravaged real economy. In April 2020, the US unemployment rate hit 14.7%, the highest on record since the 1940s. American workers endured the pandemic with less direct economic support from the federal government than their European counterparts, and were further disadvantaged by US public-health policy. US federalism shaped the response to the pandemic, with states adopting different policies towards the outbreak. In return for federal assistance, President Trump demanded public praise from Democratic state governors, excerpts of which were later used in his re-election advertisements, and sought to undermine lockdowns in such states by encouraging protests by local armed groups. Consistent with his long-held anti-immigration views, Trump temporarily restricted legal immigration, supposedly to protect American workers during the crisis. Eager to revive the faltering economy, Trump made various fantastical suggestions to address the pandemic, ranging from promoting unproven medication to musing on whether injecting bleach or exposing internal organs to UV light might cure COVID-19 patients.

Along with COVID-19's human and economic toll, developments in Democratic politics narrowed Trump's re-election prospects. Senator Bernie Sanders, the iconoclastic outsider who had run Hillary Clinton close for the Democratic nomination in 2016, enjoyed early success in the Democratic primary elections but was unable to counteract Biden's decisive lead among African Americans and women. Sanders conceded on 8 April, and Biden secured sufficient delegates to be the Democratic presidential nominee on 5 June. Biden was essentially a restoration candidate seeking to appeal to those eager to return to pre-2016 norms. Yet the COVID-19 crisis encouraged him to consider more radical policies. As of June 2020, it remained unclear how many of Sanders's voters, typically younger than most Biden voters, would support Biden in the general election. Nonetheless, Trump now faced the more testing prospect of an opponent who consistently out-polled him in key states, often by significant margins.

It seemed likely that Trump would continue his strategy of emphasising division and fomenting the culture war to mobilise his electoral base. In this light, the large-scale protests that began in May 2020 after the police killing of an unarmed black man, George Floyd, which inspired a broader movement against racial inequality, seemed likely to feature heavily in Trump's re-election strategy. The 'Battle of Lafayette Square' on 1 June, in which mainly peaceful protesters were dispersed violently in order to allow Trump to walk from the White House and pose by a nearby church with Bible in hand, was intended to appeal to his key constituencies, including the Christian right. Yet it may also have rallied opposition to the president. By that stage, such opposition included several former senior military leaders and members of Trump's administration, who warned of Trump's authoritarian and divisive style of government, and of the risk of deploying active members of the armed forces against protesters.

US foreign policy

Much of Trump's first term was marked by hostile relations with China, the most notable example being the trade war that he launched in 2018. Increased hostility towards Beijing was one of the few bipartisan trends following Trump's election. While the US and China signed a preliminary deal in January 2020 that sought to alleviate their trade dispute, the pandemic prompted a rapid re-escalation of tensions. Trump and Secretary of State Mike Pompeo took to referring to COVID-19 as the 'China virus' or 'Wuhan virus'. In May 2020, the US tightened export controls against Huawei, the Chinese technology company whose international expansion had prompted significant concern, and announced plans to cancel the visas of thousands of Chinese students in the US with alleged links to the People's Liberation Army. This re-escalation reflected long-held views of influential members of the Trump administration concerning Chinese enmity towards the US. It also reflected a more partisan calculation that making the Chinese Communist Party the focal point of US resentment for the pandemic might help Trump's re-election.

Another consistent feature of Trump's first term had been his determination to pressure Iran to concede a 'better deal' over development of its nuclear programme than the Joint Comprehensive Plan of Action (JCPOA), negotiated under President Barack Obama, from which Trump withdrew the US in 2018. Reluctant to become embroiled in a fresh Middle Eastern war, Trump did not retaliate overtly against various Iranian military provocations in the Gulf in the summer of 2019. John Bolton, Trump's pugnacious national security advisor and a long-time advocate of regime change in Tehran, left the administration in September 2019, reportedly in part owing to Trump's disapproval of Bolton's readiness to consider military solutions. The swift escalation in December–January that almost brought Washington and Tehran to war was therefore unexpected. A rocket attack in Iraq killed a US military contractor; a US retaliatory strike killed 25 Iranian-sponsored Iraqi militiamen. Washington accused Iran of orchestrating the subsequent storming of parts of the US embassy compound in Baghdad on 31 December. Three days later, a US airstrike at Baghdad airport killed General Qasem Soleimani, the leader of Iran's Revolutionary Guard Corps Quds Force, along with a senior Iraqi leader of pro-Iranian militias. While Soleimani's killing signalled a sharp departure from Trump's previous caution, Washington was keen not to escalate the confrontation further, and did not respond militarily to a limited Iranian retaliatory missile attack.

Trump has long been sceptical of multilateral agreements and commitments to allies, portraying them as equivalent to exploitation by foreigners. This outlook was manifest in Washington's absence from its traditional leadership role and reluctance to engage in multilateral cooperation during the COVID-19 crisis, as well as its withdrawal from the Intermediate-Range Nuclear Forces (INF) Treaty – announced in 2018 and completed in August 2019 – and the Open Skies Treaty, announced in May 2020. While all three of the signatories of a successor treaty to NAFTA, the United States–Mexico–Canada Agreement (USMCA), ratified it in 2019–20, the Trump administration's general suspicion towards free trade was visible in its obstruction of the functioning of the World Trade Organisation (WTO). Since 2018, Washington had prevented new judges from being appointed to the WTO's Appellate Body; when the terms of two of the three existing judges ended in December 2019, the WTO was deprived of its means to resolve disputes.

More specifically, Trump demonstrated his continued indifference, or even hostility, towards traditional notions of transatlantic solidarity. The Ukraine episode had demonstrated that Trump was willing to risk the security of a US partner fighting Russian aggression in order to attack a domestic political opponent. While reports in March 2020 that the Trump administration had attempted to purchase a German pharmaceutical firm to secure exclusive access to a potential COVID-19 vaccine were later denied by the firm itself, the fact that they were initially confirmed and condemned by the German government was symbolic of diminishing transatlantic trust and goodwill. More concretely, it was revealed in

June 2020 that the Trump administration planned to withdraw 9,500 US troops from Germany, a reduction of about 25%, and permanently limit forces there to 25,000.

Canada

Canada's general election on 21 October 2019 saw Prime Minister Justin Trudeau lose his governing majority in the House of Commons. Trudeau's popularity had been damaged by a story that had emerged earlier in 2019, which seemed to implicate him in efforts to shield a major Quebec-based company from corruption charges, along with the appearance of photographs of him as a young man in blackface, which complicated his trademark public advocacy of racial and social justice. Trudeau decided that his Liberal Party would govern as a minority government, rather than seek a formal coalition. Given that the average lifespan of a Canadian minority government is less than two years, Trudeau faced a formidable challenge to keep his government alive. Yet his approach to the COVID-19 pandemic, which, in contrast to that adopted by the US federal government, emphasised safety and robust measures against the virus, was popular. By April 2020, Trudeau's approval rating had leapt 20 points to 54%, with 62% approving of his pandemic policies. Despite this, Trudeau faced significant future challenges, not least because lower oil prices compounded the impact of COVID-19 on Canada's budget.

In foreign relations, Canada's dispute with China rumbled on. The dispute began in late 2018 when Canada detained Meng Wanzhou, a senior executive of Huawei and daughter of its founder, on a US warrant. Soon afterwards, Beijing charged two Canadians, including a former diplomat, with espionage, and imposed non-tariff barriers on various Canadian exports. In May 2020, the Supreme Court of British Columbia ruled that Meng's extradition to the US could take place. Canada's security review of Huawei's role in its 5G networks was ongoing as of the end of June 2020. The three biggest Canadian telecommunications firms signed agreements with Huawei rivals to develop their 5G networks, effectively excluding Huawei from Canadian networks.

US–China Relations
No way back?

During the first half of 2020 relations between the United States and China declined so precipitously as to raise the question of whether the two countries could re-establish any kind of modus vivendi – and, if not, what the consequences would be. The immediate cause of the decline was the COVID-19 virus, which had the effect of exacerbating already substantial tensions driven by China's increasingly assertive international posture and the United States' reaction. What looked like becoming a classic confrontation between a rising and a dominant global power had implications for the entire planet amid growing talk in Washington of a new cold war.

Relations between the US and China had never been straightforward. They could best be described by a Chinese phrase that literally translates as 'same bed, different dreams'. During the 1990s and early 2000s it had suited both parties to downplay these differences. The US was enjoying its unipolar moment of unchallenged dominance in all forms of power, while China, focused on economic development, behaved as a status quo power that was happy to adhere to the US-led global order and free-ride on the global public goods that US hegemony supplied, such as peace and stability in the Asia-Pacific region and an open trading and investment order. Particularly following China's admission to the World Trade Organisation (WTO) in 2001, much US manufacturing capacity was relocated to China to take advantage of low labour costs and tax holidays, enabling US consumers to benefit from a flood of cheap manufactured goods. The proceeds from these export sales were invested in US Treasury bonds, providing American consumers with access to cheap money and funding the US federal deficit.

The downsides of this process were the hollowing-out of domestic US manufacturing and the associated unemployment – between 1999 and 2010 the US lost six million manufacturing jobs – together with social dislocation and the distorting effect on the global financial system of a savings glut generated by China. These were overlooked until the 2008 financial crisis made them no longer possible to ignore. A much wealthier China also began to question its adherence to the US-led global order. When, in 2012, Chinese Communist Party (CCP) General Secretary Xi Jinping proclaimed his vision of the China Dream, it became evident that the US was dealing with a country that was no longer content to play second fiddle. China began to promote its own world view – a community of common destiny for mankind, a combination of Marxist and Confucian ideologies that amounts in essence to a Chinese-dominated world order – and to obtain international validation for its authoritarian model of governance. Its rapidly expanding high-tech sector began to challenge US dominance in areas such as fifth-generation mobile telephony (5G) and even the internet itself, with China promoting an alternative, top-down and state-controlled architecture. Meanwhile, China's growing military capabilities and claims to sovereignty over the South China Sea challenged US military dominance in the Asia-Pacific region.

Trade war

The dam broke in July 2018 when US President Donald Trump announced the imposition of tariffs of 25% on US$50 billion of Chinese imports, subsequently rising to US$370bn by the end of 2019. Trump's actions represented the culmination of growing exasperation, which had begun to crystallise towards the end of the Barack Obama administration, with what many perceived as unfair Chinese trading practices. The charge sheet against China included pervasive intellectual-property (IP) theft, the forced transfer of technology as a condition for US firms to operate in China, failure to live up to promises to open its markets in areas such as financial services and insurance, and levels of state subsidies to Chinese industries that far exceeded what was permitted under WTO rules.

Trump's actions caught China by surprise. They led to prolonged negotiations that culminated on 13 December 2019 in the signing of a 'phase one' trade deal. This committed China to purchasing an additional US$200bn of US agricultural produce, energy, finished goods and services over the period 2020–21. It also committed China to implementing enhanced protections for US IP and to open up its financial sector to

US companies – secure in the knowledge that US companies would have no prospect of competing with Chinese counterparts in the retail sector. Tariffs already imposed on Chinese imports remained in force pending satisfaction that the terms of the deal were being honoured. The deal did not, however, achieve a key US objective of forcing China to abandon its market-socialist economic system in favour of a market economy and open trading system.

The trade negotiation acted as a catalyst for the release of long-pent-up differences of ideology, values, and approaches to global governance and geopolitical power. Rather than contain these, the phase-one deal simply served to exacerbate them. The situation was further complicated by differences of outlook and tactics within each of the negotiating parties. Within the US camp, Trump, who had made resetting the commercial relationship with China a key plank of his election campaign and whose instincts were isolationist and protectionist, adopted a purely transactional approach, seemingly concerned only to achieve a deal that would enhance his re-election prospects. Trump's focus on the economic aspect of US–China relations was broadly supported by Trade Representative Robert Lighthizer, who had been the main force behind demands for fundamental changes to China's economic system, and to a degree by Treasury Secretary Steven Mnuchin.

By contrast, the China hawks, who included Vice President Mike Pence, Commerce Secretary Wilbur Ross, Trump's trade adviser Peter Navarro and Deputy National Security Advisor Matt Pottinger, took a much harder line that advocated restrictions on American use of Chinese technology and foreign investments, a robust US military posture in the Asia-Pacific and a partial economic decoupling from China. In speeches to the Hudson Institute in October 2018 and to the Wilson Institute a year later, Pence set out a comprehensive list of US grievances with China.

In the face of mixed, though predominantly hawkish, messages from the top levels of the administration, US government agencies concluded that it was open season on China, with each pursuing specific China-related grievances. This led former treasury secretary Hank Paulson to observe in August 2019 that the US had 'a China attitude, not a China policy'. China became one of the few issues on which a bitterly divided US Congress was able to achieve bi-partisan accord, with members striving to outdo each other in drafting anti-China legislation, most of which had no prospect of being passed during what was left of the legislative period.

But some legislation was adopted. The 2019 Hong Kong Human Rights and Democracy Act required the executive branch to confirm that Hong Kong enjoyed a high degree of autonomy. If it could not do so, the US would impose sanctions. The March 2020 Taiwan Allies International Protection and Enhancement Initiative sought to reverse China's efforts to constrict Taiwan's ability to act internationally as a separate national entity. The Uyghur Human Rights Policy Act and the Hong Kong Autonomy Act, both passed in 2020, imposed sanctions on individuals and entities perceived to have violated the rights of Xinjiang's Uighurs and Hong Kong residents respectively.

A perception took hold among US policymakers and intellectuals that the policy of engagement with China that had begun with president Richard Nixon's 1972 visit was now exhausted. Far from having produced a more liberal China integrated into the US-led international order, engagement had, it was argued, empowered the rise of an authoritarian and ideologically hostile regime. However, this perception was not uniformly shared by the US business community, in particular among technology companies which relied on China for a significant proportion of their sales of advanced semi-conductors and software, and for access to low-cost manufacturing.

For China, the trade war came as an unwelcome surprise. The Chinese party-state realised that the days of double-digit export-led growth had come to an end. It made sustained efforts to move towards an economy based on domestic demand. But Xi made clear in his 2018 speech to Davos that China remained committed to a globalised trading system. In 2019 exports still accounted for 18.4% of China's GDP. As trade negotiations progressed, some Chinese intellectual and policy circles expressed concern that China risked challenging the US before it was ready to do so, and should seek to accommodate US concerns while focusing on further reform and opening up its economy. These assertions were forcefully rebutted in China's state-controlled media. Throughout 2019 Xi made a series of speeches emphasising that China would face complex domestic and international challenges, and calling for the Chinese people to embrace a spirit of struggle – including, implicitly, against the US.

Huawei

Seen from Beijing, the US appeared determined to frustrate China's rise. The trade dispute was one aspect. Arguably more significant was the effort to contain China's technology ambitions – starting with Huawei, the telecommunications national champion seeking to become a global leader in 5G. Huawei had invested heavily in researching and developing 5G, filing numerous patents and taking an active role in international negotiations of technical standards. US companies were unable to manufacture all the components needed to put together 5G networks, nor did they feel incentivised to act as systems integrators in the way that Huawei, bolstered by substantial state subsidies, had done.

However, Huawei remained dependent for its 5G networks on advanced microprocessors, which relied in turn on US software and technology that were subject to US export controls. On 16 May 2019 Huawei was added to the US Department of Commerce Entity List, effectively preventing US companies from buying telecommunications equipment from, or selling it to, Huawei. On 15 May 2020 the US government further bore down on Huawei by barring semi-conductor manufacturers that used US technology and software in microchip design from shipping their products to Huawei without government permission. It had previously banned federal agencies from purchasing Huawei equipment, a decision that had been the subject of an ultimately unsuccessful lawsuit filed by the Chinese company in a Texas court in March 2019.

The US also launched prosecutions against Huawei for breaches of sanctions on Iran and for IP theft, and put pressure on allies to exclude Huawei from their 5G networks – including the UK, which in May 2020 reversed a decision it had taken four months earlier to allow limited Huawei participation. Huawei was not the only Chinese technology company to be targeted: the US also sanctioned Hikvision, a global leader in video-surveillance technologies whose cameras were widely installed outside US federal-agency and Defense Department buildings.

The pandemic

The COVID-19 outbreak that began in Wuhan in December 2019 proved a decisive moment in US–China relations. Initially, local officials sought to cover up, then downplay, the extent of the crisis, intimidating into silence local health workers who sought to raise the alarm. By the time the central government instituted a lockdown to contain the outbreak on 23 January 2020, a large number of Wuhan residents had already left the city for other domestic and international destinations, with the result that the virus spread and became a global pandemic. Thereafter, China's containment of the virus was remarkably successful, as were efforts to sequence the COVID-19 genome, which led to the early development of a nucleic-acid test.

The US not only failed to lead the global response to the pandemic but failed egregiously to deal with it within its own borders. In an attempt to distract attention from these shortcomings, Trump blamed China for its initial concealment of the outbreak and took to referring to COVID-19 as 'the China virus', 'the Wuhan virus' and 'kung flu'. He suggested that China should be held to account both through an independent international investigation into the origins of the virus and by seeking financial reparation for the economic dislocation it had caused. Consideration was given to stripping China of its sovereign immunity, suing the CCP rather than the Chinese state, and even to refusing to honour interest payments on Chinese holdings of US Treasury bonds, ostensibly in recompense of pre-1949 debts that the People's Republic of China government had refused to honour.

China interpreted the stumbling US response to the coronavirus as further evidence of the CCP's long-held conviction that America was a power in terminal decline. Beijing let loose its new breed of 'wolf warrior' diplomats, who responded to US attacks with a disinformation campaign suggesting that COVID-19 had originated in a US military laboratory. China also highlighted its actions in supplying medical equipment and teams of doctors to a range of countries, including some in the developed world. The result was a public slanging match, in the course of which US Secretary of State Mike Pompeo was subjected to highly personal attacks in Chinese state media.

From bad to worse

Relations deteriorated in other areas. Tit-for-tat visa denials for journalists began in February 2020 when China expelled three journalists from the *Wall Street Journal* in retaliation for an op-ed entitled 'China is the Real Sick Man of Asia'. Harkening back to the late nineteenth century, when China was routinely portrayed in Western media as irremediably backward, the article was highly neuralgic for Beijing. The US responded by requiring Chinese media organisations to register as foreign agents and to significantly reduce their staff.

Higher education was also affected. It began to suffer in 2018, when the US government reduced from five years to one year the duration of visas for Chinese graduate students wishing to study aviation, robotics and advanced manufacturing. The US also made it harder for foreign students – not just Chinese – to work in the country after graduation, and subjected university science-and-technology links with Chinese institutions to greater scrutiny. The FBI arrested and prosecuted a number of US-based Chinese – and American – academics for espionage. Bioscience and healthcare became key battlegrounds. In the words of one analyst, Luz Ding, 'since the National Institutes of Health launched a campaign to root out foreign influence on federally funded research in August 2018, universities in the US have investigated nearly 200 scientists. Many of these scientists are well-funded and established researchers who are ethnically Chinese.' In June 2020 FBI Director Christopher Wray revealed that his agency was undertaking over 2,000 investigations into economic espionage and covert political influence.

Meanwhile, it became evident that China's armed forces were exploiting US disarray to push their operational boundaries by increasing military flights over Taiwan airspace and continuing naval patrols and exercises in the South China Sea. China established new facilities on two militarised reefs in the South China Sea and intensified its exploitation of the sea's resources, establishing a record for the amount of gas extracted from the seabed in the course of a single day. In April 2020 a Vietnamese trawler was sunk following a collision with a Chinese coastguard vessel in contested waters around the Paracel Islands. From April onwards China picked fights with an increasing number of states, apparently based on the perception that, like it or not, its economic and political influence was so strong that other states would have to get used to dealing with China on its terms.

In the face of a faltering Trump re-election campaign, the Republican party decided to go all out in attacking China, leading to a further barrage of anti-Chinese comments. In May Deputy National Security Advisor Pottinger used a speech delivered in Chinese at an event commemorating the 101st anniversary of the May Fourth Movement to issue an appeal for China to adopt universal values. He emphasised the contribution of veteran Chinese diplomat P.C. Chang to the 1948 Universal Declaration of Human Rights, which, in the words of China historian John Pomfret, 'married Western belief in the primacy of the individual with Chinese concern for the greater good'. Few if any Chinese citizens would have heard the speech. But Pottinger's advocacy of universal values was anathema for Beijing and would have been seen as an unequivocal statement of intent to pursue regime change. On 18 June Secretary of State Pompeo met State Councillor Yang Jiechi in what was billed as an effort to prevent further deterioration. But the two sides merely set out their respective positions. By then it had become clear that, for the hawks in the Trump administration, the key issue was ideological.

Meanwhile, faced with the prospect of a second Trump term or a Biden presidency, Beijing appeared to decide that Trump would be the better option because of his more transactional approach and lack of ideological conviction. Within Chinese leadership circles the joke went that Trump's isolationist instincts and short-term self-interest had done more to further China's interests than China could have done itself. Beijing appeared to have calculated that a Biden presidency would be more focused and structured in its approach to China, better placed to reinvigorate US alliances and infused by a strain of moralism absent from the Trump administration. But neither Biden nor Trump as president is an attractive prospect for the CCP.

Outlook

The economic decoupling of China and the United States had been under way for some time, as rising costs and a more restrictive environment led foreign businesses – including major US technology companies such as Apple, Dell and Hewlett-Packard – to relocate some China-based manufacturing to cheaper locations. But COVID-19 revealed how much China had become a single point of failure for some global supply chains, in particular medications and medical equipment. Trump trade adviser Navarro argued that the US needed to re-establish its ability to manufacture these products, while Director of the National Economic Council Larry Kudlow suggested that the US offer to fund all relocation costs of US companies moving their operations back to the country from China. Decoupling was also taking place in the other direction. Chinese investment in the US dropped to almost zero in 2019–20, while Chinese companies listed there began to relocate to Hong Kong to avoid having to comply with US auditing rules.

By mid-2020 the two countries faced a collapse in their relations. These had fallen to their worst state since Nixon's visit to China in 1972. An American China policy based on pragmatic realism had given way to ideological animus, fundamental distrust and arguments for a complete severing of ties. Beijing, though anxious to avoid an outright confrontation, showed no readiness to rethink its efforts to achieve global pre-eminence within the time frame set by Xi. In Chinese intellectual circles there is concern that China may have shown its hand too early, thereby enabling the US to orchestrate a pushback. Such a response will likely prove effective if the US can recover its diminished willingness to exercise global leadership and a reputation for competence badly tarnished by its response to the COVID-19 pandemic.

In the words of Yuan Peng, President of the China Institutes of Contemporary International Relations, the highly respected research institute of the Chinese Ministry of State Security, 'in the next three to five years, the international scene will be a jumble of non-polarity, warring states and transitions, and the difficulty of great-power cooperation will clearly increase'. It has begun to look as if there may well be no way back for US–China relations.

Trump and the Transatlantic Relationship
How far has US policy towards Europe changed?

Speaking in the White House Rose Garden on 24 June 2020, United States President Donald Trump announced that his administration would be drawing down troops stationed in Germany and redeploying some to other European countries. Trump's guest at the event was Polish President Andrzej Duda, who faced the first round of his re-election battle in Poland just four days after their meeting. Duda used the event to underscore the strength of Polish–American relations under his leadership. Trump sought to highlight the fact that other European countries were less favoured by his administration.

Trump made clear in his remarks that the redeployment of American troops was punishment for Germany's 'delinquency' in meeting its obligation to increase its military outlays to 2% of gross domestic product. Trump stressed that Poland would receive troops because its government was willing to pay for them. 'Other places in Europe' would be in line to receive US troops as well, but only if they lived up to 'their monetary obligations' as Poland had.

Few in Europe were overly surprised by the decision, although the timing of the announcement was unexpected. Nevertheless, many worried that the move would weaken NATO's operational readiness to respond to a threat from Russia – particularly given that the bulk of the US 2nd Cavalry Regiment would be redeployed back home. A few also argued that this latest affront to German Chancellor Angela Merkel chipped away at the solidarity underpinning both the Atlantic Alliance and the transatlantic relationship.

At a minimum, the move added momentum to French President Emmanuel Macron's efforts to elaborate a coherent European security and defence identity. Since Trump attended his first NATO summit in May 2017, Merkel had openly expressed concern that the US could no longer be trusted to come to the defence of Europe. From July 2017, Macron tried to convince Merkel to work with France in building institutions and policies that would not only give Europe what he called greater 'strategic autonomy' but could also constitute a kind of 'European sovereignty' – extending beyond security and defence and into manufacturing, trade and finance. These efforts laid the foundation for a historic meeting of minds between France and Germany during the COVID-19 pandemic in May 2020.

Has the Trump administration widened a rift between the US and Europe that a future president would be unable to bridge? It is tempting to see the Trump presidency as just another in a long series of 'political differences' between the two sides of the Atlantic. This was the view expressed by NATO Secretary General Jens Stoltenberg at a press conference with Trump on the eve of the 70th anniversary NATO summit in December 2019. Stoltenberg went on to insist that 'the paradox is that despite … some political differences, we are always able to agree and unite around our core task to stay together'. This time may be different.

An imperfect past

Assessment of the significance of the changes wrought by Trump requires a baseline for comparison. US Secretary of State Mike Pompeo underscored this point in his speech entitled 'The West is Winning' to the Munich Security Conference in February 2020. Pompeo argued that 'the death of the transatlantic alliance is grossly over-exaggerated'. So long as NATO remained committed to 'national sovereignty', promoted 'free enterprise' and defended 'human rights', it would be easy to distinguish 'the West' from 'illiberal alternatives'. In this sense, Pompeo argued, the Alliance remained essentially undiminished. But this did not mean unchanged, and the baseline Pompeo set – of the West in opposition to illiberal alternatives – may now be less clearly delineated than at any time since the end of the Second World War.

When the Soviet Union ceased to be a security threat at the end of the Cold War, a range of new issues faced the Atlantic Alliance. These included the stabilisation of societies undergoing democratic transition, the rise of Islamic fundamentalism, the resurgence of Russian authoritarianism and the emergence of China as a strategic competitor. NATO itself began to enlarge. It had to adjust to a greater and more diverse membership, a broader area of operations and a wider spectrum of security threats, including cyber and hybrid warfare. Not all member states adapted at the same pace, contributed the same resources or embraced the wider agenda to the same degree.

As a result, defence spending, force readiness and the willingness to use force in NATO operations varied across member states. This generated tensions in the Alliance at critical moments, notably the bombing of Serbia in 1999, the US-led invasion of Iraq in 2003, the Russian invasion of Georgia in 2008, the NATO-led air campaign against Libya in 2011 and policy towards Syria since 2013. Such events raised important questions about whether NATO should act without UN authorisation or beyond the European theatre; how far NATO should expand or extend security guarantees to non-members; and how effectively NATO could act without US leadership or with the support of a US administration that preferred to 'lead from behind'.

Most importantly, these factors raised questions about burden-sharing within the Alliance. US defense secretary Robert Gates made this explicit in his 2011 valedictory address to NATO. Reflecting on the uneven involvement of European allies in Libya, Gates noted: 'many of those allies sitting on the sidelines do so not because they do not want to participate, but simply because they can't. The military capabilities simply aren't there.' He pointed to 'the very real possibility of collective military irrelevance'. And he cautioned about the limited patience of the American people 'to expend increasingly precious funds on behalf of nations that are apparently unwilling to devote the necessary resources … to be serious and capable partners' for the US.

Transatlantic relations have long faced challenges in economics as well as security. Difficulties arose on both sides of the Atlantic in negotiating large multilateral trade agreements such as the Uruguay Round (successfully concluded in 1994) and Doha Round (launched in 2001, so far without success) talks. In 2002 the George W. Bush administration imposed punitive steel tariffs, but later withdrew them in the face of European retaliation. The Barack Obama administration fought hard to protect American technology giants from European competition rules.

Given this history of differences between the US and Europe, how significant are the changes that the Trump administration has wrought in the style and substance of its transatlantic policy? This question suggests a very different baseline from the one used by Pompeo because it focuses not on relations between 'the West' and 'the rest', but on relations within the Alliance itself. Previous US administrations took a tough line with their European allies but often tempered this with public diplomacy. The George W. Bush administration's criticism of 'Old Europe' was balanced by its appeal to a 'New Europe' determined to place the transatlantic relationship at the centre of its foreign policy. Obama did not attend the 20th anniversary of the fall of the Berlin Wall in 2009 or the European Union–US summit in Spain in spring 2010. Nevertheless, he remained popular in Europe and used his celebrity status there to soften the blow of his administration's pivot to Asia.

Trump's new approach and priorities

By contrast, Trump showed no patience for public diplomacy. He expressed little interest in European politics or the mechanics of European integration and refused even to pay lip service to the conventions of transatlantic diplomacy. First as a presidential candidate, and later in office, Trump repeatedly questioned the necessity of the United States' Article 5 commitment to NATO. He floated the idea of negotiating bilateral trade deals with individual European countries, repeatedly accused the Europeans of unfair trading practices and openly supported the Brexit campaign.

The common threat of the COVID-19 pandemic might have been expected to encourage transatlantic cooperation. But Trump maintained rather than mitigated 'America First' policies, including towards America's closest allies. On 11 March 2020, Trump issued a proclamation letter restricting travel from the European Union to the US without consulting his European allies. Importantly, this further slighted the EU by exempting the recently departed United Kingdom, though the restrictions were later extended to include the UK. Four days later, officials of the German Ministry of Health revealed that Trump was trying to purchase the German biopharmaceutical company CureVac to ensure that any vaccine it produced would be exclusive to the US. Together, this unprecedented range of actions raised serious concerns in Europe about the Trump administration's commitment to the transatlantic relationship.

But there is another dimension to Trump's engagement with Europe. Whenever his officials talk about their foreign-policy accomplishments, they underscore the substantive change Trump has made in the pattern of American engagement with the rest of the world. In particular, they point to the number of NATO allies that have committed to increasing their defence spending following Trump's tough talk about burden-sharing. Vice President Mike Pence underscored this point in a speech in Washington DC on 3 April 2019, and Secretary of Defense Mark Esper reinforced it at the German Marshall Fund (GMF) in Brussels in October of that year. Trump himself highlighted it in his February 2020 State of the Union address, in the only part of the speech that mentioned the transatlantic relationship or even alluded to Europe. But this success is only partial. The German government's defence spending remains well below the agreed NATO target of 2% of GDP. Trump administration officials never lose an opportunity to point this out – as Trump did during his Rose Garden press conference with Polish President Duda.

The Trump administration's policy toward Iran is another area of substantive change, with dramatically different perceptions on both sides of the Atlantic. Starting in May 2018, the focus in the transatlantic relationship under Trump was on enforcing US sanctions on Iran after the United States' withdrawal from the Joint Comprehensive Plan of Action (JCPOA). The re-imposition of these sanctions, with the threat of secondary sanctions on European firms – including the financial telecommunications company SWIFT – created serious diplomatic tension.

By 2019, however, it was clear both that US sanctions would be enforced and that relations between Europe and Iran were deteriorating. This combination of factors reduced the salience of the issue, as it made Iran's withdrawal from the JCPOA more likely than a successful reset of the agreement. For many Europeans and supporters of the Obama administration's policy of engagement, the breakdown in relations with

Iran is a tragic missed opportunity. For members of the Trump administration, however, it is a diplomatic success.

The top priority for the Trump administration in 2019 and 2020 has been to persuade Europeans to focus on the threat of China. This was foreshadowed in the 2017 National Security Strategy, which identified China as a leading strategic competitor. These efforts ran alongside a series of US initiatives to block the spread of fifth-generation (5G) cellular technology by the Chinese industrial giant Huawei, strengthen the protection of intellectual-property rights and rebalance trans-Pacific trade. Thus, while the Trump administration negotiated directly with the Chinese government, officials also underscored the threat to Western values and European economic interests posed by the Chinese Communist Party (CCP).

China became a centrepiece in the transatlantic conversation in 2020. Esper devoted his entire speech at the Munich Security Conference in February to the question of what Europe should do to thwart the CCP's influence across issues ranging from security to economics. Pompeo underscored this when he spoke to the GMF's Brussels Forum the following June. In his speech he concluded that: 'The United States is not forcing Europe to choose between the free world or China's authoritarian vision. China is making that choice between freedom and democracy.' In doing so he embraced a new US–EU dialogue on China as a forum where the two sides of the Atlantic could coordinate more closely.

This focus on China does not mean Russia was forgotten. The Trump administration opened the door to sanctioning firms working on Russia's Nord Stream 2 pipeline to Germany in December 2019 with Senate approval of the National Defense Authorization Act. Initial efforts focused on those firms engaged directly in building the new infrastructure. By July 2020, however, the Trump administration, acting with bipartisan congressional support, was ready to target any firms connected to the project.

The signalling in the Trump administration's Russia policy has nevertheless been mixed. Although the sanctions on Nord Stream 2 were clearly targeted and drew bipartisan support in Congress, the administration's decision to draw down troops stationed in Germany seemed at odds with its assertions about a Russian security threat. Administration officials tried to reconcile the two issues by reiterating the claim that no president had been tougher on Russia. Such assertions have done little to resolve the confusion.

Efforts by the Trump administration to thwart Russian interference in the 2020 presidential elections have also created a sense of ambivalence – largely as a result of the president's denial of any collusion with the Russian government in 2016 or, indeed, that Russian interference influenced the outcome of that contest. At least part of Trump's argument rests on his acceptance of Russian President Vladimir Putin's denial that Russia ever interfered in the 2016 contest. He has also repeatedly insisted that any allegations to the contrary are part of an attempt either to delegitimate his electoral victory or to cover up connections between Hillary Clinton and Joe Biden and political operatives in Ukraine. Trump viewed the publication of Robert Mueller's report in March 2019, and his acquittal during the Senate impeachment trial in February 2020, as a complete vindication of these positions. Thus, despite the intelligence community's view that the Russian government was preparing to interfere in the 2020 contest, the Trump administration's response remained muted.

European reactions

Trump's reassurances that Putin was telling the truth about Russian interference in 2016 did not persuade European governments. Nor did many of the Trump administration's claims about how it had improved the functioning of the NATO Alliance or identified the strategic threat posed by China and Russia. On the contrary, many European observers regard the Trump administration's positioning toward NATO as inconsistent and have questioned what it means for the longer-term future of the Atlantic Alliance.

French and German leaders have focused increasingly on forging a more effective European partnership capable of working independently of the US if necessary. The July 2017 Franco-German summit and June 2018 Meseberg summit illustrated this. The declaration from the latter did not mention the US and began with a call to streamline and strengthen the EU as an actor in foreign and security policy. A further initiative came with the appointment of Ursula von der Leyen as the new European Commission president and the formation of what she called a 'geopolitical Commission' that is 'ambitious, strategic and assertive in the way that [it] act[s] in the world'.

Such actions do not mean that most Europeans reject America's participation in European security. Nor do they doubt that closer trading relations would be beneficial, or that China presents a growing threat. Polling evidence from the GMF's June 2020 'Transatlantic Trends' study also shows broad support for the transatlantic relationship. There is, though, stronger support for European solidarity and a more effective EU.

The question is whether this support will produce more effective institutions. The European Council may have made a first step in that direction by moving to enhance the borrowing capacity of the European Commission in order to fund Europe's economic recovery. The von der Leyen Commission is working hard to reinforce its policy priorities. Should these initiatives bear fruit, the EU could gain greater autonomy in areas as diverse as military procurement, industrial policy, technological innovation and taxation.

The future of the transatlantic relationship

Future US administrations will need to take that greater European unity into account. A more unified Europe could emerge as a better partner for the US. This was the dream that president John F. Kennedy outlined in the early 1960s and it was the hope of Obama's administration during his first term. But a more unified Europe could also prove to be a more difficult competitor in both economic performance and strategic priorities. This concern, held during the first term of the George W. Bush administration, is widely shared in the Trump administration.

At the close of June 2020, future US administrations seemed likely to find a more united European partner within a less trusting transatlantic relationship. Any decision by Trump to play even more roughly with Europe was set to be met by greater European willingness to retaliate on matters related to trade and to strike a more independent posture on defence. A potential Biden presidency looked likely to ease some tensions but unlikely to restore the transatlantic relationship to the status quo ante. Trump has changed the transatlantic relationship in substance as well as style – not just on Iran, trade relations and NATO burden-sharing, but in the larger positioning of American strategic priorities, particularly related to China. Whatever the next four years hold, that relationship will have to evolve.

CHAPTER 6

Asia

ASIA

Drivers of Strategic Change

REGIONAL SHARE OF GLOBAL POPULATION, GDP AND DEFENCE BUDGET

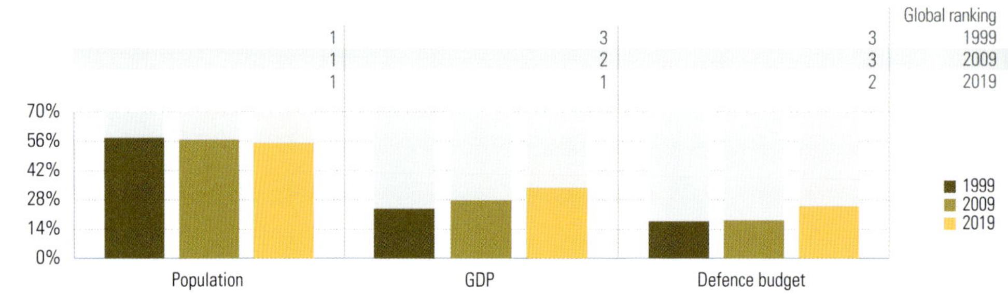

POPULATION

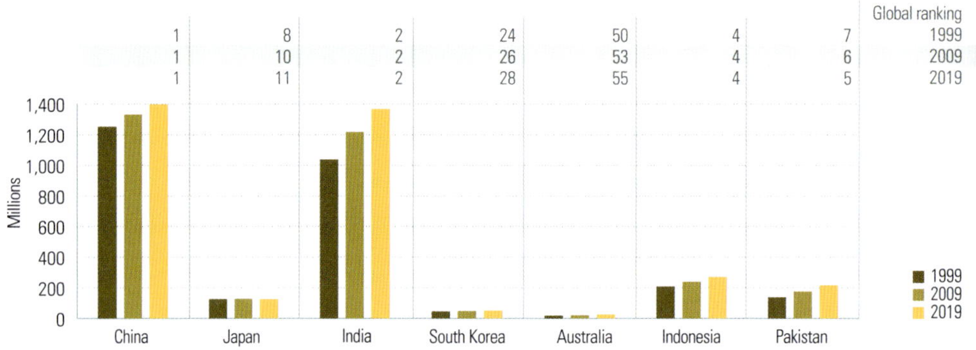

AGE STRUCTURE

(Percentage of national population)

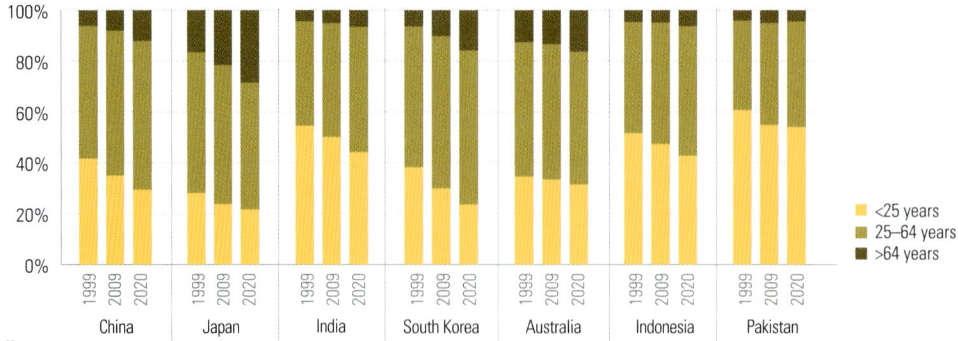

Median age	China	Japan	India	South Korea	Australia	Indonesia	Pakistan
2000	30.0	41.2	22.7	31.9	35.4	24.4	18.8
2010	35.0	44.7	25.1	38.0	36.8	27.2	20.7
2020	38.4	48.4	28.4	43.7	37.9	29.7	22.8

GDP

(Constant 2010 US dollars)

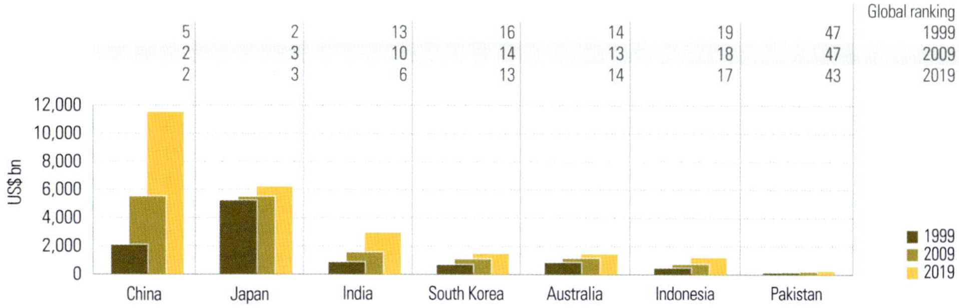

GDP PER CAPITA

(Constant 2010 US dollars)

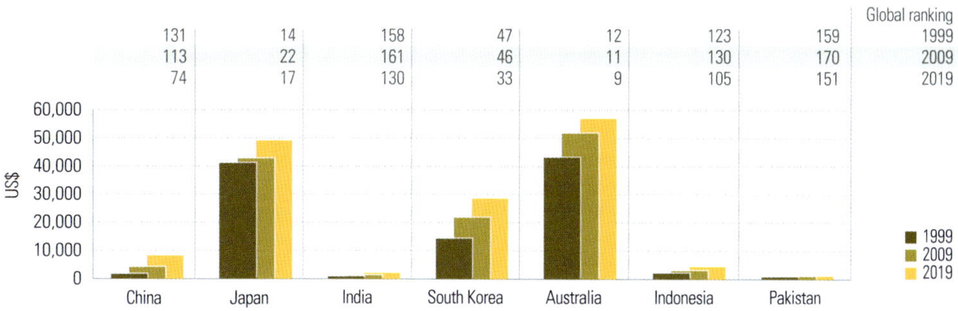

DEFENCE BUDGET

(Constant 2015 US dollars)

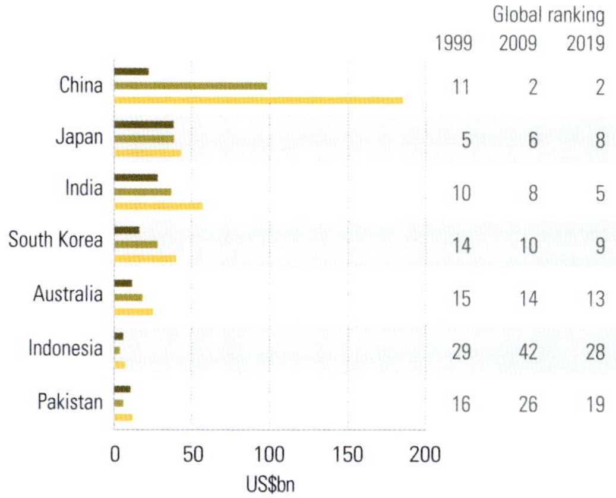

ACTIVE MILITARY PERSONNEL

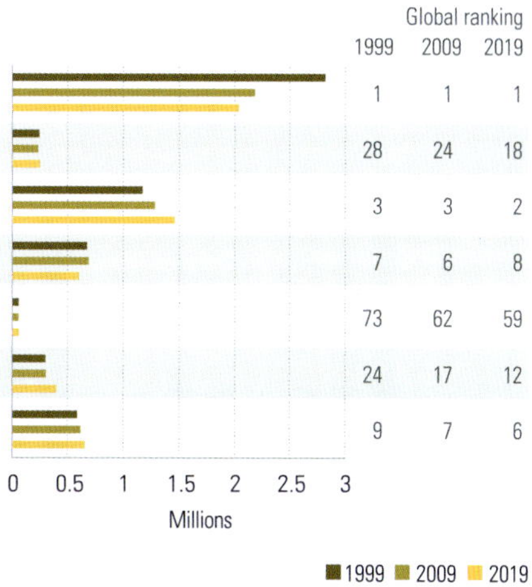

For explanation of drivers and sources, see page 7

HUMAN DEVELOPMENT INDEX (HDI)

(Score between 0 and 1, where 0 denotes a low level of development and 1 a high level of development)

Global ranking	China	Japan	India	South Korea	Australia	Indonesia	Pakistan
1999	100	15	111	29	2	95	121
2009	102	22	136	26	3	115	147
2019	85	19	129	22	6	111	152

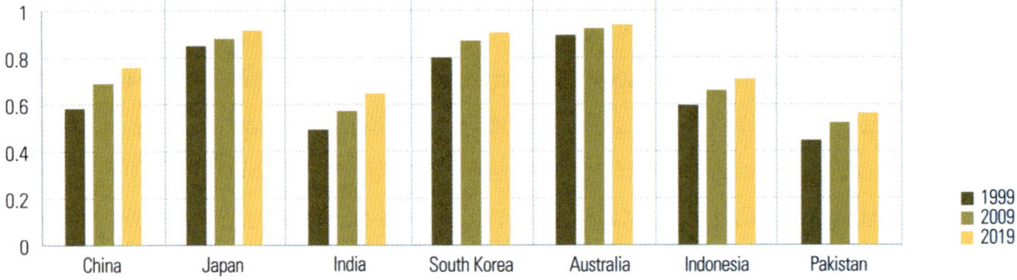

POLITICAL SYSTEM

(Score between 0 and 100, where 0 denotes no political freedom and 100 fully free)

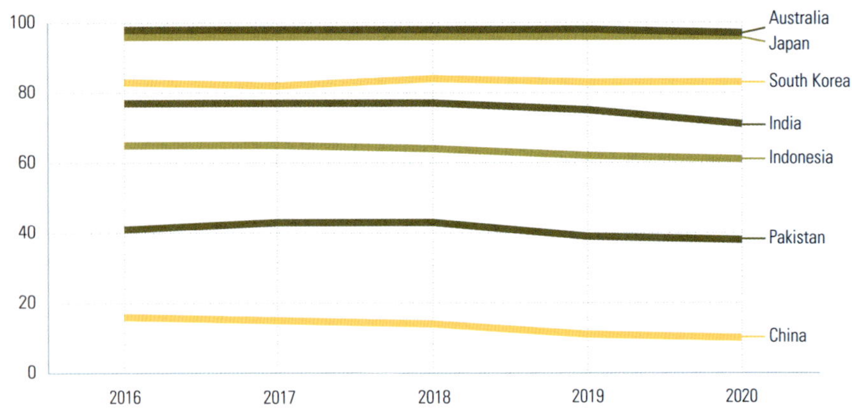

TRUST IN INSTITUTIONS

(Average level of trust)

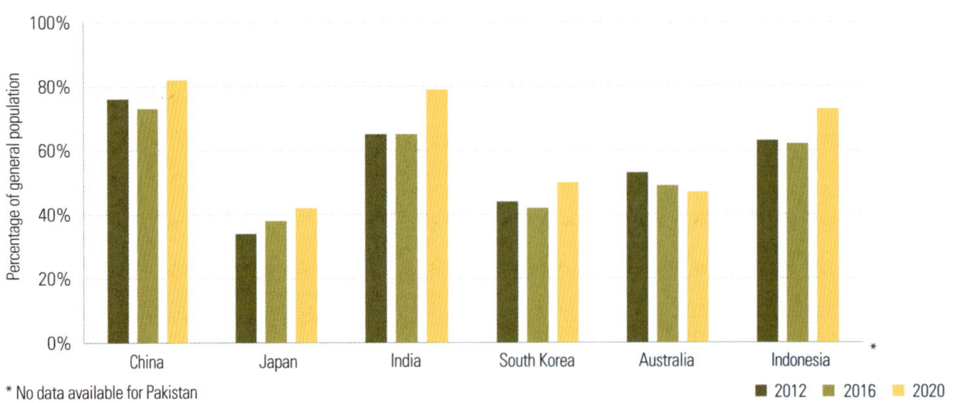

* No data available for Pakistan

For explanation of drivers and sources, see page 7

2019–20 Review

Impact of COVID-19 in Asia-Pacific

The COVID-19 pandemic struck the Asia-Pacific before any other region, impacting its states, societies and economies profoundly. From February 2020 the virus preoccupied governments throughout the region. While their responses were not without shortcomings, overall they were significantly more effective than those seen in the West. Possible reasons for this included a tradition of strong, paternalistic government in many Asian countries, a relatively high degree of social conformity (sometimes reinforced by military conscription), health systems with recent experience of dangerous, highly infectious diseases (including the regional outbreak of SARS in 2003) and possibly the regional prevalence of a relatively weak strain of the SARS-CoV-2 virus that caused the COVID-19 illness.

But the major geopolitical issues, and relations with key external partners, that had dominated security policy before the pandemic remained centrally important. It seemed clear that China's increasingly active drive to become the region's preponderant power; the Trump administration's robust policies towards not only China but also its own allies and partners; North Korea's continuing missile and nuclear-weapon programmes; and India's rivalry with Pakistan would still dominate the regional security environment once the pandemic had died down.

The COVID-19 outbreak, believed to have resulted from the transmission of a virus from animals to humans, began in China's Hubei province in late 2019. The United States, supported by some other Western countries, alleged that China mishandled and covered up the outbreak in its early weeks, facilitating its spread internationally. US President Donald Trump suggested in late April that the escape of the SARS-CoV-2 virus from a laboratory in the city of Wuhan may have started the pandemic, and that the US should seek 'substantial' compensation from Beijing for its COVID-19 policy failures.

After initial hesitation and missteps in late December 2019 and early January 2020 (including police harassment of physicians who warned of the scale and severity of the outbreak in Wuhan, its initial epicentre) the Chinese authorities responded firmly and effectively to the threat that the virus posed. In the largest quarantine effort in human history, from 23 January, the Chinese government locked down Wuhan and other cities in Hubei – a total of 57 million people – to contain the outbreak. President Xi Jinping warned on 25 January that the country faced a 'grave situation' and on 10 February launched a 'People's War' aimed at containing the virus.

China imposed constraints on the outdoor activities of more than half its population. Recorded infections with the virus numbered around 80,000 and approximately 2,900 people had died from COVID-19 there by the end of February. While some international observers distrusted China's subsequent COVID-19 figures as the virus spread internationally and wreaked havoc in other regions, it became clear that China's efforts to limit its domestic impact had been relatively successful. On 24 March, Premier Li Keqiang claimed that the authorities had controlled the outbreak inside the country.

Beijing's emphasis then shifted to preventing infected people from entering China from elsewhere and to providing support for other countries' efforts against the virus. Nevertheless, a fresh eruption of the virus in Beijing – though quickly brought under control – emphasised that the COVID-19 challenge was likely to endure beyond the short term.

At the start of 2020, COVID-19 spread quickly from China to its Asian neighbours, transmitted mainly by Chinese tourists. Other governments in the region also recorded successes in containing the spread of the virus and the numbers of fatalities. The first confirmed case of infection outside China was in Thailand on 13 January. Local transmission there followed. Surveillance and contact tracing did not prevent the first COVID-19 deaths in the country, and infections began to grow substantially. On 26 March, Prime Minister Prayuth Chan-ocha announced a state of emergency. A night-time curfew followed on 3 April along with a ban on incoming passenger flights. Overall, Thailand's response was highly effective, particularly in limiting mortality.

Some other Southeast Asian countries also performed well. The Vietnamese authorities acted promptly by preventing infected people from entering the country; locking down localities where infections were detected; closing down schools and universities; tracing and quarantining indirect as well as direct contacts of infected people; and communicating public-health messages clearly. These measures limited recorded infections to several hundred by early June. By that stage, no one had died from the virus in Vietnam. In Singapore, the government's consistent emphasis on contact tracing severely limited the spread of infection in the general community. But the poor living conditions, and initially inadequate medical access, of low-paid foreign workers (mainly in the construction sector) provided fertile ground for an explosion in overall infections to over 45,000 by early July. Yet the city-state's high-quality medical care ensured that only 26 people had died in Singapore from COVID-19 by then.

Elsewhere in the region the picture was less rosy. In the Philippines, despite a strictly enforced lockdown from mid-March in Metro Manila, Luzon island and elsewhere, the initial denial by the administration of Rodrigo Duterte that COVID-19 posed a serious threat contributed to the spread of the virus and one of the worst national death tolls in Southeast Asia. But Indonesia was hardest hit in the region, and the situation there remained serious in early July 2020. Fearing the economic impact of measures to restrict the virus, President Joko Widodo's government waited until 15 March to call for social-distancing measures in the most-affected parts of the country. These measures were relaxed from late May onwards. But it was evident that, while the spread of the virus was under control in provinces such as West Java and Bali, it continued to spread in other regions such as East Java. Low rates of testing and contact tracing meant the full extent of the challenge remained unclear.

In Northeast Asia, the picture was also mixed. North Korea implemented rigorous measures to prevent the spread of COVID-19. It suspended international rail and air travel, restricted road and sea travel, imposed severe quarantine restrictions, closed schools and restaurants, and cancelled public events. Pyongyang claimed that these measures had succeeded in keeping the country free of the virus. In South Korea, the scale of the COVID-19 challenge became apparent from late February with the appearance of a cluster of almost 4,500 infections linked to a church in Daegu city. Seoul's initially complacent response shifted as the authorities implemented a massive programme of testing for the virus, accompanied by stringent efforts to isolate those infected and trace and quarantine their contacts. South Korea reaped considerable international kudos for its effective management of COVID-19, which was achieved without large-scale lockdowns. However, further outbreaks during May and June emphasised the persistence of the challenge.

In Japan, some observers claimed that the government's relatively slow response was partly due to concerns that recognising the extent of COVID-19 might impact adversely on international participation in the politically and economically important 32nd Summer Olympics scheduled for July–August. On 24 March, Japan announced that the Games would be postponed by one year. The authorities began implementing a range of measures (involving school closures and restrictions on incoming travellers), which they bolstered following the outbreak of a 'second wave' of infections, mainly from Europe, in late March. Overall, Japan's response was broadly effective. Without lockdowns and with only minimal testing for infection and non-digital contact tracing, the national death toll from COVID-19 was a fraction of that in many Western countries.

The most widely praised response in Northeast Asia was that of Taiwan. From the end of December, it systematically implemented a wide range of measures to prevent the spread of the virus, including screening of incoming air travellers and case tracing. This approach was so successful in limiting infections and deaths that New Zealand's Prime Minister Jacinda Ardern modelled her government's COVID-19 strategy on Taiwan's. New Zealand itself received international plaudits for the timeliness and success of its measures.

While many countries in the region successfully contained COVID-19 infections and mortality, far-reaching economic repercussions of the health crisis were unavoidable. Travel bans, lockdowns and other social restrictions had a huge impact on economic activity throughout the Asia-Pacific, leading to a rapid contraction in growth prospects in 2020.

At the same time, governments needed to provide financial support to individuals and businesses. With government revenue falling but expenditure rising, national budgets quickly came under serious stress. Some governments, notably Singapore's, could draw on national financial reserves, but many others did not have this option. Even in the short term, the crisis impacted some regional states' defence spending. By June, Indonesia, South Korea and Thailand had all announced cuts in their planned military budgets for 2020 as resources were redirected to crisis-relief measures.

Worsening geopolitical outlook

The COVID-19 crisis had significant geopolitical ramifications, essentially by accentuating existing regional patterns and trends. Most importantly, a major dispute erupted in April 2020 between China on the one hand and the US and some of its allies – notably Australia – on the other over China's supposed responsibility for the global spread of COVID-19 due to its initially inadequate and secretive response. This became a new strand in a broader political and economic conflict that, during the first half of 2020, increasingly came to resemble an incipient cold war between China and the West. Other sources of tensions included the US drive to recalibrate trade relations; widespread concern in the West over the potential security threats posed by the adoption of Chinese 5G technology; pushback against Chinese efforts at political interference; Western perceptions of a growing threat posed by China's rising defence spending and increasing military capability; and fears over the threat to freedom of navigation posed by China's creeping efforts to assert control over the South China Sea.

The widening Sino-US rift disconcerted other Asian states. China's economy, on which they had become increasingly dependent, would be severely undermined if it were 'decoupled' from those of the US and other Western powers. Moreover, as Singapore Prime Minister Lee Hsien Loong wrote in June, 'Asia-Pacific countries do not wish to be forced to choose between the United States and China'. Intensifying geopolitical rivalry between the US and China made such a prospect seem less theoretical than it had in the past. Whether the two major powers could manage their differences and cooperate was, in Lee's words, 'a fundamental issue of our time'.

Competition intensified in the region. Beijing seemed steadily more intent on asserting strategic primacy in its maritime littoral. The US government, not to mention many Western and some Asian analysts (particularly in Japan), saw evidence that China was taking advantage of other governments' preoccupation with the COVID-19 pandemic to gain strategic advantage throughout what the US, its allies and other states including India and Indonesia increasingly referred to as the 'Indo-Pacific region'.

In the South China Sea, Chinese activities continued, apparently uninterrupted by COVID-19, during the first half of 2020. China maintained pressure on its main South China Sea adversary, Vietnam. In early April, a Chinese Coast Guard (CCG) vessel sank a Vietnamese fishing boat near the Paracel Islands. Significantly, Indonesia and Malaysia found they were now effectively 'front-line states' facing intrusions into their exclusive economic zone (EEZ) of maritime jurisdiction from CCG and maritime militia vessels using facilities on Chinese-occupied features in the Spratly Islands as forward bases. China also kept up pressure in the East China Sea. In early May, CCG ships harassed a Japanese fishing boat inside the territorial waters of the Senkaku/Diaoyu Islands (administered by Tokyo, but claimed by Beijing), leading to a stand-off with the Japanese Coast Guard, which reported in late June that it had observed Chinese government vessels in waters close to the Senkaku/Diaoyu Islands for 70 consecutive days, a new record.

China also intensified its attempts to intimidate Taiwan, where President Tsai Ing-wen of the Democratic Progressive Party – which, while stopping short of favouring *de jure* independence, strongly asserts the island's quasi-national status – was re-elected with a resounding 57% of the vote in January. From early February onwards, Chinese combat aircraft flew close to or within Taiwan's Air Defence Identification Zone numerous times, including on four consecutive days in mid-June. During April, at a time when COVID-19 had undermined the operational readiness of the US Navy's Seventh Fleet in the region as crew members reportedly fell sick on 40 warships (including the aircraft carrier USS *Theodore Roosevelt*), the Chinese aircraft carrier *Liaoning* and its accompanying strike group sailed close to Taiwan as part of a month-long deployment that some observers interpreted as a demonstration of Beijing's growing capacity to project power not only to Taiwan but to Japan and Southeast Asia, particularly Vietnam.

China also dramatically escalated its efforts to assert dominance over Hong Kong, where large-scale and widely supported protests against Beijing's tightening political and jurisdictional control had attracted attention worldwide since June 2019. In apparently direct contradiction to the principle of 'One Country, Two Systems' that had provided the framework for relations between Beijing and the former British colony since it reverted to Chinese rule in 1997, in June 2020 the Standing Committee of China's National People's Congress enacted a national-security law for Hong Kong. The new law sought to prevent foreign interference (which supposedly 'harmed the rule of law and threatened national sovereignty, security and development'), criminalised acts that 'threaten national security', such as 'subversion and secession', and allowed China's national-security organisations to establish branches in Hong Kong 'when needed'. This legislation further damaged China's already fraying relations with Western countries. In response, the United Kingdom announced that it would allow Hong Kong residents born before the end of British rule in 1997 and their dependents to settle in the UK, and would provide a path to British citizenship for them.

US engagement remained the most important anchor for the existing regional order, which many supporters in Asia and beyond characterised as a 'Free and Open Indo-Pacific'. While the COVID-19 pandemic continued to prevent bilateral and multilateral exercises involving US forces in the region, by June the US Navy had largely recovered from the impact of the pandemic on its operational readiness and sent three carrier strike groups to the Indo-Pacific region for the first time since 2017 – a move widely interpreted as a demonstration of the strength of continuing US military commitment in the face of China's geopolitical assertiveness.

The US became increasingly conscious of the political and economic, as well as military, dimensions of its intensifying struggle with Beijing for influence across the region. A strong bilateral consensus in the US Congress supported this. In December 2019, Congress authorised US$2.5 billion in funding to implement the Asia Reassurance Initiative Act (ARIA) of 2018. ARIA had established initiatives to promote democracy, human rights and transparency across the region; mandated reports from the US government on North Korea, counter-terrorism, engagement with the Association of Southeast Asian Nations (ASEAN) and on other subjects; and mandated the US administration to develop strategies for trilateral security cooperation with Japan and South Korea, and diplomatic coordination with Indo-Pacific allies. Reflecting acute awareness of China's efforts to exert influence through infrastructure projects built under its Belt and Road Initiative (BRI), in January 2020 the US International Development Finance Corporation, an agency created by the Better Utilization of Investments Leading to Development (BUILD) Act in 2018, began operations and assumed responsibility for channelling US commitment to overseas development financing. As a result of the BUILD Act, overall US capacity for such funding has more than doubled to US$60bn.

Multilateralism and policy coordination with allies and partners, emphasised by ARIA, were often absent in the Trump administration's policy execution. Indeed, the president's continuing pressure on US allies and partners to boost their financial and military contributions to their relationships with the US – and notably on Japan and South Korea to pay more for the presence of US forces on their soil – appeared to risk undermining the security ties upon which America's regional role depended. Nevertheless, the two most important US allies in the region – Australia and Japan – continued to develop regional policies that were closely congruent with Washington's. Both allies' foreign and defence policies reflected growing concern over what they, like the US, viewed as the major challenges posed by China's growing defence spending, military capabilities and strategic activism.

In September 2019, Tokyo's Defense White Paper for the first time explicitly identified China, not North Korea, as posing the most serious military threat to Japan. It argued that infrastructure built through Beijing's BRI investments could be used to 'promote the activities of the PLA [People's Liberation Army] in the Indian Ocean, Pacific Ocean and elsewhere'. In Australia, concern over what was perceived as a growing multidimensional threat from China spurred tougher policies. In April 2020, Canberra questioned China's transparency in relation to the initial COVID-19 outbreak and demanded an international investigation into the origins of the SARS-CoV-2 virus and how it had spread. In June, responding to concerns over undue Chinese economic and technological influence, Australia's government announced pending legislation for a new screening process for foreign investment in the telecommunications, energy, technology and defence-manufacturing sectors. A Defence Strategic Update in mid-2020 referred to Australia's strategic environment having deteriorated more rapidly than anticipated when the country's latest Defence White Paper was published in 2016, and set out plans to modify the national defence posture in favour of developing a more powerful and self-reliant military deterrent. While direct references to China were sparse, the focus of Canberra's concern was nevertheless clear. In response to the BRI challenge, Australia and Japan continued to develop their collaboration with the US through a Trilateral Partnership for Infrastructure Investment in the Indo-Pacific that was agreed in 2018.

South Korea, the other major US ally in the region, remained focused primarily on the immediate challenge of North Korea. Pyongyang had resumed testing ballistic missiles in May 2019 after a pause of almost 18 months and had continued its nuclear-weapons development. Meanwhile, the diplomatic process failed to advance after a third summit between Trump and North Korean leader Kim Jong-un in June 2019. In October 2019, North Korea's delegation walked out of working-level talks with the US in Stockholm on ending its missile and nuclear programmes. A key obstacle to progress was disagreement over the meaning of the term 'denuclearisation'. Inter-Korean diplomacy, which since 2018 had appeared to provide a vital complement to the parallel US–North Korea negotiations, also stalled as the US insisted that the South Korean administration of President Moon Jae-in adhere to Washington's 'maximum pressure' strategy of applying economic sanctions against Pyongyang. This dashed hopes for the restoration of cooperative projects on the peninsula.

In the second half of 2019, North Korea resumed its hostility towards Seoul. In a stark demonstration of the breakdown in relations, in mid-June 2020 the North Korean government used high explosives to demolish the inter-Korean liaison office in Kaesong on its side of the border. While this, and the threat of further escalation, appeared intended to coerce the US and South Korea into lifting economic sanctions, the response from Seoul was the opposite of conciliatory: Moon's spokesman warned that South Korea would 'no longer tolerate irrational remarks and actions by the North'.

At a time when close policy coordination between Seoul and Tokyo regarding North Korea was necessary and important, relations between South Korea and Japan remained frosty. Following a decision by South Korea's Supreme Court in October 2018 that plaintiffs could sue Japanese companies over their use of Korean slave labour in factories during the Second World War, in July 2019 Tokyo – citing national-security concerns related to North Korea – restricted exports of important chemicals used in South Korea's semiconductor industry. In August, Seoul retaliated by giving notice that it would withdraw from the trilateral General Security of Military Information Agreement (GSOMIA) with Japan and the US to share intelligence on North Korea. In the same month, South Korea also staged a major naval and marine exercise around the Dokdo/Takeshima Islands, which are occupied and administered by South Korea but claimed by Japan. In September, Seoul removed Japan from a 'white list' of countries with fast-track trade status.

The two sides subsequently attempted to moderate their dispute, largely as a result of US pressure. In November, Seoul said it would remain in the GSOMIA (while reserving the option to pull out at any time). In February 2020, the two countries held talks intended to improve relations but made little progress towards reconciliation. Within days of the February talks, Japan launched a complaint against South Korea in the World Trade Organization (WTO) over Seoul's subsidies for its shipbuilding industry. In June, Seoul restated its own claim to the WTO that

Japan's export controls were politically motivated. It seemed increasingly unlikely that this bilateral dispute could be significantly tempered under the current leaderships of Japan and South Korea. This suggested that a resolution was unlikely before 2022 at the earliest.

Despite the deteriorating regional security environment, most Southeast Asian governments remained primarily focused on managing domestic concerns. The COVID-19 pandemic only accentuated this, notably in Indonesia and Malaysia, which faced stronger Chinese pressure on their maritime interests. In Indonesia, the government of President Widodo (also known as 'Jokowi'), who had been re-elected in May 2019, continued to prioritise economic considerations in its foreign policy, notably the importance of Chinese investment in a national infrastructure programme. In late 2019, incursions by Chinese fishing vessels, supported by coastguard ships, into waters close to Indonesia's Natuna Islands led Jakarta to strengthen its military deployments in the area. Despite this, it seemed unlikely that Indonesia under Jokowi's leadership would revert to the more active regional and international role that it played under his predecessor, Susilo Bambang Yudhoyono (2004–14). Rather, Jokowi seemed likely to prioritise the balancing of factions and personalities within his cabinet, domestic economic reforms, and planning for the relocation of Indonesia's capital city to East Kalimantan province, which was announced in August 2019.

Malaysian politics was no less focused on domestic matters. In February the fragile Pakatan Harapan (Alliance of Hope) coalition, led by veteran politician Prime Minister Mahathir Mohamad, was unseated following infighting within and between its constituent parties, and Mahathir's refusal to set a date for ceding the premiership to designated successor Anwar Ibrahim. The government under Prime Minister Muhyiddin Yassin that replaced it enjoyed only a narrow majority in parliament. Constant domestic political wrangling, combined with the need to manage COVID-19, meant that Muhyiddin's government had little capacity to engage in regional geopolitics. The month-long intrusion of a Chinese survey ship into its EEZ in April–May 2020 prompted the US and Australia to stage a joint naval exercise nearby, and Washington to call on China to cease its 'bullying behaviour'. Meanwhile, in late April, Malaysia cautiously called for South China Sea disputes to be resolved peacefully.

Singapore and Vietnam demonstrated greater resolve than their Southeast Asian neighbours in responding actively to the deteriorating regional strategic environment. In the face of rapidly deteriorating relations between China and the US, Beijing's geopolitical assertiveness and Trumpian unilateralism, Singapore's government continued to argue that Sino-US accommodation, adherence to international law, and multilateral approaches to trade agreements and global challenges remained vitally important. In September 2019, Singapore Prime Minister Lee and Trump signed an amendment that extended to 2035 a Memorandum of Understanding regarding US use of military facilities in Singapore.

In November 2019, Vietnam assumed the chairmanship of ASEAN. This provided an important opportunity to use multilateral means to shape the regional environment in support of Hanoi's objectives. These included reinforcing unity among ASEAN's diverse member states and making progress towards concluding an ASEAN–China Code of Conduct to constrain Beijing's assertiveness in the South China Sea. Both sides had already agreed this should be achieved by 2021. But COVID-19 prevented the physical meetings of principals and the associated informal personal interactions between them that were vital to progress on the Code of Conduct, and at the same time severely distracted the attention of ASEAN as well as its member states. In mid-April, a virtual special ASEAN summit focused on formulating collective responses to the pandemic (and which proposed establishing a 'COVID-19 ASEAN Response Fund', among other measures) substituted for the normal, wide-ranging ASEAN summit, which was postponed to late June and then also staged virtually.

In the meantime, Vietnam responded to Chinese pressure in other ways. In November 2019, Deputy Foreign Minister Le Hoai Trung suggested that Hanoi might take China to an international court over its claims in the South China Sea, as the Philippines had in 2013. In April 2020, Vietnam publicly protested a series of Chinese moves, claiming that Beijing's establishment of two new administrative districts for the

Paracel and Zhongsha islands, and for the Spratly Islands, 'seriously violated Vietnam's sovereignty'. By May, it was widely rumoured that Vietnam might seek to extend its role as ASEAN chair by another year, so as to develop the political and security agenda that COVID-19 had side-lined. However, while Brunei (next in line to be chair) was apparently agreeable to an extension for Vietnam, given that the ASEAN Charter prescribes that the chair 'shall rotate annually' it was unclear whether this would be possible.

Impact of COVID-19 in South Asia

In South Asia, the COVID-19 pandemic severely disrupted economies and tested governance and public-health infrastructure. Densely populated urban areas, poverty and poor sanitation, weak public-health systems, and low testing rates by Asian standards challenged the region's ability to limit infections and deaths. On 24 March, Indian Prime Minister Narendra Modi ordered the world's largest lockdown of 1.3bn people with only four hours' notice. State borders were closed, and transport services suspended, causing considerable hardship for some 10m migrant workers travelling across the country to return home. The 75-day lockdown was gradually eased from 8 June. But by the end of June, India looked set to overtake Russia as the country with the third-largest number of COVID-19 cases, even though it had the eighth-highest number of fatalities. With a peak expected weeks later, India looked likely to have the highest number of COVID-19 infections after the US, and to be among the top three countries in terms of fatalities, along with the US and Brazil.

In Pakistan, Prime Minister Imran Khan imposed a six-week country-wide lockdown from 26 March to 9 May, despite initial concerns over its impact on the economy. The army was deployed to enforce lockdown measures, and the country's powerful Inter-Services Intelligence (ISI) directorate deployed surveillance technology (normally used to locate militants) as part of the country's 'track and trace' system for suspected COVID-19 cases. In Bangladesh, the spread of COVID-19 in the country's Rohingya refugee camps caused particular concern. The pandemic prompted the first summit (albeit a virtual one) in six years of the eight-member South Asian Association for Regional Cooperation (SAARC), the main regional forum, on 15 March. Amid India–Pakistan sparring over Kashmir, the main outcome was a voluntary COVID-19 Emergency Fund of some US$21.6m.

South Asian economies were badly hit. The World Bank expected India's economy to contract by 3.2% in FY2020–21, making it likely that the country would suffer its first annual recession since 1980. In 2019, India's economy had grown by 4.2%, its slowest growth in 11 years. More than 120m people reportedly became unemployed as a result of COVID-19. Pakistan's economy, having contracted the previous year, was expected to contract further, resulting in the World Bank predicting growth of only 0.18%.

COVID-19 also impacted Afghanistan's fragile road map to peace. Widespread infection among Afghan security forces led to the suspension of training programmes and joint operations with US forces, the latter primarily in eastern Afghanistan against the Islamic State, also known as ISIS or ISIL. With the Trump administration aiming to withdraw most US troops in 2020, on 29 February it signed an unprecedented peace agreement with the Taliban intended to pave the way for an intra-Afghan dialogue. Violence had intensified across the country in the months leading up to the signing, primarily due to an increase in attacks by the Taliban. Despite the agreement, the Taliban continued its campaign of violence, and political disputes within the Afghan government also continued.

Kashmir tensions

On 5 August, the Indian government announced the controversial decision to end the 'semi-autonomous' constitutional status of the Indian province of Jammu and Kashmir that had been in place for nearly 70 years. Although the government stated that its actions were an internal matter and did not necessitate any change to India's external boundaries, the move led to tensions with Pakistan, China and Nepal.

The most vociferous opposition to India's actions in Kashmir came from Pakistan, which disputed India's control of the province and had historically laid territorial claim to parts of it. India released a new map of its territorial boundaries on 2 November 2019, which also resulted in a diplomatic rift with Nepal: Kathmandu objected to the disputed area of Kalapani, at the tri-junction of India, Nepal and China, being shown within India's borders. This was exacerbated by India's inauguration of a road passing through Kalapani to the Lipulekh pass on the Line of Actual Control (LAC – the non-demarcated land border between China and India) on 8 May. Nepal issued a new map on 20 May showing, for the first time, the disputed areas of Kalapani, Lipulekh and Limpiyadhura as part of Nepalese territory.

Although Modi and Chinese President Xi held their second 'informal' summit in as many years in Chennai, India, in October 2019, tensions in the eastern Ladakh sector along the LAC flared up in May and June 2020. A violent clash on 15 June in the Galwan Valley led to the deaths of 20 Indian (and an unknown but possibly larger number of Chinese) soldiers, the first fatalities on the LAC in 45 years. A slow-moving disengagement process had begun by 30 June, but bilateral tensions remained high and the eventual de-escalation process was expected to take months. India's response options remained primarily diplomatic and political, not military.

The motives for the clash were not clear on either side. For China, factors may have included the perception of Indian construction of border roads and upgrades of transport infrastructure (bridges, tunnels and the world's highest airfields) as an attempt to alter the LAC in India's favour; an increased appetite for 'strategic opportunism' while India was distracted by its COVID-19 response; concern over India's deepening of relations with the US to counter Chinese influence in the Indo-Pacific; challenging India's unilateral decision to change the status quo in Kashmir on 5 August through the change of Ladakh's administrative status to a union territory (controlled by India, but with some Chinese territorial claims); and countering New Delhi's rhetorical reiteration over its claim to Aksai Chin (the Chinese-controlled sector of Kashmir that India claims is part of Ladakh) and Gilgit-Baltistan (controlled by Pakistan), the latter being a key region for the China–Pakistan Economic Corridor (CPEC).

Indian motivations for these skirmishes were likely to have included the countering of Chinese intrusion and assertive patrolling within its own claimed area of the LAC; and defending against a Chinese attempt to alter the LAC in China's favour through territorial control.

India's new citizenship law

In another controversial move, India passed its Citizenship (Amendment) Act (CAA) in parliament on 11 December 2019. The CAA amended the Citizenship Act of 1955 to fast-track citizenship for religious minorities (specifically Hindus, Sikhs, Buddhists, Jains, Parsis and Christians) from Afghanistan, Bangladesh and Pakistan who had entered India prior to 2015. Although this was aimed at giving stateless people citizenship rights, it attracted strong criticism because religion had never been used as an explicit criterion for citizenship under Indian law. Furthermore, since its provisions did not extend to Muslims, there was concern that, together with a prospective Indian National Register of Citizens (NRC), the CAA would lead to the marginalisation of India's Muslims. Large-scale protests against the CAA across India resulted in numerous arrests and 84 deaths between December 2019 and February 2020, including 53 deaths during the Hindu–Muslim communal riots in northeast Delhi in February 2020. The COVID-19 lockdown, with strictures in place to prohibit gatherings of four or more people in public places, stalled further protests.

India deepened its bilateral relationship with the US. Following the first India–Japan–US–Australia foreign-ministerial quadrilateral dialogue in September 2019, India and the US held their first tri-service military exercise in November. In December, India appointed its first Chief of Defence Staff. Trump's state visit in late February 2020 symbolised the countries' enhanced bilateral relationship and the importance of defence cooperation. US defence sales to India since 2008 reached US$20bn by June 2020.

In Pakistan, Khan approved a three-year extension for Chief of Army Staff Gen. Qamar Javed Bajwa, who was set to retire in November 2019, citing a tense 'regional security environment'. Bajwa's extension came as Pakistan was expected to remain on the 'grey list' of the intergovernmental Financial Action Task Force (FATF) until October 2020 for failing to combat terrorist financing. By February, the FATF noted that Pakistan had addressed half of the 27 action items it required, while in June Pakistan's interior minister said that Pakistan had met most of the FATF's conditions. Hafiz Saeed, the chief of the Lashkar-e-Taiba (LeT) terror group who had been in prison since February, was moved to house arrest in April due to the COVID-19 pandemic, but the founder of the Jaysh-e-Mohammad (JeM) terror group, Masood Azhar, remained free and resided in Pakistan.

In Sri Lanka's presidential election of 16 November 2019, Gotabaya Rajapaksa of the Sri Lanka People's Front defeated Sajith Premadasa – a minister in the previous coalition government – by 52% to 42% of votes. This was the result of strong anti-incumbency sentiment over the slowing economy, compounded by the previous government's failure to prevent the April 2019 terrorist attacks.

Japan's Free and Open Indo-Pacific Vision at the Crossroads

Will it endure after Abe?

In August 2016, Japanese Prime Minister Abe Shinzo announced his 'Free and Open Indo-Pacific (FOIP)' vision at the Sixth Tokyo International Conference on African Development (TICAD-VI). The vision has three pillars: promotion of the rule of law, freedom of navigation and free trade; pursuit of economic prosperity by improving connectivity and strengthening economic partnerships; and commitment to peace and stability.

FOIP is the Abe government's central diplomatic strategy to respond to the rise of China in Southeast Asia and India, regions strategically important to Japan. After Donald Trump became US president, Japan also sought to use FOIP to retain America's commitment to the region at a time when 'America First' populist protectionism risked creating a leadership vacuum.

The FOIP vision has so far been vague but has come to play a central part in Abe's goal of providing a 'proactive contribution to peace'. Since its launch, annual white papers from a range of ministries have included FOIP-related policy achievements. FOIP also features in policy speeches given by Abe and senior government officials. Japan's allies and regional partners have also recognised the three pillars of FOIP and the idea of the 'Indo-Pacific' as a geographical concept. But since early 2020, the public-health and economic crises caused by the COVID-19 pandemic have challenged the fundamental principles of FOIP. The intensifying US–China economic, political and security rivalry also risks destabilising the regional order.

To what extent has Japan achieved the goals of FOIP? Will FOIP remain a viable Japanese strategic vision in 2021 and beyond? This will largely depend on Japan's political and economic capacities in the post-COVID era, its ability to institutionalise the FOIP concept, and the external factors shaping the geostrategic environment in the Indo-Pacific region.

Milestones of Japan's Indo-Pacific policy

Japan's main FOIP priority so far has been to enhance the rules-based open trade and economic order. In March 2018, it played a leadership role in concluding the Comprehensive and Progressive Agreement for Trans-Pacific Partnership (CPTPP), the world's third-largest free-trade zone, which now includes Canada and ten other countries in the Asia-Pacific region. This agreement commits members to raise industrial standards and abide by e-commerce rules, with the implicit goal of encouraging China eventually to abide by these rules.

Japan has also fostered economic rule-making by promoting the 'quality infrastructure' investment standard. Unlike some China-funded development projects under the Belt and Road Initiative (BRI), this standard includes a due-diligence process to ensure the transparency and sustainability of projects. The culmination of this effort was the G20 Leaders' Summit in Osaka in June 2019, when Japan secured endorsements of the G20 Principles for Quality Infrastructure Investment from all participating countries, including China. In September 2019 Japan and the European Union signed the Partnership on Sustainable Connectivity and Quality Infrastructure that embodied these principles.

The FOIP concept has also helped to maintain US focus on the region. From 2017, senior officials in the Trump administration began to echo Japan's vision of a free and open order in the Indo-Pacific. The most advanced form of cooperation emerged in the economic-connectivity vision for the region. In 2018, Japan and the US released a joint statement prioritising the development of energy, infrastructure and digital connectivity through senior-level coordination, cooperation between public and private financial institutions, and standard-setting. Examples of successful projects include private-sector cooperation for the Japan–Guam–Australia fibre-optic submarine cable system project. In November 2019, Japan and the US, together with Australia, launched the Blue Dot Network scheme to mobilise public and private investment for infrastructure projects meeting the quality standards endorsed at the G20 summit.

Japan and the US have also worked towards a common vision of digital connectivity. The United States' withdrawal from the Trans-Pacific Partnership (TPP) agreement in 2017 was a major blow to Japan's effort to create a digital partnership with like-minded countries that would support e-commerce rule-making. Despite this, in March 2019 the two countries set up a bilateral working-level policy group, the Japan–US Strategic Digital Economy Partnership (JUSDEP), to coordinate on regional policies around e-commerce rule-making, cyber-security capacity-building and promotion of open, reliable and interoperable digital-infrastructure development from 5G to smart-city projects. Japan succeeded in bringing the US back into a regional framework on e-commerce rules with the signing of the US–Japan Digital Trade Agreement in October 2019.

On the security front, the Trump administration recognised the importance of enhancing security architecture in the Indo-Pacific region. The 2017 US National Security Strategy was the first official document to refer to the concept of the Indo-Pacific region, and the US Pacific Command was renamed the US Indo-Pacific Command accordingly. The 2018 Defense Strategy noted the importance of a 'free and open Indo-Pacific region' and stated that the US would 'strengthen our alliances and partnerships in the Indo-Pacific to a networked security architecture capable of deterring aggression, maintaining stability, and ensuring free access to common domains ... to preserve the free and open international system'. At the 2019 IISS Shangri-La Dialogue in Singapore, acting US secretary of defense Patrick M. Shanahan further articulated the Indo-Pacific as the 'priority theatre' for the US, and reassured regional countries concerned about the challenges to rules-based maritime order amid the rise of China.

Japan's push for the Indo-Pacific is not new. Abe originally launched the idea of connecting the seas of the Indian Ocean with the South China Sea and the Pacific Ocean in his 2007 speech at the Indian parliament, entitled 'Confluence of the Two Seas'. He attempted to operationalise security cooperation by establishing in 2007 the Quadrilateral Security Dialogue (also known as 'the Quad') with Australia, India and the US.

Like the US, Australia and India have included elements of FOIP in their policy statements and speeches. Australia's 2017 Foreign Policy White Paper set out the goal of promoting 'an open, inclusive and prosperous Indo-Pacific region'. At the 2018 IISS Shangri-La Dialogue, Indian Prime Minister Narendra Modi articulated his country's Indo-Pacific vision for the first time, using the same geographical scope outlined by Abe in 2016: from the 'shores of Africa to that of the Americas'. These documents and statements marked a departure from the previous security-oriented approach of the Quad in emphasising the role of diplomatic, as well as military, relations with Southeast Asia.

The Abe administration has expanded and promoted FOIP to regional and global like-minded partners beyond the Quad countries too. In 2018, Indonesia also launched its Indo-Pacific Cooperation Concept as part of its effort to develop a non-aligned stance in the US–China strategic rivalry. This concept eventually served as the foundation of the 'Outlook on the Indo-Pacific' released by the Association of Southeast Asian Nations (ASEAN) in June 2019. The United Kingdom and France, Europe's major maritime powers, have also accepted the principles of FOIP in the Indo-Pacific region. They have enhanced ties with Japan and deepened their strategic posture to assert freedom of navigation in the region. In 2017, the prime ministers of Japan and the UK agreed on a Joint Declaration on Security Cooperation, and by 2020 had hosted bilateral exercises between all branches of the military. The potential deployment of the Royal Navy's new HMS *Queen Elizabeth* carrier in the contested waters of the Indo-Pacific is also a sign of the high priority that the UK places on security in the region. French President Emmanuel Macron and Abe have accelerated talks to enhance security and stability in the Indo-Pacific since Abe's visit to France in October 2018. France is also taking steps to expand strategic partnerships with India, Vietnam and Australia.

The impact of COVID-19 on FOIP

Despite these achievements, the outbreak of the COVID-19 pandemic and the subsequent travel bans, supply-chain disruption and strict lockdown measures by Japan and other countries in the region immediately affected the fundamental goals of FOIP. Strict containment measures and restrictions severely hit economic growth. By June, Japan had imposed travel bans on 146 countries. It was the only G7 country to apply the same re-entry rules to foreign residents and foreign visitors. Suspension of the production of value-added manufacturing goods such as semiconductors and automobiles also hurt the economy and prompted the Japanese government and companies to shift production from China back to Japan, to enhance economic security. In April 2020, the government allocated JPY220 billion (approximately US$2.2bn) to help Japanese manufacturers achieve this. The government also set aside JPY23.5bn (approximately US$222.5 million) to diversify its supply chain to other regional countries, but this budget was limited compared to the amount allocated to bring production back to Japan.

In Southeast Asia, the relatively weaker medical infrastructure and smaller capacity of wards and intensive-care units in hospitals prompted governments to impose stronger measures to contain the virus by restricting economic and social activities. Cooperation on health and medical supplies, such as masks, ventilators and protective equipment, was limited, and most countries did not have enough to meet domestic needs. These responses challenged the 'connectivity' goals of FOIP, and seriously affected the regional economy.

The COVID-19 outbreak also challenged the FOIP goal of maintaining an open and transparent information space in the region. The Chinese government faced widespread international criticism over its alleged under-reporting of case and mortality statistics, censorship of medical researchers and the lack of accountability in its handling of its domestic COVID-19 outbreak. Japan, the US and other liberal-democratic partners also viewed China's extensive use of digital and artificial-intelligence technologies to track and contain the virus as challenges to human rights and liberal values. The success of democracies in the region, such as Taiwan and South Korea, in managing the trade-off between liberal values and the use of digital technologies to combat COVID-19 offered an alternative model to China's. However, these two nations have not yet adopted Japan's FOIP concept.

Moreover, when regional cooperation was most needed to meet the challenges the pandemic posed, neither Japan nor the US had the capacity to initiate it. On 7 April, a rapid rise in COVID-19 cases led Abe to announce a state of emergency in seven prefectures, which was expanded nationwide on 15 April. His administration faced criticism for its handling of the pandemic, and its net approval rate turned negative. But even in this difficult situation, Japan continued to promote the FOIP vision by attending the virtual Special ASEAN Plus Three Summit on Coronavirus Disease 2019 on 14 April, convened at the request of Vietnam. At the meeting, Abe implored regional states to embrace free, 'transparent' and 'timely' information-sharing and proposed the establishment of an 'ASEAN centre for emerging diseases, and public health emergencies'. However, the domestic political environment, and potential future shortage of medical supplies, did not allow the administration to promote widely its financial, medical and economic support to the region. The lack of active US diplomacy also proved problematic. America became the worst-affected country by infected cases and number of deaths. It retreated from multilateralism, pulling out of the World Health Organization and failing to convene the G7 Leaders' Summit as the 2020 chair. It even failed to cooperate bilaterally with Japan, one of its closest allies. Three Trump–Abe phone calls by the end of May led to agreement only on basic coordination of information-sharing on vaccine development.

China's growing confidence and assertiveness

While Japan, the US and the rest of the Indo-Pacific struggled to manage the pandemic, China's relatively quick recovery enabled it to assert political, economic and military influence. From March, Chinese leader Xi Jinping and senior officials repeatedly announced the intention to create a 'Health Silk Road' to offer masks and medical supplies to pandemic-stricken countries, with the aim of painting the country in a more positive light. Furthermore, China increasingly acted in ways that undermined all three pillars of FOIP. Such activities faced backlash from other countries in the region, and further intensified major-power competition with the US.

In the security realm, China intensified its challenge to the rules-based maritime order. In April 2020, it sank a Vietnamese fishing boat in the South China Sea, heightening tensions between the two countries. In the East China Sea, China also increased its presence and activities near the Japanese-administered Senkaku/Diaoyu Islands, which are also claimed by China. From April 2020, Chinese patrol vessels remained in the contiguous zone around the Senkaku/Diaoyu Islands for more than 100 consecutive days. The Japanese government also detected the activity of Chinese submarines near Japan's Amami-Oshima in the first island chain.

In the economic realm, China's projection of geo-economic power – the ability to pursue strategic interests through economic means – undermined the rules-based economic order in the region. China threatened to retaliate economically against countries that described the pandemic in ways it considered unfavourable to its image. A notable example was its suspension of beef imports from Australia in May 2020 in retaliation against the Australian government's call for an independent inquiry into the origins of the pandemic.

This assertiveness prompted other countries to review their ties with China and strengthen their links with regional strategic partners. The strongest reaction came from the US. After the National Security Council released its new strategy document on China in May 2020, senior officials repeatedly made statements advocating a fundamental shift in US policy towards China to confront its growing assertiveness. The US strengthened its export controls on Chinese telecommunications giant Huawei and urged like-minded countries to align with US policies. Its push for economic and technological decoupling from China risked further disruption of regional and global supply chains.

India and Australia also enhanced security ties among the Quad countries. On 4 June 2020, the two countries signed the awaited Mutual Logistics Support Agreement to allow mutual access to military bases for logistics support to strengthen interoperability. The two governments also signed the Defence Science and Technology Implementing Agreement to improve collaboration for defence and technology research and development. On the same day, the heads of both states held a virtual summit to make a Joint Declaration on a Shared Vision for Maritime Cooperation in the Indo-Pacific, which reiterated the two countries' 'enduring interest in a free, open, inclusive and rules-based Indo-Pacific region'.

Since the two governments signed the agreements, there has been growing momentum in the Indian policy community and media to invite Australia to join the *Malabar* naval exercise conducted by the US, Japan and India. Such a move would indicate a significant change in Indian policy. India has long been cautious about inviting Australia, despite the latter's request, in order to avoid the perception that it was promoting security cooperation against China in the Quad framework.

Australia is also expected to conclude a Reciprocal Access Agreement with Japan in July 2020. This will define a legal framework for Australian and Japanese troops when temporarily stationed on each other's territories for joint exercises. This will be an important step for the two countries to deepen their relations as quasi-allies. These efforts are expected to strengthen security ties between the Quad countries to support a rules-based maritime order in the Indo-Pacific, with the potential to involve ASEAN partners in the longer term.

Prospects for FOIP after Abe

The COVID-19 pandemic challenged the fundamental FOIP principle of connectivity. As countries shifted their policy focus and resources to tackle domestic public and health crises, neither Japan nor the US was capable of leading an immediate multilateral response to the crises in the Indo-Pacific region. Bilateral, multilateral and most regional dialogues were largely put on hold after the outbreak of the pandemic. As a result of the weak regional response, China extended its geopolitical and geo-economic influence in the region. However, China's increasing confidence and assertiveness also served as a catalyst for Japan's like-minded regional partners to speed up their multilateral coalition-building to counter Chinese activities and uphold the rules-based liberal order.

To make the FOIP vision resilient to these crises, Japan should use existing regional multilateral institutions to discuss pressing issues on pandemic preparedness, economic security and digital governance. Annual regional forums expected to take place in late 2020 – such as the ASEAN Regional Forum, East Asia Summit and the Asia-Pacific Economic Cooperation meeting – offer major opportunities for Japan to bring in the US and other like-minded countries and gain wider support for its idea of establishing an ASEAN Centre for emerging diseases and public-health emergencies. It is also critical that discussions take place on cooperation on trade schemes to stock and maintain the flow of masks, ventilators, protective equipment and medicine to provide an alternative to China's Health Silk Road development.

Such an effort will require Japan and the US to take a leadership role to maintain and enhance the FOIP. Two major uncertainties could challenge the viability of the FOIP vision in 2020 and beyond. The first is US commitment. As of June 2020, the significant fall in GDP, growing unemployment and rapid surge of reported COVID-19 cases and death rates all cast doubt on a quick US recovery. Unless the domestic situation improves, it will be difficult for the US to take initiatives to launch new projects or accelerate security cooperation to enhance and maintain the FOIP vision.

The second uncertainty is Japan's political and economic capacity to sustain a proactive diplomatic, economic and security role in the region. The high number of COVID-19 cases, and grim economic and fiscal outlook for 2021, may undermine Japan's ability to resource its regional diplomatic goals. Nor is it clear that Japan will retain the political will to pursue a coherent vision of regional order. In November 2019, Abe became the longest-serving Japanese prime minister in the history of constitutional government in terms of totals days in office. His long tenure has underpinned Japan's recent proactive diplomacy. However, the term of his current administration is set to expire in autumn 2021. As of June 2020, it was unclear if Abe would serve a fourth term – and, if not, whether a post-Abe government would continue to pursue his FOIP vision.

Hong Kong
A triumph of experience over hope?

Hong Kong was the test-case for the 'One Country, Two Systems' concept devised by veteran Chinese Communist Party (CCP) leader Deng Xiaoping to achieve one of the CCP's core aims, national reunification. This involved the recovery of the colonies of Hong Kong and Macao, as well as Taiwan, which, since 1949, has been to all intents and purposes an independent state but is seen by Beijing as a rebel province.

Apart from fomenting an abortive uprising there during the Cultural Revolution, China had shown itself in no hurry to regain control of Hong Kong from the British, though it always maintained that it would do so in due course. But the need to offer certainty about the future of commercial leases that would fall due after 1997 – when the lease on Hong Kong's New Territories would expire – led to a period of intensive negotiation. The result was the 1984 Sino-British Joint Declaration, an international treaty deposited at the United Nations. This set out in extensive detail how what would become the Hong Kong Special Administrative Region (SAR) would be governed, and what freedoms its residents would enjoy once sovereignty reverted to China.

These freedoms went beyond anything available on the Chinese mainland. Hong Kong's colonial-era administration and institutions were left intact, with the colonial governor being replaced by an elected chief executive with ministerial rank. Hong Kong retained its own Legislative Council; its own common-law system; its own currency, pegged to the US dollar; and the rights of freedom of expression and free assembly. Hong Kong also had its own mini constitution in the form of the Basic Law, drafted by Beijing. China opened a liaison office in Hong Kong and installed a small People's Liberation Army garrison that maintained a low profile.

From the start, China's readiness to abide by the terms of the Joint Declaration evoked concern. But Hong Kong was initially left to its own devices. At the time of the 1997 handover, Hong Kong's GDP was roughly 20% of China's, and it played a vital role in China's development as a source of foreign capital and technology. A successful recovery of

Hong Kong was good both for China's international image and, perhaps more importantly, for its aspirations to recover sovereignty over Taiwan peacefully. But as China's economy underwent a rapid expansion following its accession in 2001 to the World Trade Organisation, its relationship with Hong Kong began to change. By 2020 Hong Kong's GDP was only 3% of China's. Peaceful reunification with Taiwan seemed an ever-more-remote prospect. As China became stronger and more assertive, its tolerance of Hong Kong's exceptional status began to fade. By 2019 what had previously been an asset was coming to be seen in Beijing as a political and security liability.

Rising tensions

There had always been a gap in expectations between Beijing and Hong Kong. This first manifested itself in 2003 when Hong Kong secretary for security Regina Ip attempted to introduce legislation on subversion and secession in accordance with Article 23 of the Basic Law, which stated that 'Hong Kong shall enact laws on its own to prohibit any act of treason, secession, sedition, subversion against the Central People's Government'. In the face of large-scale peaceful demonstrations, this legislation was shelved, and was not subsequently addressed again until 2020.

Concern about China's readiness to respect the high degree of autonomy afforded under the terms of the Joint Declaration began to emerge in 2014 with a dispute over the method for electing Hong Kong's chief executive. Beijing, which had accepted that this would eventually be achieved through universal suffrage, insisted on a list of candidates that it had pre-approved. This gave rise to protests under the label of Occupy Central, more informally known as the Umbrella Movement as the protesters used umbrellas to protect themselves against police pepper spray. The protests fizzled out and the movement's leaders were sentenced to brief terms of imprisonment for public-order offences. Beijing did not give ground on its insistence that candidates for the post of chief executive had to be vetted.

Thereafter a series of small but significant events raised further questions about Beijing's continuing commitment to the Joint Declaration.

The first occurred in 2015, when the founder and four employees of the Hong Kong publisher Causeway Bay Books were detained in China. The company's Gui Minhai, who was a Swedish citizen, was rendered to China from Thailand, while the others were abducted from Hong Kong. The five detainees made televised confessions that one detainee, on his return to Hong Kong, confirmed had been scripted by the Chinese authorities. This episode was denounced by the United Kingdom's then foreign secretary, Philip Hammond, as 'a serious breach of the Sino-British Joint Declaration'.

In 2016 Xiao Jianhua, a wealthy mainland Chinese financier who had assisted many of China's political elite in moving their money to Hong Kong, was abducted by Chinese security agents from an apartment in the Four Seasons Hotel, where he had been based for the preceding three years. And in 2019 it emerged that China's Ministry of Public Security, the police ministry that also exercises some domestic-security functions, had conducted extensive surveillance operations in Hong Kong against Western journalists investigating Chinese links to Malaysia's 1Malaysia Development Board scandal.

The 2019 demonstrations

A growing sense of unease among many Hong Kong residents about China's intentions erupted into a full-blown crisis in February 2019, when the SAR government proposed legislation that would enable criminal suspects to be handed over to jurisdictions with which Hong Kong had no extradition agreement, including mainland China. The proposal gave rise to fears that critics of China might be extradited there for political crimes and face certain punishment. The proposed legislation led to large-scale peaceful demonstrations. But Hong Kong Chief Executive Carrie Lam, a career civil servant rather than a politician, pressed ahead with a second reading of the bill. At that point, the demonstrations escalated and began to turn violent, with protesters breaking into and defacing the Legislative Council Complex and subsequently occupying Hong Kong's airport.

Faced with a dramatic deterioration in public order and ever-more violent confrontations between police and demonstrators, Lam first announced a suspension of the extradition bill and then pronounced it to be dead – though Beijing made clear that the bill could not formally be withdrawn. But that was not enough to satisfy the protesters, who issued five demands: the formal withdrawal of the extradition bill, the resignation of Lam, the withdrawal of the designation of the protests as riots, an independent inquiry into police conduct and an amnesty for all those arrested in the protests. Lam reportedly tendered her resignation to Beijing and had it rejected. Admitting that she was serving 'two masters' and had little room for manoeuvre, Lam, under direction from Beijing, rejected all five demands, with the result that violent demonstrations continued.

Although both Beijing and the SAR government sought to play down the level of support for the protests, polls suggested that this was consistently high, at around 60% – though the number of respondents who professed themselves willing to take part was somewhat lower. The demonstrators defied categorisation. Those in the front line were mostly young and well educated, and appeared to be motivated first and foremost by fear of a China that had become progressively more authoritarian and intolerant of criticism. They enjoyed widespread support from older residents, many of whom provided services such as a 'school bus' network of getaway vehicles to enable protesters to evade the police. The protests, organised through social media such as Telegram, were by design leaderless. Their tactics followed the injunction of the late martial-arts film star Bruce Lee to 'be like water', meaning that their deployments should be fluid, flexible and agile. As the protests grew in intensity, a protest culture evolved with slogans such as 'Recover Hong Kong, Revolution for Our Time' and an anthem, 'Glory to Hong Kong'. Some protesters called for independence – a red line for Beijing that it could not contemplate accepting in any form.

Beyond the immediate fear of an overbearing China, the protests also reflected frustrations with a Hong Kong administration seen as remote, out of touch and dependent on a wealthy elite of pro-Beijing business interests. Dealing with the protests demanded political skills which the Hong Kong government, made up of civil servants, simply did not have.

Rather than seeking to engage with the protesters, they relied on the police to deal with the demonstrations.

It quickly became clear that Beijing had been caught unawares by the protests and failed to understand their causes. China's official press ascribed the outbursts to economic inequalities and an absence of affordable housing, which they laid at the door of the wealthy Hong Kong business leaders who had been their staunchest supporters. As the protests grew more violent, Beijing blamed what it described as acts of terrorism on 'foreign black hands' seeking to instigate a colour revolution. A propaganda campaign was launched both within China and among Chinese diaspora communities with overt messaging on Twitter and Facebook aimed at discrediting and disparaging the protesters. This was amplified by bots using Twitter accounts originally designed to promote pornographic sites and purchased on the dark web.

In Hong Kong itself, Beijing sought to rally its traditional supporters, many of whom had been wavering, while putting pressure on the private sector to dissociate itself from the protesters. Failure to see which way the wind was blowing led on 4 September 2019 to the resignation of the chairman and CEO of Hong Kong's flagship airline, Cathay Pacific. But fears that Beijing might deploy its own gendarmerie, the People's Armed Police, went unrealised. Beijing instead focused on expressions of support for the Hong Kong police, and provided it with intelligence.

Legislation and international repercussions

Beijing's discomfiture with the protests was exacerbated in November 2019, when local elections resulted in pro-democracy candidates taking control of 17 of the 18 district councils and tripling their number of seats, largely at the expense of veteran pro-Beijing politicians. In late January 2020, the protests were brought to a temporary close not by the actions of the Hong Kong or Beijing governments, but by the COVID-19 pandemic. Protest groups repurposed themselves to provide community support for managing the outbreak. In November 2019 Beijing had dismissed head of the Hong Kong Liaison Office Wang Zhimin and director of the Hong Kong and Macao Affairs Office Zhang Xiaoming, who took the blame for failing to read the situation correctly and misinforming Beijing. They were replaced by Luo Huining and Xia Baolong respectively, both enforcers rather than conciliators, in a move indicative of the shape of things to come.

Beijing also made it clear that Hong Kong had to pass anti-subversion legislation as mandated in its own constitution. Then, in the face of gridlock in Hong Kong's Legislative Council due to persistent filibustering by pro-democracy legislators, on 21 May 2020 Beijing announced that its own legislature, the National People's Congress, would pass this law.

This decision led to an international outcry amid allegations that it was a clear breach of the Joint Declaration. In 2019 United States legislators had passed the Hong Kong Human Rights and Democracy Act, which required the US government to certify annually that Hong Kong retained a high degree of autonomy. Hong Kong's continued enjoyment of privileged US treatment would depend on this – specifically, the provisions of the 1992 Hong Kong Policy Act that treated Hong Kong as a separate customs territory and guaranteed the peg to the US dollar. On 27 May 2020 US Secretary of State Mike Pompeo testified to Congress that this high degree of autonomy no longer existed, potentially paving the way for these privileges to be revoked. President Donald Trump announced that he would end the United States' special relationship with Hong Kong, a move that would 'affect the full range of agreements … from our extradition treaty to our export controls on dual-use technologies and more, with few exceptions'. He did not go into detail on what measures his administration would impose, though the US government subsequently announced on 29 June 2020 that it would cease defence exports to Hong Kong and restrict transfers of sensitive dual-use technologies, as it does with China. It also announced visa restrictions on CCP officials.

UK Prime Minister Boris Johnson promised that, if the new legislation went ahead, the UK would change its immigration system to enable the three million Hong Kong residents born before 1997 to apply for British National (Overseas), or BNO, passports that would grant the right to move to the UK and pursue a path towards citizenship. The BNO passport was a device conceived by the UK to incentivise Hong Kong's elite to remain in the territory after the handover by offering an escape if things went wrong. Beijing had from the outset been unhappy about the passports and reacted forcefully to the Johnson initiative by warning the UK not to interfere in China's internal affairs.

It quickly became evident that Beijing would not be deterred by these responses and that the relevant legislation would be fast-tracked through China's legislative process. The new law, which came into effect on 30 June 2020, covers secession, subversion of state power, terrorist activities and 'collusion with foreign or external forces to endanger national security'. This latter provision is of particular concern because China's definition of national security is far more encompassing than is true for liberal democracies. Beijing has sought to provide reassurance that the legislation will only affect a small minority of cases. But there is a widespread and well-founded fear that not only would criticism of Beijing become criminalised – in what seemed like a Freudian slip, Chief Executive Lam said that Hong Kong people could speak freely 'for the time being' – but that private-sector commercial decisions that Beijing saw as inimical to its interests might be criminalised too. The legislation, worded like all Chinese laws in vague and general terms, has extraterritorial application through Article 38, which stipulates that the law applied to acts outside Hong Kong by those who were not Hong Kong residents.

Under the legislation a new Office of the National Security Commissioner of the People's Republic of China on the Hong Kong SAR is to be established to analyse the national-security situation in Hong Kong and offer advice on strategies and policies. Hong Kong is required to set up a Commission to Safeguard National Security, to be chaired by the chief executive and include a national-security adviser appointed by Beijing. Dedicated units are to be created in the Department of Justice and the police to deal with national-security cases. Judges will be appointed by the chief executive to hear cases, with the exception of those particularly egregious cases that will be heard in the mainland. China's intelligence and security agencies, which have been operating covertly in Hong Kong for many years under a variety of official and non-official guises, are authorised to operate in Hong Kong as required. Full details were still lacking at the time of the law's passage, but the general direction of travel was clear: the SAR is expected to take action, but China will decide what action will be taken and against whom. Beijing will not be deterred by the protests this move is bound to elicit.

The Chinese government has come to view the recent protests in Hong Kong as a cancer that has to be cut out lest it spread to the mainland. The protests came at a time when China's leadership was confronting a range of domestic and international challenges that included an economic slowdown, a dramatic deterioration in relations with the US and an international reputation tarnished by the initial mishandling of the COVID-19 pandemic. The protests also had a significant adverse effect on cross-straits relations, enabling localist Tsai Ing-wen, who had been trailing in the polls, to win a second term as president of Taiwan with a convincing majority. As a consequence, on 19 June 2020 the Chinese-nationalist opposition Kuomintang party published new proposals for relations with Beijing. They made no mention of the 1992 Consensus, a device whereby both China and Taiwan agreed that there was only one China while holding their separate views on what that China should be. These moves would appear to put peaceful reunification under the One Country, Two Systems formula beyond reach.

Beijing's gambit

China's leadership is in an angry, vengeful and paranoid mood and is determined to bring Hong Kong to heel at whatever cost. Beijing will hope to preserve Hong Kong's status as a major global financial centre and to continue to benefit from Hong Kong's ability to access US technology that cannot be exported directly to China. And it will calculate, quite possibly correctly, that a US government in disarray, led by a president whose approach to China has been ambivalent and inconsistent, will stop short of invoking the nuclear option of withdrawing support for the Hong Kong dollar peg. That would leave the currency open to a damaging speculative assault that would undermine confidence in Hong Kong's economy and its credibility as an international financial centre.

But Beijing appeared confident it would be able to ride out the storm because of the international draw of its economy and markets. In the worst-case scenario, Beijing would be ready to bail Hong Kong out and could use its own holdings of US dollars to guarantee the peg. Nor is it overly concerned about the risk posed by local demonstrators. The new law has already had a chilling effect on pro-democracy groups, many of which disbanded and sought to erase their digital footprints. Some have moved to Taiwan, which has offered them asylum. Beijing appeared to be calculating that making a high-profile example of early offenders would intimidate the population into compliance, on the time-honoured Chinese principle of 'killing the chicken to frighten the monkeys'.

Hong Kong's exceptional status will become more a matter of form than substance. Beijing can be expected to eliminate not just all vestiges of British colonial rule but all the aspects of Hong Kong that set it apart from mainland China. Hong Kong will become just another Chinese city, merged into the Greater Bay Area together with Macao and nine cities in southern Guangdong province. The CCP has always prioritised stability over all other considerations. In this respect the likely outcome for Hong Kong can be seen as a triumph of experience over hope.

Changes in Jammu and Kashmir

What are the strategic implications?

In a sudden and controversial move on 5 August 2019, the Indian government announced its intention to revoke the special 'semi-autonomous' constitutional status of the Indian province of Jammu and Kashmir that had been in place for nearly 70 years. This marked a major unilateral change to the status of a province that for decades has been a source of tension between three nuclear powers – India, Pakistan and China. Although New Delhi stated that its actions were an internal matter and did not necessitate any change to India's external borders, it heightened existing tensions with Pakistan and China, strained relations with Nepal and raised the risk of regional instability.

The decision was announced in parliament by powerful Home Minister Amit Shah through the submission of legislation abrogating Articles 370 (of 17 October 1949) and 35A (of 14 May 1954 – part of Article 370) of the Indian constitution, which had provided special status to the province, and dividing the province into two separate union territories – Jammu and Kashmir, and Ladakh – to be governed directly by the central government. Article 370 had allowed the province to have its own constitution and flag, and to establish its own laws (except on defence, foreign affairs, finance and communications), while Article 35A had given its permanent residents special privileges, including prohibiting other Indian citizens from purchasing land or property in the province.

The province had been under president's rule since 20 December 2018 and its local legislature dissolved, so the parliament in New Delhi took up the issue immediately. With the ruling Bharatiya Janata Party (BJP) able to ensure a two-thirds majority in both the Lok Sabha (the lower house) and the Rajya Sabha (the upper house), and enjoying overwhelming support throughout the country (including from non-Kashmiri Muslims), parliament duly revoked Articles 370 and 35A on 6 August 2019. The presidential order of 9 August 2019 declared that all provisions of the Indian constitution and parliamentary laws would now apply to Jammu and Kashmir and Ladakh. President's rule ended on 31 October 2019 and two separate union territories were formally established, depicted in a new map of India's territorial boundaries released on 2 November. Jammu and Kashmir has a Muslim-majority population with a large Hindu minority, while Ladakh has a Buddhist-majority population.

These legislative changes were accompanied by a strict lockdown of Jammu and Kashmir from 5 August to prevent any violent mass protests (and civilian loss of life by security personnel) and terror attacks originating locally as well as from across the Line of Control (LoC) in Azad Jammu and Kashmir (Pakistan-controlled Kashmir). An additional 46,000 Indian paramilitary and army personnel were deployed to Jammu and Kashmir, which is fighting a decades-long anti-India insurgency. A curfew was put in place, with executive orders prohibiting gatherings of more than four people. Three former chief ministers of Jammu and Kashmir were placed under house arrest and some 3,500 Kashmiris were arrested on a preventive basis in August. Internet, mobile and landline connectivity services were suspended. The internet shutdown lasted for 175 days, making it the longest ever imposed in a democratic country. Nonetheless, several local protests still took place and nearly 3,000 additional arrests were made. A local human-rights group recorded 33 civilian deaths in Jammu and Kashmir between 5 August and 31 December.

Amid heightened international media criticism, the Indian government began easing the restrictions. On 14 October, call and text-messaging services were restored for postpaid mobile phones, and these services resumed on all mobile phones on 18 January 2020. Mobile internet services were gradually reinstated from 25 January, initially allowing access to 301 central-government-approved websites (albeit with slow connections) and then to social-media websites on 4 March. Broadband internet services were partially restored the following day.

By early November 5,000 of the 6,300 people arrested had been released. Father and son Farooq and Omar Abdullah, both former chief ministers, were released in March 2020, though another former chief minister, Mehbooba Mufti, remained in detention. The Indian government was keen to showcase a return to near normality in the region, including carefully scripted visits for groups of Delhi-based foreign envoys. But stringent laws remained in place prohibiting protests and allowing for preventive arrests and detention for up to two years without trial. Some political activists remained in detention (even though most had been released by June), curbs on media reporting continued and slow 2G internet speeds were the norm. Local political parties (other than the BJP) remained inactive. Restrictions on economic activities severely affected livelihoods, and the local population faced hardships. Schools were closed in August and did not reopen, except for a ten-day period in March. The situation was exacerbated by the 75-day countrywide lockdown from the end of March until early June to counter the COVID-19 pandemic.

Despite the presence of a larger number of security personnel than usual, attacks on civilians and security personnel continued, and skirmishes between militants and security personnel were reported. At least 119 militants were killed between January and July 2020, along with several security personnel. But there were no attacks on the scale of the suicide-bomb attack in Pulwama in Jammu and Kashmir in February 2019, which killed 40 Indian paramilitary personnel and resulted in a retaliatory Indian airstrike against a training camp of the terror group Jaysh-e-Mohammad (JeM) in Balakot in Pakistan's Khyber Pakhtunkhwa province. Pakistan's counter-response led to the shooting down of an Indian MiG-21 fighter aircraft. Heightened bilateral tensions had eased in March 2019 after Pakistan returned the downed pilot to India.

India's rationale

The BJP's aspiration to revoke Jammu and Kashmir's special status had been reflected in its 2014 and 2019 general-election manifestos. Its 2014 manifesto stated that 'Jammu and Kashmir was, is and shall remain an integral part of the Union of India', and that the BJP was committed to the abrogation of Article 370. This was reiterated in 2019, with the additional proviso that the party was committed to annulling Article 35A because it was an 'obstacle in the development of the state'. Yet despite its single-party majority in the 2014 general election, the

Narendra Modi-led central government ignored this aspect of the manifesto during its first five-year term. Other priorities may have taken precedence, in particular providing the country with effective governance, development and sustained economic growth, and initiating the resumption of an official-level peace dialogue with the Pakistani prime minister, Nawaz Sharif, which had been stalled since December 2012. But with Modi's landslide general-election victory in 2019 (the first time in nearly 50 years that an incumbent government had won an absolute majority and an increased mandate), the appointment of a powerful new home minister and heightened tensions with Pakistan under Prime Minister Imran Khan, the second Modi government sensed an opportunity to deliver early on its electoral pledge on Kashmir.

The rationale for the decision was that Articles 370 and 35A were a 'temporary provision' that had prevented the full merger of Jammu and Kashmir with India, rather than being the basis of its merger (it was a Hindu-ruled Muslim-majority princely state with independent status at the time of the partition of British India). Moreover, on 8 August, Modi stated that the special status had led to 'secessionism, terrorism, nepotism and widespread corruption on a large scale'. In effect, the government argued that revoking the special status would make Jammu and Kashmir a 'true part of India' and signal an end to any possibility of the secession of Kashmir. To this effect, on 20 May 2020 the central government formalised the repeal of 29 laws and the amendment of 109 others governing the province. Crucially, in line with national provisions, recruitment to all local-government posts was expanded from only 'permanent residents' to those having 'domicile' in the union territory (including non-Kashmiris). Two of the main Kashmiri political parties criticised this move as an attempt to change the region's demography.

Several mainstream opposition political parties also challenged the change in Jammu and Kashmir's status. They argued that the central government could not change the internal boundaries of the province without consulting its legislative assembly; that this was the first time that a province had been turned into a union territory; and that the Jammu and Kashmir (Reorganisation) Bill 2019 set a constitutional precedent to revoke Article 371 of the Indian constitution, which provides special provisions to 11 other Indian provinces.

Regional opposition

The most vociferous opposition came from Pakistan, which disputed India's control of the province and has historically laid territorial claim to parts of it. Islamabad stated that India's unilateral step was illegal and a violation of UN Security Council (UNSC) resolutions because Kashmir was an 'internationally recognised disputed territory'. It described India's actions as an 'illegal annexation'. Pakistan also described the Kashmir dispute as an 'international nuclear flashpoint'. The Indian central government stated that its actions were an internal matter and did not necessitate any change to India's external boundaries, including the LoC or the International Border with Pakistan.

Pakistan also described human-rights violations during the lockdown as 'war crimes'. Khan was scathing in his opposition, comparing Modi to Adolf Hitler and saying that the Indian government was 'inspired by Nazi ideology', and suggesting that India's actions in Kashmir amounted to 'ethnic cleansing'. At his address at the UN General Assembly (UNGA) in September 2019, Khan warned of a 'bloodbath' with violent protests after the Indian government lifted restrictions in Kashmir. Pakistan called for the restrictions in Jammu and Kashmir to be lifted in view of COVID-19, with nearly 7,500 cases and just over 100 deaths in the union territory reported by 30 June. Khan also condemned India's controversial Citizenship (Amendment) Act (CAA) of 11 December 2019 as part of Modi's 'Hindu supremacist agenda'. The CAA offered Indian citizenship to minority communities of six religions (excluding Muslims) from Afghanistan, Bangladesh and Pakistan who were fleeing religious persecution.

Diplomatically, Pakistan responded on 7 August with a series of punitive measures against India. These included the downgrading of bilateral relations through the expulsion of the Indian high commissioner in Islamabad and a sustained pause on deploying its high commissioner-designate to New Delhi. Bilateral trade (albeit a paltry US$2.6 billion in FY2018–19) was also suspended and all agreements were reviewed.

Subsequently, Pakistan suspended cross-border train and bus services and banned all cultural exchanges with India, including the screening of Indian films in Pakistan.

Pakistan also urged China – its strategic partner and India's larger northern neighbour, with which New Delhi has strained relations – to strongly oppose India's decision. Although India emphasised that Ladakh's new administrative status as a union territory was an internal matter with 'no implication for India's external boundaries or the Line of Actual Control (LAC) with China', on 6 August China strongly opposed this as being 'unacceptable' and as 'undermining its territorial sovereignty'. Shah's statement in parliament on 6 August that the region of Kashmir, including Pakistan-controlled Kashmir and Aksai Chin (the Chinese-controlled sector of Kashmir that India claims is part of Ladakh), was an 'integral part of India' may have provoked the robust Chinese response. China warned India that it should 'refrain from taking actions that will unilaterally change the status quo and escalate tensions'. This was likely a factor in the subsequent Sino-Indian clashes in May and June in the eastern Ladakh sector along the LAC. A violent clash on 15 June in the Galwan Valley led to the deaths of 20 Indian (and an unknown but possibly larger number of Chinese) soldiers, the first fatalities on the LAC in 45 years.

Pakistan's most significant diplomatic achievement regarding Kashmir took place at the UNSC with China's support. On 16 August 2019, on China's initiative, the UNSC held an informal closed-door meeting on Kashmir, the first special meeting on the issue for nearly 50 years. But the UNSC members in attendance failed to agree a joint press statement, leaving the Chinese and Pakistani envoys to brief the media. The Chinese envoy urged India and Pakistan to refrain from taking 'unilateral action' in Kashmir, while the Pakistani envoy stated that the meeting 'nullified India's claim that Jammu and Kashmir is an internal matter for India'. A second UNSC discussion on Kashmir, scheduled for December, was postponed. At a UNSC meeting on 15 January 2020, the other four permanent members, led by the United States, stated that the Kashmir dispute was a bilateral matter for India and Pakistan to resolve, and should not be raised at the UNSC.

India's release of its new map on 2 November 2019 resulted in a diplomatic rift with Nepal, with Kathmandu objecting to the disputed area of Kalapani – located at the tri-junction of India, Nepal and China – being shown within India's borders. Nepal subsequently issued its own new map, which was formally included in its constitution on 18 June. For the first time, this showed the disputed areas of Kalapani, Lipulekh and Limpiyadhura as part of Nepalese territory.

International criticism: limited impact

Mounting international media and official criticism of India affected its standing as the world's largest democracy. On 8 August, UN Secretary-General António Guterres called for 'maximum restraint', expressing concern that the restrictions in Kashmir could exacerbate the human-rights situation in the region. The General Secretariat of the 57-member Organisation of Islamic Cooperation (OIC) on 31 August 2019 reaffirmed UNSC resolutions on the internationally recognised status of the Jammu and Kashmir dispute and called for restrictions to be lifted. Malaysia and Turkey were particularly critical of India's decision. India subsequently restricted imports of palm oil from Malaysia (India is the world's largest importer of the product) and announced that it would cut imports of oil and steel products from Turkey.

In January 2020, the US Department of State's Bureau of South and Central Asian Affairs made a rare statement about India's actions in Kashmir, expressing concern about the 'detention of political leaders and residents, and internet restrictions'. The following month, the European Union called on India to lift the remaining restrictions. In April, the annual report of the US Commission on International Religious Freedom listed India as a country of particular concern for the first time since 2004.

However, India's key partners, including four of the five permanent UNSC members (France, Russia, the UK and the US), officially reiterated that Kashmir was a bilateral issue between India and Pakistan. In February, the UK Minister for Asia and the Pacific stated in parliament that the UK government's 'longstanding position is that it is for India and Pakistan to find a lasting political resolution on Kashmir, taking into account the wishes of the Kashmiri people ... it is not for the UK to prescribe a solution or to act as a mediator'.

Saudi Arabia and the United Arab Emirates also supported India bilaterally, even though multilaterally they agreed with the OIC critique. When Khan raised the Kashmir dispute with US President Donald Trump on the sidelines of the UNGA on 23 September 2019, Trump once again offered to mediate a peaceful resolution between the two countries, as long as both agreed to this. While Pakistan was in favour, India firmly rebuffed the offer, having consistently maintained that no third party should have a role in mediating bilateral disputes. Pakistan also failed to secure the support of enough states to submit a resolution on Kashmir at the 47-member UN Human Rights Council in September. Although India's actions in Kashmir resulted in greater international attention and scrutiny, Indian diplomacy successfully ensured that it remained a bilateral issue with Pakistan and avoided international mediation.

Increased prospects for regional instability

India's amendments to the status of Jammu and Kashmir exacerbated the already heightened tensions with Pakistan following the Pulwama–Balakot crisis. This resulted in sharpened public rhetoric and insults by both countries in international fora, and a doubling of ceasefire violations by Pakistan across the LoC amid a reduction in terror attacks and casualties in Kashmir.

Raising the rhetoric against India, Pakistan's Foreign Minister Shah Mehmood Qureshi warned on 5 August 2019 that India's Kashmir decision was a 'dangerous game' that would have 'serious consequences for regional peace and stability'. The following day, Pakistan's Chief of Army Staff Gen. Qamar Javed Bajwa added that the army was 'prepared and shall go to any extent' to support the people of Kashmir and their 'right to self-determination'. India responded by warning of a repeat of its February 2019 Balakot strike in response to another significant terror attack on Indian territory, while asserting its physical claim over Pakistan-controlled Kashmir and Gilgit-Baltistan. On 4 September,

Pakistan's military spokesman Maj.-Gen. Asif Ghafoor outlined Pakistan's additional response options, which included peacetime competition, conventional warlike operations, sub-conventional warfare and conventional warfare, and raised the possibility of nuclear escalation.

Nonetheless, Pakistan's military response to India remained limited. It largely took the form of 'sub-conventional warfare' through the intensification of shelling across the LoC. India claimed that Pakistan violated the 2003 ceasefire across the LoC 3,289 times in 2019, the highest-ever number of violations in a year; of these more than 1,500 incidents took place after 5 August. Pakistan claimed that India's intensified cross-LoC shelling killed 59 Pakistani civilians in 2019. Both India and Pakistan have regularly blamed each other for ceasefire violations across the LoC, stating that their respective actions were retaliation for the other side's violation of the ceasefire agreement.

Attacks against civilian and security personnel in Jammu and Kashmir continued. In March India stated that 79 terrorist incidents had taken place in the six months since August 2019, which it claimed were supported by Pakistan's military intelligence agency. The Indian security forces highlighted the renewed role of the Lashkar-e-Taiba (LeT) terror group (responsible for the November 2008 Mumbai attacks) in these attacks through its new offshoot, the Resistance Front. This raised the prospect of another significant terrorist attack in India with worrying implications for regional stability.

The situation worsened in early May 2020, when India accused Pakistan of supporting infiltration and terrorism in Kashmir at the same time as Indian and Chinese troops were involved in skirmishes and clashes along the LAC. Khan denied these accusations, saying that they were diverting world attention from 'Indian state terrorism' in Kashmir and were a pretext for a 'false flag' operation to target Pakistan. Bilateral tensions were further strained in June 2020 with India's expulsion of two Pakistani diplomats on accusations of espionage; the subsequent harassment of two Indian diplomats in Pakistan; and India's decision on 23 June to halve the number of Pakistani diplomatic staff in India, leading to the withdrawal of 39 Pakistani diplomats from India and 28 Indian diplomats from Pakistan by 30 June. Khan blamed India for a terror attack on the Karachi stock exchange on 29 June.

The increased tensions and tit-for-tat accusations took place in the absence of an official peace dialogue between New Delhi and Islamabad, which looked increasingly unlikely to resume. India stated that it would not engage in bilateral talks until Pakistan-based terrorist groups stopped conducting attacks in India. New Delhi insisted that, following the change in status of Jammu and Kashmir, there could no longer be a Kashmir dimension to such talks, other than to discuss the return to India of Pakistan-controlled Kashmir and Gilgit-Baltistan (both of which India continued to claim in its new official map). Rhetorically, India also reiterated its claim over Gilgit-Baltistan. Pakistan stated that it would not be party to any official talks with India until New Delhi reversed its 5 August decision on Kashmir.

The absence of official dialogue, 'backchannel' links or a crisis communication mechanism between the two countries could fuel misunderstanding and misperceptions. Another significant Pakistan-supported terror attack on India would raise the risk of the conflict escalating into a new crisis. Aggressive Indian and Pakistani conventional postures following the Pulwama–Balakot incidents, and the nature of their asymmetrical nuclear doctrines, are also fraught with risks. At the end of June 2020, the situation remained strained, and Kashmir a potential flashpoint.

Violent incidents in Kashmir

● **Major recent terrorist incidents**

29 October 2019
Kulgam district: six civilians killed in a Hizbul Mujahideen attack.

1 January 2020
Nowshera sector of Rajouri district: two security-forces personnel and three militants killed.

5 April 2020
Keran sector of Kupwara district: five security-forces personnel and five militants killed as the latter attempted to cross the Line of Control.

18 April 2020
Sopore in Baramulla district: three security-forces personnel killed following attack by the Resistance Front.

22 April 2020
Shopian district: four militants of Ansar Ghazwat-ul-Hind killed.

2 May 2020
Handwara area in Kupwara district: five security-forces personnel and two militants killed following a hostage-rescue operation.

4 May 2020
Handwara area in Kupwara district: three security-forces personnel and one civilian killed.

7 June 2020
Shopian district: five Hizbul Mujahideen militants killed.

10 June 2020
Shopian district: five Hizbul Mujahideen and Lashkar-e-Taiba militants killed.

18–19 June 2020
Shopian and Pulwama districts: eight Lashkar-e-Taiba militants killed in overnight encounters.

21 June 2020
Srinagar and Kulgam districts: four Islamic State in Jammu and Kashmir/ Hizbul Mujahideen militants killed in two separate encounters.

Ceasefire violations across the Line of Control

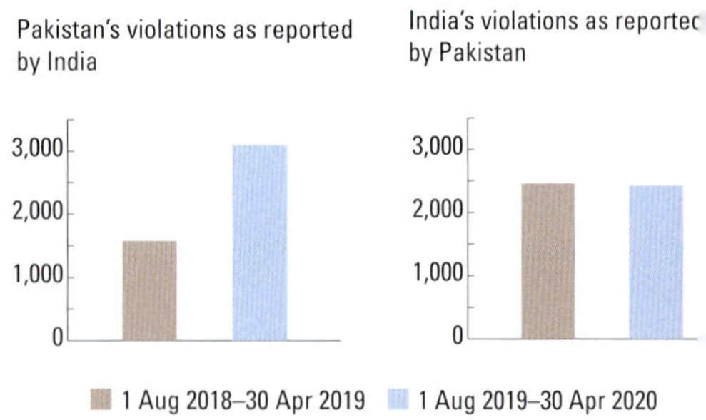

Note: map accurate as of June 2020

Sources: Government of India; Government of Pakistan; IISS; South Asia Terrorism Portal

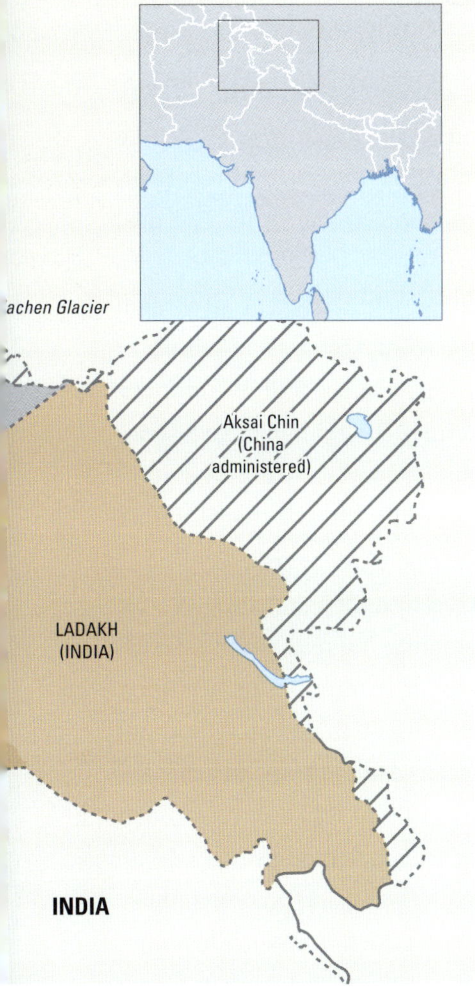

Civilians, security forces and terrorists killed in Jammu and Kashmir, India

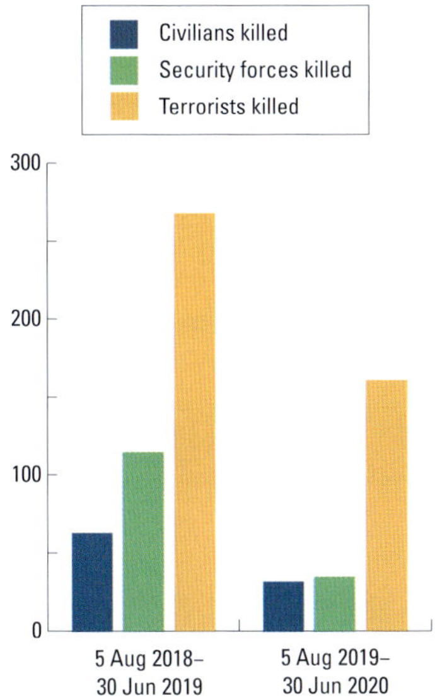

Terrorist attacks in Kashmir

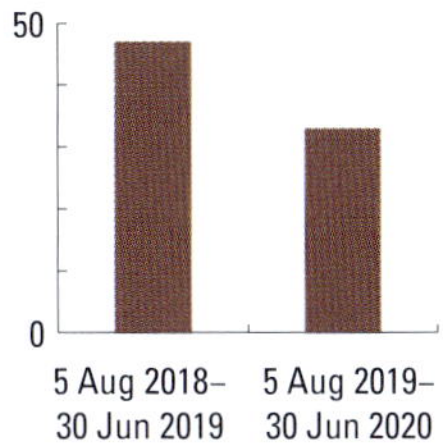

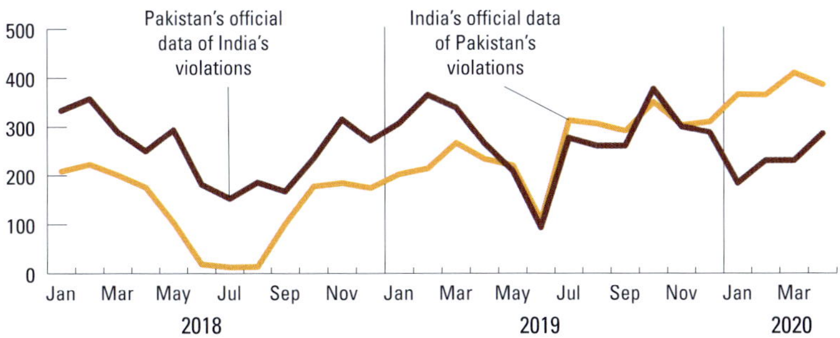

The US–Taliban Deal

Will Afghanistan's peace process work?

On 29 February 2020, the United States government and the Afghan Taliban (which calls itself the 'Islamic Emirate of Afghanistan') signed a historic peace agreement. This followed multiple rounds of direct bilateral talks that had begun 17 months earlier in October 2018. The deal aimed to end the longest war in US history.

The agreement laid out a timetable for the withdrawal of all foreign troops (including around 12,000 US troops) from Afghanistan within 14 months, with an initial reduction of US troops to 8,600 and five military bases to be vacated by mid-July 2020. In return, the Taliban gave a series of security commitments, including guarantees that the Afghan territory it controlled would not be used by foreign militants and that it would not pose a security threat to the US and the rest of the world.

Under the deal, the Taliban was expected to fight against the Islamic State, also known as ISIS or ISIL, and dismantle the militant infrastructure of other international violent extremist groups in its territory, including al-Qaeda. The deal also stipulated that the Taliban would not conduct attacks against foreign forces based in Afghanistan, and that the US would not attack Taliban positions unless there was a serious threat to Afghan forces, large towns or military bases. Although the Taliban largely refrained from attacking US/NATO forces and major military bases, and from carrying out large-scale attacks (including suicide bombings and car bombings) in major cities, the Taliban and Afghan forces continued fighting each other in rural areas, resulting in the killing and injuring of many fighters and civilians.

The deal was signed in Doha by US Special Representative for Afghanistan Reconciliation Zalmay Khalilzad and Mullah Abdul Ghani Baradar, a co-founder of the Taliban and the group's deputy leader for political affairs who also heads its Qatar-based political office. The signing ceremony was attended by US Secretary of State Mike Pompeo, several Taliban leaders, and representatives from around two dozen countries and international organisations, among them the foreign ministers of Pakistan, Turkey and Uzbekistan. On the same day, US Secretary of Defense Mark Esper signed a joint declaration with the Afghan government in Kabul about the need for negotiations with the Taliban and to combat terrorism.

Although the US kept Kabul abreast of developments in the US–Taliban talks, the Afghan government was not part of these negotiations. The government in Kabul was excluded from the Doha agreement because of the Taliban's long-standing position of not recognising it as a legitimate entity and true representative of the Afghan people. However, the simultaneous US–Taliban deal created a framework for direct peace talks between the Taliban and representatives of the Afghan government and factions supporting the Kabul government and the current political system in the country.

US President Donald Trump hailed the US–Taliban deal as a major achievement and spoke by telephone to Abdul Ghani Baradar on 3 March 2020. The half-hour call was the first direct exchange between a US president and the insurgent leadership since the creation of the Taliban in 1994. On 10 March the United Nations Security Council (UNSC) unanimously backed the deal – a rare endorsement of an agreement with a militant group – and called it 'an important step towards ending the war and opening the door to intra-Afghan negotiations'.

Challenges and obstacles

Despite an upbeat start, the US–Taliban deal faced several obstacles and challenges. These have caused delays, and key deadlines have already been missed. Moreover, within days of the signing, the Taliban and the US accused each other of violating certain principles of the agreement, particularly the 'accepted level' of violence, the exchange of prisoners and the nature of the Taliban's alleged links with al-Qaeda. While both sides implemented various parts of the deal, these accusations and counter-accusations continued.

On 10 June, the head of the US Central Command, Marine Corps Gen. Kenneth F. McKenzie Jr, said that the Taliban had not yet met the conditions required for a complete US troop withdrawal from Afghanistan by

May 2021 as envisioned in the deal. He acknowledged the Taliban's military campaign against the ISIS branch in the Afghanistan–Pakistan region (commonly known as the Islamic State–Khorasan group (IS–K) or the Islamic State–Khorasan Province (ISIS–KP)) and its role in the elimination of the latter group in the east of Afghanistan. However, McKenzie added that it was 'less clear to me that they will take the same action against al-Qaeda'. The next day, the Taliban published a strong rebuttal on its website stressing that it was committed to full implementation of the US–Taliban deal and 'would never allow anyone to use the Afghan territory against any other country including the US'. Despite these obstacles and accusations, both the Taliban and the US were keen to save the deal from collapsing.

But mutual accusations of violations, and delays in the implementation of certain parts of the deal (mainly the Afghan government dragging out the release of its prisoners and delays in starting the talks between the government and the Taliban), meant that reaching a political settlement in Afghanistan would take even longer than envisioned in the deal. US and Afghan officials have criticised the Taliban for working against the spirit of the peace deal by continuing widespread attacks on Afghan forces, mainly in rural areas. Meanwhile, the Taliban insisted that it was fully committed to the implementation of the deal and accused the US – and by extension the Afghan government – of not fulfilling their commitments, in particular the release of the Taliban prisoners from Afghan jails as promised in the US–Taliban deal. However, it seemed that both the Afghan government and the Taliban were trying to increase pressure and gain maximum leverage ahead of the imminent intra-Afghan talks – which had been expected to begin on 10 March – in which a general ceasefire was expected to be a top agenda item.

The most immediate and significant sticking point was the exchange of prisoners between the warring sides. The US–Taliban deal called for the Afghan government to release 5,000 Taliban prisoners and for the Taliban to free 1,000 government and military personnel within nine days ahead of the intra-Afghan negotiations. Although the lists of Taliban prisoners had been prepared before the US–Taliban deal was signed, and the US had shared the issue with the Afghan government, Afghan

President Ashraf Ghani insisted that his government had not committed to the release of thousands of Taliban prisoners. In addition to mentioning legal, technical and logistical problems, he linked the release of the Taliban prisoners to a ceasefire by the insurgent group and its willingness to enter into direct talks with his government.

While the US pressed Ghani to implement this key part of the agreement in order to open the door for the intra-Afghan dialogue, his domestic opponents accused him of using the issue of prisoner release as a bargaining chip to extract political concessions from the US government at a time when he was engaged in a bitter row with his rival presidential contender Abdullah Abdullah. Although the Afghan electoral commission had declared Ghani the winner of the September 2019 presidential election, Abdullah contested the result and accused Ghani of electoral fraud. As US efforts to resolve the political crisis failed, Abdullah and Ghani both held inauguration ceremonies in Kabul and took simultaneous oaths for the presidency on 9 March 2020. Weeks after sustained encouragement and pressure by the international community, especially the US, and mediation by Afghan political figures including former president Hamid Karzai, the two rivals reached an agreement. Ghani was confirmed as president, while Abdullah was given the leading role in the peace process and future negotiations with the Taliban as the chairman of the High Council for National Reconciliation. His supporters were also given 50% of the ministerial positions in the new cabinet, as well as several other government positions.

The Taliban accused Ghani of deliberately creating obstacles to the implementation of the US–Taliban deal in order to 'prolong his power'. The group stressed that the prisoner exchange was a prerequisite for commencing the intra-Afghan dialogue, which was a key part of the US–Taliban deal. After stalling for weeks, the first phase of prisoner swaps eventually began in April and unfolded slowly amid increasing tension and intensifying violence between the Afghan government and the Taliban. The prisoner exchange was meant to be a confidence-building measure before the start of the intra-Afghan talks. By the end of June 2020 the exchange had not been completed, the intra-Afghan dialogue had not begun and no date had been fixed for talks to start. The continuing delay and mutual accusations created friction in the lead-up to the planned intra-Afghan dialogue and further delayed the creation of a sustainable road map for peace in the country.

In addition to the delay in the exchange of prisoners, and the presidential-election crisis, two other major factors contributed to the delay in starting the intra-Afghan talks. These were the absence of an 'inclusive' negotiating team from Kabul and the formation of the High Council for National Reconciliation. These two bodies were not formed early enough, mainly due to differences between Ghani and his political rival Abdullah. The formation of the negotiation team was finally announced on 26 March, but the High Council for National Reconciliation was yet to be formed by the end of June.

Meanwhile, the withdrawal of US troops from Afghanistan, which began immediately after the signing of the deal, continued on schedule. The US–Taliban deal stated that the US troop presence would drop from about 12,000 troops to 8,600 within 135 days of its signing (that is, by mid-July 2020). McKenzie revealed on 19 June that the US had indeed reduced its troop presence in Afghanistan to 8,600. Driven by several factors, including cost reduction and the US presidential-election timetable, Trump repeatedly expressed his desire to withdraw US forces from Afghanistan as soon as possible. This would enable him to claim a foreign-policy success before his re-election campaign in autumn 2020. Ending the US war in Afghanistan had been one of his campaign pledges in 2015–16. Although the Doha agreement made the foreign-troop withdrawal contingent mainly on the Taliban's compliance with anti-terrorism commitments, it was unclear whether the US would wait until the intra-Afghan dialogue produced a desirable outcome. Furthermore, the COVID-19 pandemic complicated US force operations and partnership with Afghan counterparts.

Some officials in the Trump administration hinted that US troops might be withdrawn from Afghanistan even faster than envisaged in the US–Taliban deal. However, the security risks of a precipitate withdrawal from Afghanistan could outweigh Trump's pre-election interest in completing this withdrawal. His urgency and unpredictability have left the Afghan government in a vulnerable position. US politicians warned that a rushed and premature withdrawal would risk losing the gains achieved in Afghanistan over nearly two decades, not only in counter-terrorism but also in building Afghan institutions. Some Afghan officials and politicians seemed to hope that the US would slow down or even reverse withdrawal until the outcome of the US presidential election was clear.

The US–Taliban agreement was the first phase of the Afghan peace process. Given the nature of the conflict in Afghanistan and the divergent political agendas of the two main local belligerent parties (the Afghan government and the Taliban), the second phase, dubbed the 'intra-Afghan negotiations', will undoubtedly be more complicated and difficult. The gulf between the two parties on major issues – including a durable ceasefire, security-sector reforms, civic and political rights of citizens, and the future political system – will test their negotiation skills and flexibility. The Afghan government wants to preserve the republic and its democratic system, while the Taliban might want to reimpose the Emirate. Reconciling the wishes and desires of the two sides to achieve a consensus will not be easy. The intra-Afghan dialogue will also test the unity of the Taliban when the group is expected to present a more specific vision for the future of Afghanistan.

The timeline of the intra-Afghan negotiations also remained a major unknown. Remarks by Ghani and his close allies indicated that he wanted to complete his second five-year term, which would mean he remained in office until 2024. But the Taliban seemed keen on the creation of a caretaker/interim neutral or national unity government before a more permanent set-up is established as one of the outcomes of the intra-Afghan negotiations. A coalition of relevant regional and international players is thus needed to influence the Afghan negotiating parties and help them reach a settlement sooner rather than later.

The final challenge for the Afghan peace process is the implementation and monitoring of any future agreement and the creation of verifiable enforcement mechanisms. The UNSC or a group of key regional and international state actors and international organisations could fulfil this role.

Supporters and spoilers

Pakistan's readiness to play a constructive role and to dismantle the sanctuaries and activities of the Afghan Taliban on its soil – the existence of which is usually denied by the country's officials – is widely seen as a key factor for the success of the Afghan peace process. Following intense pressure by the Trump administration, Pakistan has repeatedly assured the US of its cooperation in the peace process and of its willingness to use its influence over the Taliban in a positive manner. Although US pressure had substantially decreased over the past two years, some US and Afghan officials remained sceptical of Pakistan's short- and long-term intentions.

A quarterly report by the US Department of Defense to Congress, released in mid-May 2020, noted that Pakistan continued to harbour the Afghan Taliban. In a visible attempt to show Pakistan's support for the Afghan peace process, Pakistan's powerful army chief Gen. Qamar Javed Bajwa visited Afghanistan on 9 June for meetings with Afghan leaders, during which both sides reiterated Pakistan's support for the process. The Afghan presidential-office statement quoted Bajwa as saying that Pakistan 'supports independence, and the republican and democratic [political system] in Afghanistan'. US officials have repeatedly stated that peace in Afghanistan offers an unprecedented opportunity to advance security, connectivity and development for the region. Yet it was unclear when and how Pakistan would move against the reported Taliban sanctuaries on its soil, and whether it would take action against the group during the intra-Afghan negotiations (which appears unlikely) or afterwards. However, there were hopes, at least in the White House, that this time around, Pakistan would follow through on its promises.

Pakistan is not the only neighbour to retain an active interest in Afghanistan. Despite the economic hardship and international political and financial pressures it faces, Iran has expanded its regional influence by establishing ties with the Taliban. The constructive role of these two immediate neighbouring countries as well as other key regional actors such as China, India and Russia is of paramount importance for the success of the Afghan peace process.

Peace in Afghanistan is a shared global interest. The ongoing conflict has had a negative impact on many countries in the form of increased insecurity, militancy, narcotics and refugees. Failure of the peace process could exacerbate the already precarious situation in the country and its surrounding region. Another round of chaos in Afghanistan not only carries the risk of transregional militant groups such as al-Qaeda and ISIS filling the security vacuum; it would probably result in the emergence of new and even more dangerous violent extremist outfits.

The peace talks seek to end not only the war that began in 2001, but also four decades of conflict since the communist coup in Afghanistan in 1978 and the Soviet invasion the following year. The Afghan peace process has entered a sensitive and crucial phase. However, this current opportunity – and the local, regional and international consensus of support – could easily be squandered if the situation is not handled with care. A framework involving the region and key international players is needed to coordinate efforts and deter and prevent spoilers and war profiteers from undermining the Afghan peace process.

Potential spoilers include regional and international violent extremist groups such as ISIS and militant outfits that have traditionally focused on fighting in the Indian-administered Kashmir. They can be expected to increase their attacks and propaganda both in Afghanistan and the wider region to create doubts and tensions among key actors and prevent the long-awaited political settlement.

Such a framework could be led or convened by the UN, which should ensure that the relevant states are on the same page and allay misperceptions and mutual suspicion about intentions. While the political future of Afghanistan will be largely determined by the Afghans themselves as part of the intra-Afghan dialogue, the behaviour of key regional states (many of which have interfered in Afghanistan, mainly by supporting their local Afghan allies or various Afghan groups and factions) will be equally important for the success of the peace process and the implementation of a final agreement.

China's Strategic Ambitions in the South Pacific
Past the peak?

For most of the past 60 years the South Pacific has been geopolitically benign. In the last three years this has been upended. Spurred largely by China's renewed engagement in the Pacific, the region has roared back onto the global stage and to the centre of Australian foreign-policy thinking. Australia has organised monthly state visits, expanded its aid to the region and reshaped its foreign-policy bureaucracy to step up its relationship with its immediate neighbourhood. Like-minded nations like Japan, New Zealand, the United Kingdom and the United States are similarly engaged with 'resets', 'elevations' and 'uplifts' in relations with the region.

But just how significant is China's presence in the region? What are China's ambitions for its renewed engagement, and what are the major risks of this ambition for the West? How will China's engagement in the South Pacific play out in the year ahead?

As Pacific nations become increasingly cautious of China's renewed presence, and the West refocuses its attention on the Pacific, China will find it more difficult to build influence in the Pacific in 2020–21. This will test the resolve and depth of its ambitions in the South Pacific. Those ambitions will likely contract as China, and the rest of the world, deal with the severe economic hangover of the COVID-19 pandemic.

The Pacific Islands

The idyllic image of the Pacific Islands as a paradise of palm trees and beaches masks a region facing profound, and in some cases existential, developmental challenges. With a cumulative population of under 13 million, most of which reside in Australia's former colony Papua New Guinea (PNG), these 14 nations span over 15% of the world's surface, making up what Pacific leaders have come to call a vast oceanic continent, or 'blue Pacific'.

The dual challenges of small size and remoteness make conventional economic growth pathways in most of these countries very difficult, if not impossible. With some of the highest population growth rates in the world, and the real and imminent threat of climate change, the Pacific is one of the most fragile regions in the world, and the most aid-dependent.

During the Second World War these islands had great geostrategic importance. From Guadalcanal in Solomon Islands to Kwajalein in Marshall Islands, control of them was critical for maintaining logistical supply lines and military force projection. Since 1945, the Pacific has been a relatively benign geopolitical environment. While the gradual decolonisation of the region has led to a reduced UK presence, the remaining key actors – Australia, France, Japan, New Zealand and the US – have remained a largely fixed presence. Nonetheless, over the years the Pacific Islands have been more concerned about receiving too little attention than too much. This benign neglect from Australia, the South Pacific's primary aid, trade and investment partner, was felt particularly acutely. While money was always available, attention, particularly at the political level, was often found wanting.

China's growing presence in the Pacific

China's growing presence in the Pacific Islands region in recent years has shifted these dynamics. China has long had a presence in the region, and significant pockets of ethnic Chinese have resided there for more than a century. Julius Chan, the second prime minister of PNG and still a member of parliament, is half ethnic Chinese. Some of the oldest trading houses across the Pacific region originate from China's Fujian province, similar to many parts of Southeast Asia and beyond.

In addition to these historical economic links, the China–Taiwan divide is nowhere more stark than in the Pacific. As of 2020, four of Taiwan's remaining 15 diplomatic allies were Pacific Island nations, following the loss of its two largest allies in the region, Kiribati and Solomon Islands, in 2019. This vying for diplomatic alliances has resulted in some degree of political attention and chequebook diplomacy in the Pacific by both China and Taiwan for decades.

Since 2006, however, China has significantly scaled up trade, aid, diplomacy and commercial activity in the Pacific region and is emerging as an increasingly important trading partner. For all Pacific countries, China is a growing export market for the region's commodities – lumber, raw minerals, oil and gas, and fish. It is also a growing source of imports. By 2018, China had emerged as a top-three two-way trading partner for most Pacific nations, according to the IMF. If PNG is excluded, Pacific exports to China and imports from China have overtaken trade with Australia in the last five years.

Chinese engagement has been especially visible in foreign aid. China provided close to US$1.5 billion in foreign aid to the Pacific Islands region through a mixture of grants and loans between 2006 and 2017.

In a short time, China has become the third-largest donor to the Pacific, the most aid-dependent region in the world. Australia and New Zealand remain the top two donors, together accounting for 54% of all aid flows to the region in 2011–17.

Figure 1: Two-way trade with China as percentage of total trade in major Pacific markets, 2000–18

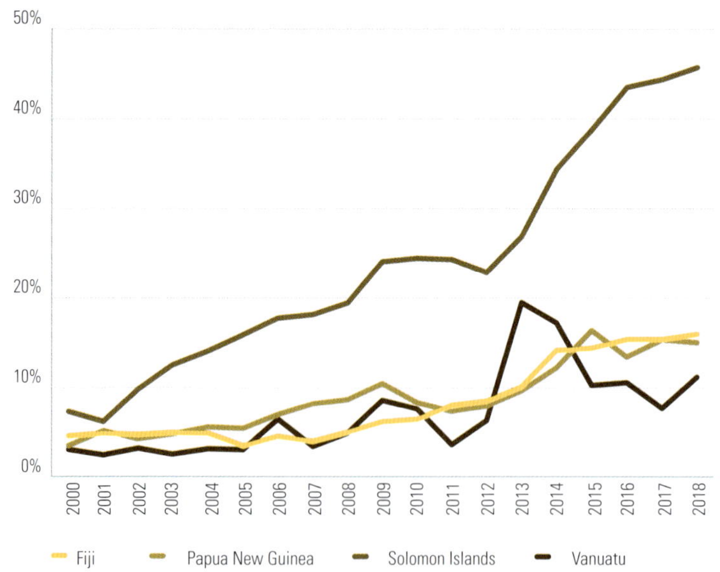

Source: IMF

Figure 2: Chinese foreign aid to the Pacific, 2007–17*

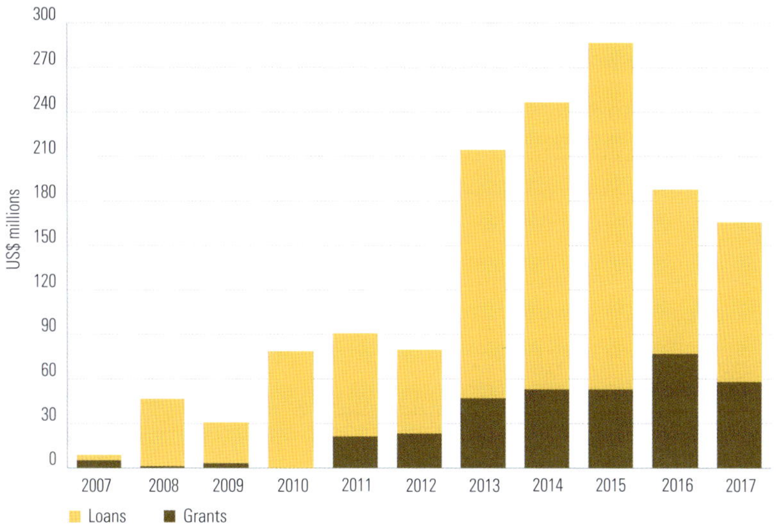

* Because most Chinese aid to the Pacific comes in the form of concessional loans, year-to-year disbursements can be uneven as large loans are dispersed

Source: Lowy Institute Pacific Aid Map ©IISS

Figure 3: Major donors' aid as percentage of total aid to the Pacific, 2011–17

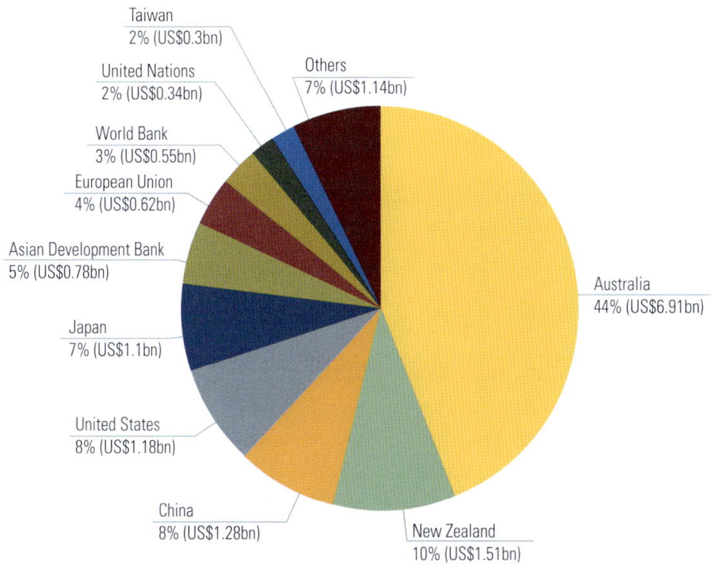

Source: Lowy Institute Pacific Aid Map ©IISS

Alongside growing aid and trade ties, China has strengthened its diplomatic links with the Pacific, and with political elites in particular. In the five years between Chinese President Xi Jinping's bilateral visits to Fiji in 2014 and PNG in 2018, Pacific leaders took part in more two-way Guest of Government visits with China (to Beijing or into the region) than with Australia. The respect of equivalence built much goodwill for China in the region, in sharp contrast to the perceived neglect Pacific leaders felt from Australia's politicians.

China solidified its diplomatic links with the Pacific region at the Asia-Pacific Economic Cooperation (APEC) Leaders' Summit in Port Moresby, PNG, in 2018. During the summit, Xi held a special meeting with Pacific leaders who have diplomatic ties with Beijing, at which they all signed up to his flagship Belt and Road Initiative. China achieved an even greater victory in September 2019, when Kiribati and Solomon Islands, Taiwan's two largest diplomatic allies in the region, both defected in the same week.

But while China has emerged as a formidable player in the Pacific Islands region, it has by no means become dominant. Its engagement, however, stands out from that of other nations in several ways, particularly in terms of its foreign aid.

China engages almost exclusively in concessional lending – cheap loans that will eventually be paid back. Three-quarters of the finance that China dispersed to the Pacific in 2011–17 came in the form of loans. These loans, and the types of projects they finance, have led to calls that China is engaged in 'debt-trap diplomacy' in the Pacific – using debt as leverage to secure strategic concessions. An extensive assessment of debt dynamics in the Pacific carried out by the Lowy Institute in 2019 concluded that 'China has not been the primary driver behind rising debt risks in the Pacific, although a continuation of business as usual would risk future debt problems in several countries'.

China also focuses its aid on large infrastructure projects that are exclusively implemented by Chinese state-owned enterprises (SOEs) and almost exclusively use Chinese labourers. Such projects are now found in every major capital. These SOEs have put down deep roots, competing in commercial activity across the board and winning commercially tendered contracts from government tenders and multilateral agencies. Chinese SOEs have become the dominant player in construction activity in many countries. According to China's own investment statistics, Chinese construction activity in the region was worth US$958m in 2017, almost six times greater than its foreign-aid activities.

A significant influx of Chinese migrants mirrors growing Chinese commercial activity. Arriving to work on construction sites or fishing vessels, many entrepreneurial Chinese citizens have been enticed by the commercial opportunities that the high-cost and low-competition economies of the Pacific present. While official statistics are limited, anecdotal evidence suggests that no significant town in the Pacific lacks a shop run by a Chinese migrant selling cheap Chinese goods. These entrepreneurs, while motivated by commerce and not directly under the influence of the Chinese state, serve to increase China's presence in the Pacific.

Reasons for concern

China's deepening presence in the Pacific is in many ways a positive thing for the region. Chinese businesses are providing competitively priced goods that are clearly in demand. China's development assistance is helping the region address some of its acute developmental challenges. The presence of a new and formidable player is forcing traditional partners to step up and is giving Pacific nations more influence and agency in investment decisions. China insists that its growing engagement in the Pacific is benign and simply reflects its growing global influence.

The pace at which China's presence is increasing, however, is a source of growing concern in Canberra, Tokyo, Washington DC and Wellington. China's strategic interests in the Pacific are a subject of debate. The Chinese do not articulate these, at least not in public. And while there is a growing community of Chinese academics focused on the region, there is far less coverage of the Pacific in China than in most other parts of the world.

Many see China's influence in the Pacific Islands as growing more by accident than by design. Regardless of whether China's capability or intent came first, its interests in the Pacific are likely multifaceted. The Pacific is rich in natural resources. Half of China's tropical-log imports come from PNG and Solomon Islands. SOEs such as Zijin Mining Group and Metallurgical Corporation of China have large stakes in nickel and gold mines in PNG. The Pacific's well-regulated fisheries are home to more than 50% of the annual global tuna catch.

The West's greatest concern is that China is trying to use its new-found influence – through diplomacy, debt, trade and elite capture – to establish a military base in the Pacific region. A military base would serve multiple objectives for China. The People's Liberation Army has regional, if not global, force-projection ambitions that require bases. A Chinese military base as little as 2,000 km from Australia's eastern coast would also force a wedge between Australia and its traditional strategic anchor, the US. While the strategic benefit and size of such a facility for China is questionable – the logistics of maintaining supply routes alone would be challenging – it would have a profound psychological impact on Australia. It would accelerate a military build-up, confirm fears about China's strategic intent in the region, and force Australia into a rapid and isolated military scale-up. The Australian Department of Defence's 2020 Defence Strategic Update and the accompanying Force Structure Plan indicated that Australia was already heading in this direction. A Chinese military base in the region would rapidly accelerate that discussion.

Some parts of the Pacific are more strategically significant than others. A military base in Kiribati or Federated States of Micronesia, for example, would have far greater strategic relevance than one in Tonga because of the proximity of these countries to US interests in Guam and Hawaii, enabling China to project force and gather intelligence more effectively. The challenges of securing concessions from the Pacific for a military base mean that China will have to be opportunistic in this pursuit. This led to reported Chinese military interest in funding and building a wharf in Luganville, Vanuatu in 2018; in acquiring a defunct naval base in Manus Island, PNG in 2019, which Australia and the US are now restoring, largely to keep China out; and more recent interests in Kiribati and Solomon Islands following diplomatic recognition in 2019.

For Australia and its allies, the stakes are high. But the probability of a Chinese military base being established in the Pacific is low, for two reasons. Firstly, the Pacific states are not merely pawns on a geostrategic chessboard. They have agency, wield it liberally, and have no interest in further militarisation of a region where only Fiji, PNG and Tonga retain standing armies. It is one thing for China to ask a Pacific nation to side with it on a UN vote on the South China Sea, in which the Pacific has no interest. It is a completely different matter for a country to cede perceived sovereignty in handing over territory for the exclusive use of a foreign military. Despite Australia's influence in the region, and strong military institutional links with those Pacific countries that maintain armed forces, it does not operate any military bases in the region. The US maintains military bases in Federated States of Micronesia and Marshall Islands.

The second reason why a Chinese military base is unlikely is that the West is now paying attention and scrambling to engage. With the largest footprint in the region, Australia is scaling up its aid spending and diplomatic networks. Crucially, it has significantly boosted its political engagement with the Pacific. In 2019 alone, Prime Minister Scott Morrison made more bilateral visits to the Pacific in one year than any of his recent counterparts did in their entire terms in office.

New Zealand has also deepened its engagement. With 20% of its population retaining Pacific heritage, it enjoys a clear cultural advantage in doing so. The US, too, has shown a growing interest. North Pacific states, in particular, have seen the US reaffirm its commitments to maintain their Compact of Free Association relationships. This was cemented by the leaders of Federated States of Micronesia, Marshall Islands and Palau when they met with US President Donald Trump in May 2019 – the first time that leaders of these countries had been hosted in the White House in their history.

Despite these factors, the acute vulnerability and relative size of these nations, and the potential for China to put on the table an 'offer they couldn't refuse', mean that vigilance will be required.

Recent and future developments

China has rapidly built its presence and influence in the Pacific in the early twenty-first century. But in the past three years it has found it harder to do so for two reasons.

The first is that Western countries, and Australia in particular, are taking much more notice and pushing back. For example, in the past three years Australia has prevented a proposed Huawei-funded submarine cable for Solomon Islands; announced a US-partnered naval base on PNG's Manus Island; and outshone China at the 2018 APEC Leaders' Summit with a massive quadrilateral PNG electrification project.

The second and more crucial point is that Pacific countries are themselves pushing back against China, particularly on Chinese debt. As the debt-repayment holidays of major infrastructure initiatives expire, Pacific leaders are left with large bills for often overpriced and inferior projects. No Pacific country except Vanuatu has accepted a new loan from China since 2018. Leaders are calling for more grants, or at a minimum budget-support loans. These come at a higher cost to China, which has been wary of providing them. Calls for debt forgiveness or renegotiation continue to rise as the region is economically damaged by COVID-19. Resentment of the new Chinese diasporas and fears of 'economic colonialism' are also growing.

Finding it more difficult to operate with its existing Pacific partners, in 2019 China focused on securing diplomatic recognition from Kiribati, which has an exclusive economic zone the breadth of Australia, and with Solomon Islands, the third-largest Pacific island nation. This served to broaden the strategic playing field for the West. Chinese influence in countries that recognise Taiwan is limited and cannot have a military component. In this regard, Kiribati's switch of recognition was especially significant, because its outer atolls are only 2,000 km from Hawaii. This will stretch the West's focus in the region. Beijing's push for recognition by Kiribati and Solomon Islands was additionally part of a broader (unsuccessful) bid to put pressure on the pro-autonomy Tsai Ing-wen government ahead of the January 2020 elections in Taiwan. China also hosted a high-level conference with the Pacific in Samoa in October 2019 but brought no new major funding announcements.

The COVID-19 pandemic adds a new dimension for China as it faces greater pushback in the Pacific. The needs of the Pacific region are greater – it will be poorer and more unstable as it faces a potential 10% economic contraction. China will face significant pressures to be more fiscally disciplined as it manages its own economic recovery. It has already seized some of the initiative by hosting digital meetings with leaders across the Pacific to discuss COVID-19 and by sending much-needed personal protective equipment and other supplies. But this support pales in comparison with what Australia has provided to date and will provide in the future. The pandemic has also reminded countries in the region – especially those dependent on tourism – of their economic reliance on Australia and New Zealand. A mooted trans-Pacific travel bubble will give Australia and New Zealand even more opportunity to strengthen their ties with the Pacific.

While China has far greater resources to bring to bear, Australia and New Zealand have far greater resolve. The COVID-19 pandemic provides China with an opportunity to expand its influence even further in the Pacific. But the cost, and the heightened competition with major regional powers this would trigger, may now prove too high a price.

Pacific Island states

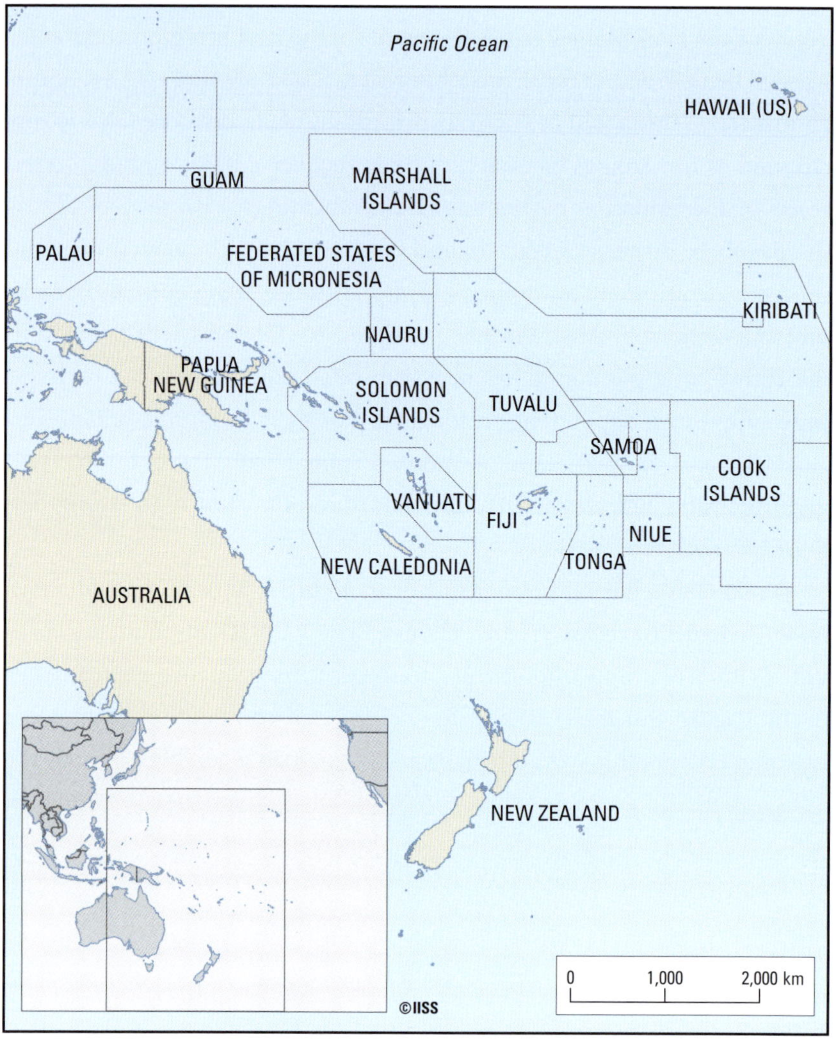

CHAPTER 7

Europe

EUROPE

Drivers of Strategic Change

REGIONAL SHARE OF GLOBAL POPULATION, GDP AND DEFENCE BUDGET

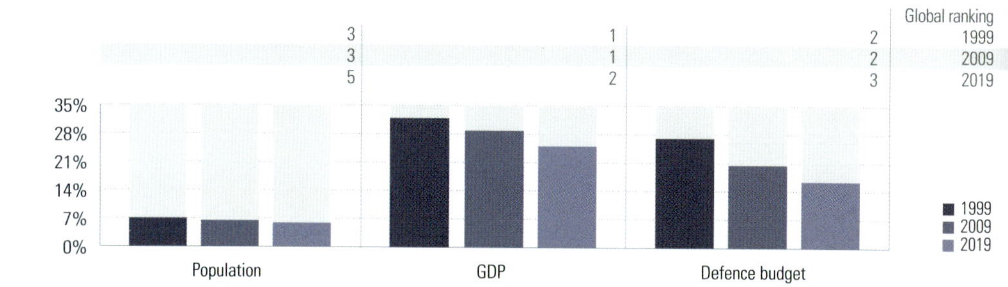

POPULATION

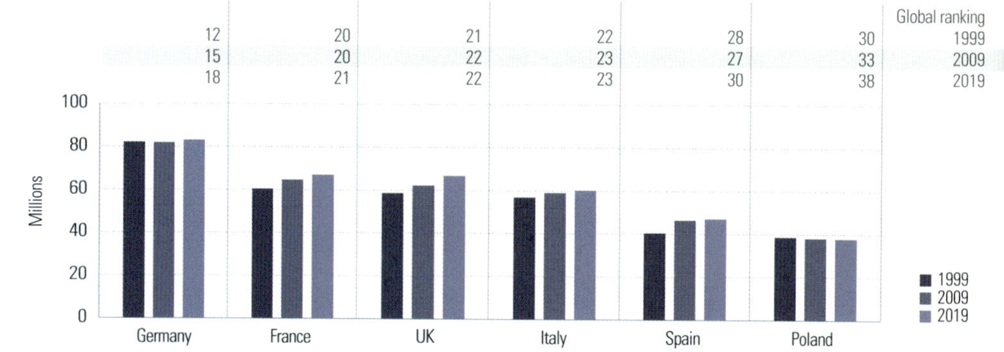

AGE STRUCTURE
(Percentage of national population)

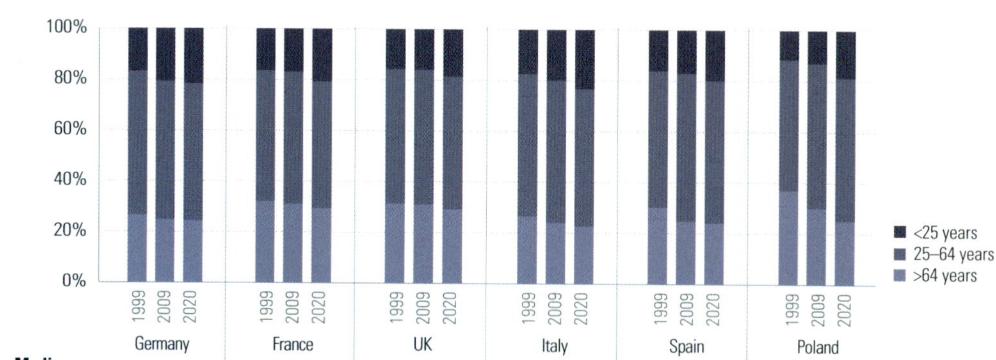

Median age	Germany	France	UK	Italy	Spain	Poland
2000	40.1	37.7	37.6	40.3	37.6	35.0
2010	44.3	40.1	39.5	43.5	40.0	38.1
2020	45.7	42.3	40.5	47.3	44.9	41.8

GDP

(Constant 2010 US dollars)

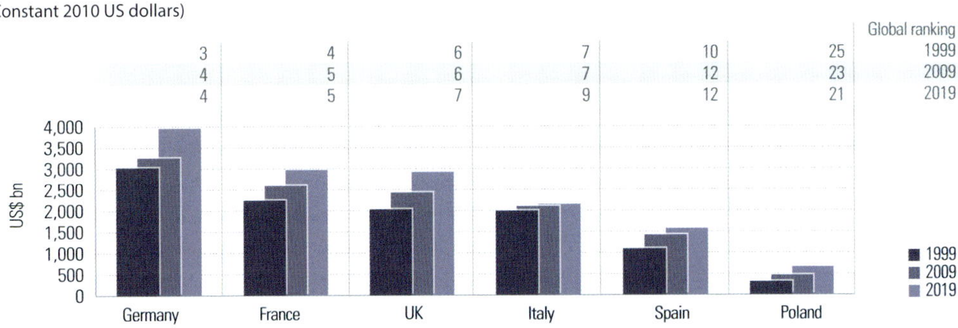

GDP PER CAPITA

(Constant 2010 US dollars)

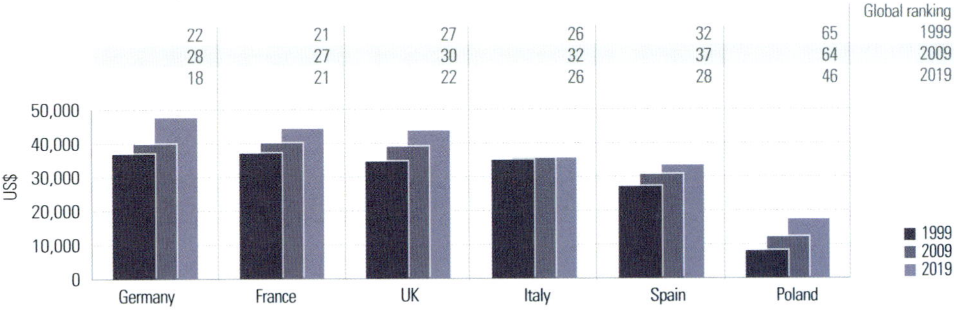

DEFENCE BUDGET

(Constant 2015 US dollars)

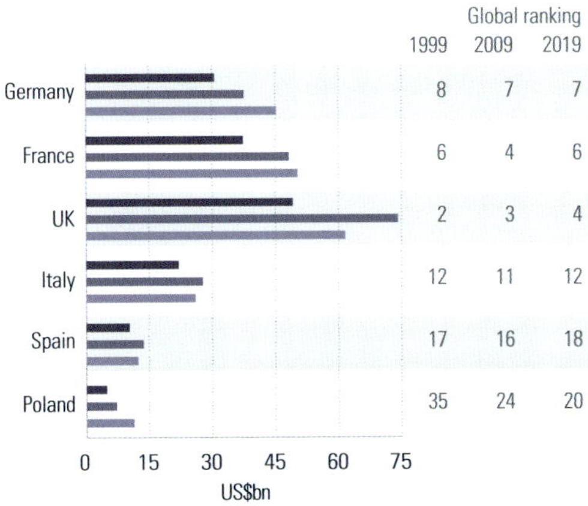

ACTIVE MILITARY PERSONNEL

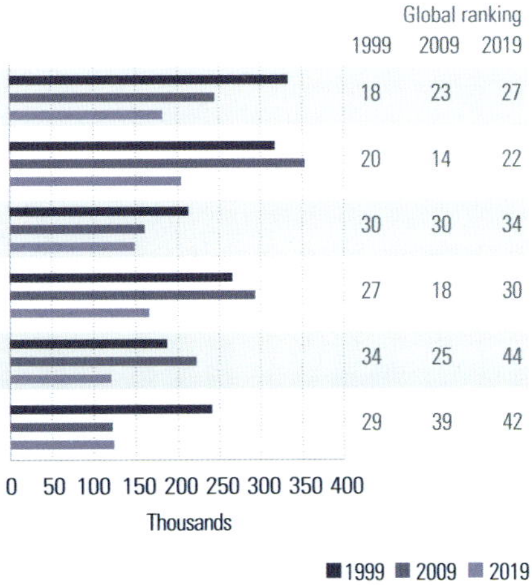

For explanation of drivers and sources, see page 7

HUMAN DEVELOPMENT INDEX (HDI)

(Score between 0 and 1, where 0 denotes a low level of development and 1 a high level of development)

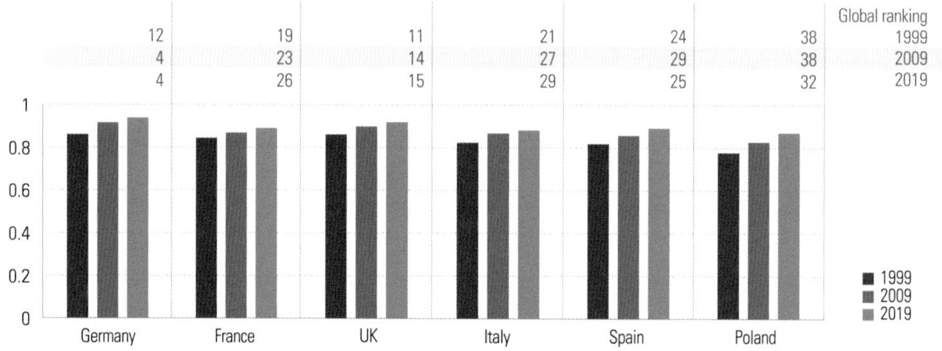

POLITICAL SYSTEM

(Score between 0 and 100, where 0 denotes no political freedom and 100 fully free)

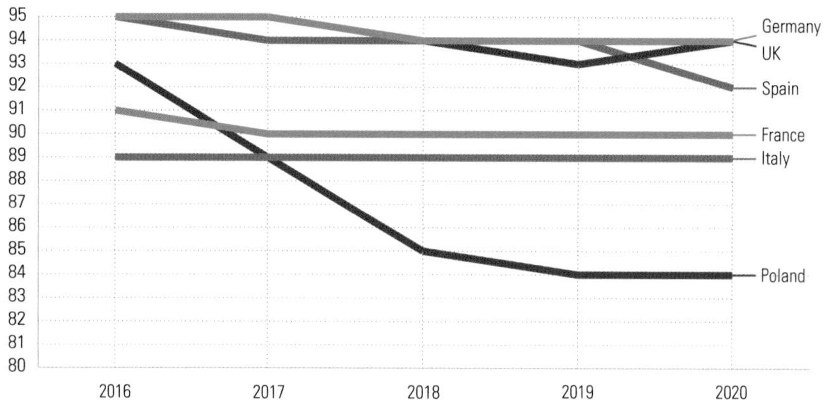

TRUST IN INSTITUTIONS

(Average level of trust)

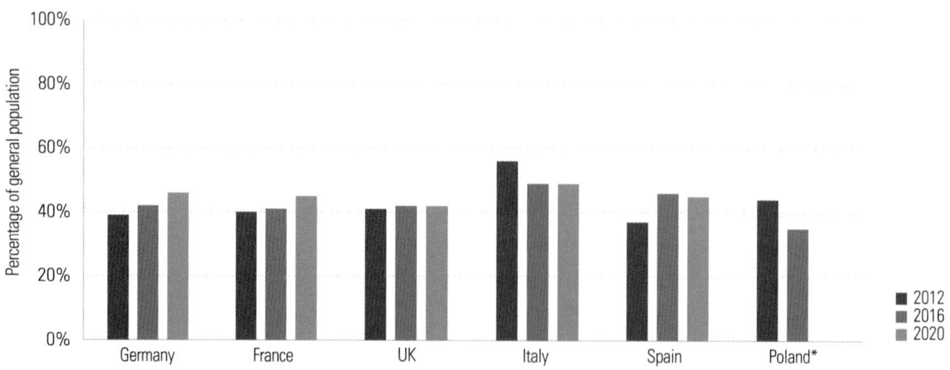

* No 2020 data available for Poland

For explanation of drivers and sources, see page 7

2019–20 Review

Europe's fault lines came into sharp relief in the year to mid-2020: north against south over reforms to eurozone governance; west against east over respect for the EU's democratic norms; federalists versus proponents of greater national autonomy over the choice of a new European Commission president; and the exit from the European Union of the United Kingdom, the bloc's third-largest economy and one of its two major military powers. The EU also faced external challenges. The transatlantic alliance came under further pressure, and China's increasingly assertive diplomacy exposed the limitations of the EU's approach to it. Finally, the COVID-19 pandemic presented the EU with an unprecedented test. Its response could prove pivotal for the future of the eurozone and hence the EU as a whole.

French calls in March 2019 for the establishment of a eurozone budget and more solidarity between economically stronger northern members and weaker, more indebted, southern ones initially came to little. Germany was sceptical, and some small northern countries, such as Austria and the Netherlands, were staunchly opposed. In May 2020, the European Commission called for an emergency €750 billion (US$852bn) recovery fund – named the 'European Recovery Instrument' – to help the EU economy overcome the downturn caused by COVID-19. But member states failed to agree on how much of the money should take the form of grants as opposed to loans, or on how to allocate the money. Moreover, the continuing negotiations over the EU's latest long-term budget (the Multiannual Financial Framework–MFF), covering the seven years from 2021 to 2027, laid bare tensions between member states that are net contributors to the budget and those that are net beneficiaries.

Germany was central to finding solutions to these interlocking challenges. The COVID-19 outbreak hit some member states harder than others, and some of the worst affected – notably Italy and Spain – had little budgetary firepower to counter the slump in economic activity. Faced with worsening economic and political divisions, Germany shifted its stance, paving the way for a groundbreaking joint Franco-German proposal in May 2020 for an EU-level fiscal-stimulus programme to be financed by debt issued by the EU and backed jointly by the 27 member states. Although intended as a temporary measure, German acceptance of 'eurobonds' – and therefore the sharing of risk across member states – was a turning point and enabled the Commission to propose its recovery fund. In June, Germany also indicated that it would accept the Commission's proposal for a 42% increase in its annual contribution to the EU budget, from €31bn (US$35.2bn) to €44bn (US$50bn).

While the member states were yet to agree on the exact terms of the recovery fund by the end of June 2020, Franco-German backing gave it strong momentum. Moreover, no government wanted to be held responsible for scuppering a deal. Failure to unify in the face of an economic shock could leave investors doubting whether the eurozone would ever find the political will to jointly issue debt. No currency union has survived in the long term without such a capacity. Failure could seriously strain relations between and within member states.

Italy remained the most serious potential flashpoint. The country's political situation stabilised over the course of the year to June 2020, with a coalition of right-wing separatist, anti-EU and anti-establishment parties making way for a broad coalition including the country's pro-EU Democratic Party (PD). But Italy's economic situation was bleak. The European Commission estimated that the Italian economy would shrink by almost 10% in 2020, leaving it smaller than when the euro was launched in 1999. The risk was that a weak recovery would exacerbate Italy's indebtedness, leading to further austerity, in turn boosting support for populist political forces.

Spain's economy was in better shape than Italy's, but it too faced an uncertain economic outlook. Its crucial tourism industry was hit hard by the pandemic and, like Italy, Spain had less scope to bolster its economy than its northern neighbours. After five general elections in as many years, in January 2020 a governing coalition was agreed between the Spanish Socialist Workers' Party (PSOE) and Podemos, a left-wing populist party formed in 2014. While pro-EU, this government was more outspoken than previous ones about the need for more burden sharing within the eurozone. Much as with Italy, a return to austerity could be politically destabilising.

The German acceptance of eurobonds and the resulting boost to the Franco-German axis was timely for French President Emmanuel Macron. He faced fierce domestic criticism for economic reforms that allegedly favoured the better-off at the cost of those on lower incomes. His assertive calls for a strategic review of NATO and an end to EU enlargement had also alienated other European leaders.

As COVID-19 spread across Europe, EU member states moved quickly to close their borders, while the EU waived budget rules and effectively suspended competition and state-aid regulation as governments intervened to halt a wave of bankruptcies or to prevent firms being bought up cheaply. Borders were reopened by the end of June 2020. EU fiscal rules governing deficit spending had to be suspended because member states needed to support their economies through the crisis. Problems will arise for the Commission if countries oppose the reimposition of these rules. The suspension of competition and state-aid rules could also pose a challenge. Any attempt by member states to hold on to these new-found freedoms could threaten the web of political trade-offs that underpin the EU's single market. This in turn could make it hard for the Commission to win support for its wider agenda, including a green deal to accelerate the transition to a low-carbon economy, or efforts to tackle tax avoidance.

The example of state aid illustrates this point. Governments of bigger and stronger economies have more firepower. While Germany comprises about one-quarter of the EU economy, it has accounted for half of all EU state aid to firms since the outbreak of COVID-19. Since smaller economies cannot provide comparable support, there was a risk that they would double down to protect policies that give their economies an advantage, such as favourable tax regimes, or would defend energy-intensive industries by opposing policies to reduce carbon emissions. For this reason, it was crucial for the Commission that member states agreed to its proposed recovery fund – as it needed to demonstrate that it does more than police the single market.

The east–west divide

Tensions between older and newer EU member states posed a further challenge. Many older members, and the EU Commission, were critical of countries, Hungary and Poland in particular, that received substantial EU funding while flouting democratic norms such as judicial independence and a free press. For their part, many newer members resented what they saw as their second-class status in the EU. They viewed EU funds as the quid pro quo for opening their markets to more established and competitive Western firms; the migration of many of their young and best-educated people; and the dominance of their economies by Western European firms.

Most of the newer members faced big cuts in their funding under the 2021–27 MFF, partly because they were now wealthier and partly because the UK's exit had left a hole in the EU's budget. Most contentious, however, was the EU's drive to make funding conditional on adherence to democratic norms and environmentally sustainable policies. A decision to withhold funds could be decided by qualified majority voting rather than unanimity, giving the Commission significant leverage. Many newer member states saw this as a further threat to their already limited national autonomy.

There is no obvious solution to this east–west divide. Indeed, the revitalised Franco-German axis essential to strengthening the eurozone could aggravate east–west tensions by reinforcing the perception of a hierarchy of states within the EU. German Minister of Defence Ursula von der Leyen took over from Donald Tusk as president of the European Commission on 1 December 2019. Strictly speaking, the job should have gone to the leader of the European People's Party, after it emerged as the largest grouping in the May 2019 EU parliamentary election. But this would have meant appointing Manfred Weber, whose right-wing Bavarian Christian Social Union had a record of accommodating right-wing populists such as Hungary's Fidesz party. Weber was unacceptable to France and others, while Germany was adamant the role should go to a German. A committed federalist from the liberal wing of Germany's Christian Democratic Union, von der Leyen's appointment reassured supporters of closer integration. But her apparent attempt to assuage the conservative right backfired when the name she gave to the Commission's new migration portfolio, 'Protecting the European Way of Life', was criticised for having far-right connotations. It was subsequently renamed 'Promoting the European Way of Life'.

France and Germany's inability to agree common positions had hamstrung the Commission's attempts to push forward an integrationist agenda for years. On the face of it, the choice of a German federalist as president, together with Franco-German backing for its ambitious recovery fund and the UK's departure from the EU, promised to reverse the Commission's loss of influence and open the way for closer integration. But this was far from certain, not least because of the political and economic fallout from the pandemic.

Brexit uncertainties

The UK's formal exit on 31 January 2020 was a major event in the history of the EU. The resounding election victory for the UK's Conservative Party in December 2019 had ended any chance of reversing Brexit, of the UK remaining in the single market as a member of the European Economic Area, or of it developing a relationship with the EU comparable to that enjoyed by Switzerland.

The UK would remain in the single market until the end of 2020 but lost its voting rights. In June, the UK government confirmed that it would not seek an extension to this transitional status, despite little progress on negotiating the country's future relationship with the EU. This lack of progress was partly the result of COVID-19, which diverted political leaders' attention, but mainly reflected the gap in negotiating positions. The UK wanted a far-reaching free-trade deal with the EU, while being free to set its own regulatory standards. The EU ruled this out, arguing that access to the single market without compliance with its rules of membership would give the UK an unfair competitive advantage. While the EU might give a little ground in the coming months, its fundamental position will not change.

The future UK–EU relationship is likely to take one of two forms: a minimal trade deal of the kind the EU has with Canada, or no deal at all

– with trade reverting to World Trade Organization (WTO) rules. Even if the former could be agreed, it is unlikely it could be implemented by the beginning of 2021 – the UK in particular has done too little to build the necessary customs checks. This would make a further transition period necessary. No deal – the only feasible alternative – would be messy. The bilateral trade relationship would be by far the most complex carried out under WTO rules. In addition to practical difficulties – the lack of necessary border infrastructure, and the absence of regulatory agreements on everything from product standards to freight – the imposition of tariffs would seriously disrupt trade. While the average EU tariff on imports of non-agricultural goods was only 2.8% in June 2020, it was as much as 10% on cars and car parts. Average tariffs on agricultural goods were close to 12%. Trading under WTO rules would also mean a physical border between Northern Ireland and the Republic of Ireland, putting the UK in breach of the Good Friday Agreement – an international treaty – and threatening peace in the province.

Either outcome – a minimal trade deal or no deal – would do much economic damage, especially to the UK. But no deal would also risk doing lasting harm to bilateral ties, making it harder to maintain close relations in other areas, such as security and defence. There was no obvious silver lining for the EU; the UK was an awkward member at times but not a major obstacle to lasting solutions to eurozone or east–west tensions. Brexit could make it easier for the EU to develop an independent defence capability, something the UK always resisted on the grounds that it could undermine NATO. But other obstacles remained, including low military spending across the EU and the lack of a strategic international world view among most member states.

External relations

EU–US relations deteriorated further in 2019–20, with tensions over trade, security and climate policy reinforcing each other. The US administration was frustrated by the failure of EU governments to raise defence spending, as well as with the US trade deficit with the EU. European confidence in US leadership – already shaken by President Donald

Trump's hostility to European integration and readiness to describe the EU as a 'foe' – was further hit by the US handling of the COVID-19 crisis, including an apparent attempt to buy exclusive rights to a vaccine being developed in Germany. A Pew Research Center poll in May 2020 suggested that Germans were now split equally on whether they would prefer closer relations with China or the US.

A host of issues remained unresolved – from disagreement over the origins of the trade imbalance, to aircraft subsidies, car tariffs and the future of the WTO. But it was the taxation of global technology firms that could be the trigger for a further worsening of transatlantic trade relations. In June, the US walked out of multilateral talks aimed at agreeing a framework for taxing global digital firms, most of which are US-based. At the same time, the US labelled attempts by European governments to tax the profits of digital firms made in their jurisdictions as discrimination against US digital firms, threatening retaliatory action against French exports to the US. France indicated that it would go ahead with its tax, raising the prospect that other member states might follow. While Trump's combative approach to Europe has aggravated trade tensions, they will not disappear under a different president.

The same goes for tensions over security. Trump's threat to pull US forces out of Europe unless European countries boost their military spending has come to little. Even the June 2020 announcement of a sharp reduction in the number of US troops stationed in Germany was followed by a boost to the numbers deployed in Poland, one of the few EU member states that has met the NATO target of spending 2% of GDP on defence. Germany is increasing defence spending, but gradually and from a low base; it stood at 1.2% of GDP in 2019. With the US becoming increasingly focused on China, frustration at low levels of European military spending will persist, especially if European countries deepen their ties with China or position themselves as neutral in a conflict between the US and China.

Closer EU–China ties were, however, far from inevitable. While China was attempting to exploit perceived US weakness to expand its influence in Europe, its assertiveness showed signs of backfiring. China's belligerent rhetoric, its threats to retaliate if European governments exclude Chinese telecoms firm Huawei from their 5G networks, and growing evidence of the scale of China's domestic-surveillance architecture have shaken European governments. While some have resisted US pressure to exclude Huawei, they have tried to reduce the firm's involvement in ways that avoid outright confrontation with China.

But there were obstacles to the unified EU approach that would prevent China from singling out individual member states for intimidation, as it did to Sweden in November 2019, when it threatened the country with 'bad consequences' for criticising its human-rights record. China subsequently cancelled a trade mission to Sweden and vocally criticised its handling of COVID-19. Germany, which accounted for almost half of EU exports to China, was particularly cautious of criticising China for fear of damaging its commercial interests. China's attempts to sow European division by cultivating ties with Central and Eastern European countries (the so-called '17+1') have also had some success. Greece and Hungary in particular were taking an increasingly pro-Beijing line.

The readiness of some EU governments to put commercial interests above strategic ones was highlighted in April 2020 when the EU's European External Action Service (EEAS) was pressured into toning down its analysis of China's global disinformation campaign. In the draft version of the EEAS report into the activities of foreign powers, the EEAS had bracketed China with Russia as the worst perpetrators, but in the final version published in April 2020, criticism of China (though not of Russia) was much more muted.

Germany's Foreign-policy Challenges
How is the EU's largest state navigating global change?

November 2020 marks the 15th anniversary of Angela Merkel's inauguration as chancellor of Germany. She is the longest-serving current leader of any European democracy. Among major European democracies, only her predecessor, Helmut Kohl, has ever served for longer.

Kohl's 16-year tenure as chancellor included the end of the Cold War and the reunification of Germany. When Merkel came to power in 2005, Germany was still known as 'the sick man of Europe'. Her time in office has been marked by impressive growth in the country's economic stature, even as an accompanying policy preference for restraint over greater international engagement has helped curtail the rise of Germany's geopolitical influence. Developments in the first half of 2020 shifted the narrative of Merkel's legacy and highlighted the policy challenges she will likely bequeath to a successor following federal elections in autumn 2021.

German leadership within the European Union

The year 2020 was always going to shine a spotlight on German leadership in Europe, thanks to the country's incumbency of the six-month rotating presidency of the Council of the EU from July to December. The COVID-19 pandemic further magnified this dynamic.

Prior to the shift to crisis management and resilience that COVID-19 forced on Europe, the Chancellery had identified three priorities for Germany's EU presidency. Firstly, it would broker a deal on the EU's next long-term budget (its Multiannual Financial Framework, or MFF). Secondly, it would host the first-ever summit between all leaders of EU member states and Chinese President Xi Jinping, ideally signing there an EU–China Comprehensive Investment Agreement that has been under negotiation since 2013. Thirdly, it would push initiatives to speed up Europe's conversion to a greener economy. With the UK's transition agreement with the EU due to expire on 31 December 2020, the Chancellery was also braced to play a key role in setting the framework for future EU–UK cooperation. The pandemic did not remove these substantive policy challenges, but it added plenty more.

In June 2020, the EU–China summit that Berlin had planned for September was postponed. Perhaps, as the Merkel circle briefed, COVID-19 was the sole reason for this decision, but the summit had become increasingly controversial. With no investment agreement ready for signature, disquiet about Beijing's initial handling of the outbreak and growing alarm over China's behaviour in Hong Kong and elsewhere, the summit risked turning into a photo opportunity for Xi and a display of disunity for Europe. In a rare speech on foreign and security policy just a week prior to the summit's postponement, Merkel noted serious policy differences with China. But she argued this was no reason not to talk or cooperate where possible. China was too big to be isolated, and dialogue became more, not less, important as US–China tensions increased. These remarks implied that the planned summit was likely to be rescheduled, especially if progress could be made on an investment agreement.

This push to maintain dialogue did not, however, stop Germany from tightening its foreign-investment laws, with COVID-19 heightening concerns about predatory Chinese investment in sensitive sectors. Germany is at the forefront of EU work to develop new tools to review and, if necessary, block takeovers of European companies by state-owned and state-supported foreign enterprises. The country is also likely to push for an updated understanding of European 'strategic autonomy' in light of vulnerabilities that COVID-19 exposed, including in medical supply chains. A rethink on EU–China relations is under way, albeit with predictable caution in the Chancellery. In May 2020, the CEO of the influential Axel Springer publishing house, Matthias Döpfner, published an article arguing for Europe to draw 'a line in the sand' with China, and in particular to 'fully assert the principle of true reciprocity'. Several members of Merkel's Christian Democratic Union (CDU) party support such a line.

As negotiations over an EU–UK free-trade agreement intensify, the UK's predilection to look first to Merkel and Germany to play the mediating role is likely to be reinforced by the country's EU presidency. Germany will not agree to anything that risks a back door for the UK into the EU single market and is braced for hard negotiations on the issue of a level playing field. But Merkel's desire to keep the UK as a close strategic partner will ensure that Berlin works hard behind the scenes to bring a deal together.

Merkel has called COVID-19 the 'toughest challenge' in the EU's history. She has been clear that, for Germany, the EU's well-being is 'not just a question of solidarity' but 'a matter of self-interest'. As the pandemic tested the internal cohesion of the eurozone, in May 2020 the chancellor agreed a proposal with French President Emmanuel Macron that broke deeply ingrained German conservative taboos. This proposal – for an EU rescue fund that would see the provision of €500 billion (US$560bn) in targeted grants to blighted member states as part of the pandemic recovery plan – was surprisingly ambitious. While this did not represent a conversion to the idea of a fiscal union, it seemed that Germany had learned lessons from its controversial management of the Greek sovereign-debt crisis. Details of the proposed Recovery and Resilience Facility would require considerable negotiation among the EU-27. But Merkel was leading again and, unlike during the 2015 migration crisis, she appeared to be bringing her party and her country with her.

But Germany is not always the EU's mediator of tensions and broker of deals; at times it can also be a more disruptive influence. For example, a May 2020 ruling by the German Constitutional Court – which gave the European Central Bank three months to provide adequate justification for a 2015 bond-buying programme that it would otherwise declare illegal under German law – appeared to pose a serious challenge to the EU's legal order. Contentious negotiations on the broader MFF package also risked undermining Germany's attempts to sustain greater EU dynamism and solidarity.

Facing up to international tensions

Germany's central foreign-policy preoccupation in the years ahead will be navigating US–China tensions. The Chancellery is anxious to avoid Germany and Europe being caught in the middle of a new cold war. In the first half of 2020, Berlin was concerned that the US election campaign would push both contenders into still-more hawkish positions on China. A Biden presidency would seem likely to help those in Europe, including within Germany, who supported a more assertive approach, as the US would be seen as a more reasonable and reliable ally. However, the extent of the bipartisan shift in the US to a more hawkish approach towards China meant that an increase in transatlantic tensions over China policy felt increasingly inevitable. While the US contemplated rolling back 40 years of ever-closer economic relations with China, Germany was well aware that pressure on Europe to 'decouple' from China would both challenge its interests and test EU unity.

Broader German–US relations have faced growing strain throughout Donald Trump's presidency. Time and again Germany has highlighted its readiness to ignore the US foreign-policy lead: on Iran and the Nord Stream 2 gas pipeline, defence spending and possibly – still – on Huawei. In an example of growing transatlantic drift, the Körber Stiftung reported that 36% of German respondents to its 2020 survey on German foreign policy thought it more important to have close relations with China than the US (up from 24% in 2019), while 37% put greater value on closer relations with the US (down from 50% in 2019). Within Germany, few criticised Merkel's rejection of Trump's invitation to attend a face-to-face meeting of G7 leaders in June. The Trump administration's subsequent announcement of its intention to withdraw nearly 9,500 of its 34,500 troops permanently stationed in Germany highlighted the new low to which bilateral relations had fallen. In Germany, suspicion swirled that the announcement was driven primarily by Trump's domestic political agenda, and even petty revenge for Merkel's disruption of his G7 plans. Alternative suggestions that the decision had more to do with Trump's continued dissatisfaction with the level of Germany's defence expenditure received little consideration.

Germany faces further foreign-policy challenges closer to home, including the management of the EU's, and its own, relations with Russia. Merkel has been a critical influence in levying and sustaining EU sanctions on Russia following the annexation of Crimea and the fomenting of conflict in eastern Ukraine. But Macron has appeared keen to chart a different path, arguing that 'pushing Russia from Europe' is a 'profound strategic mistake'. There is a risk that more Russia-friendly forces within both Germany and the wider EU will see Merkel's retirement as an opportunity to push for the easing of EU sanctions, regardless of progress on the Minsk accords. Meanwhile, Merkel has undermined her position on Russia, and Germany's credibility as an ally to some of its Central and Eastern European partners, through her persistent defence of Nord Stream 2.

More generally, multilateralism – the mainstay of German foreign policy – continues to fall out of fashion fast. Throughout 2020, many of the global institutions that Germany has long supported, notably the World Health Organization, World Trade Organization and United Nations, were further weakened. Efforts by German Foreign Minister Heiko Maas to summon an 'alliance of multilateralism' did not advance beyond the rhetorical. The difficulties Germany faced in defending these institutions exposed the limitations of its leadership.

Defence-policy implications

As Germany headed into its worst recession for at least 70 years, defence-spending commitments were put at risk by renegotiation of budgetary allocations for 2021 and beyond. This made it likely that funding earmarked for future projects, and potentially also for research and development, would be cut. Such pressure was regrettable given that Germany's commitment to defence spending had been growing in credibility, albeit belatedly and from a notably low base. Indeed, in 2019 Germany recorded the second-largest defence-budget percentage increase in real terms among the world's top 15 states by military expenditure. However, recognition of Germany's deteriorating strategic environment should help limit the damage, at least until after the federal elections in autumn 2021, after which a new government coalition will need to negotiate its own budgetary agreements. Meanwhile, declining GDP will help ease the pressure on Germany's NATO defence-spending commitments. While the 2% obligation will continue to go unfulfilled, the target that Germany set for itself in 2019

– for 1.5% of GDP to be allocated to defence spending by 2025 – looked set to be reached in 2020.

Even before the pandemic, German Defence Minister Annegret Kramp-Karrenbauer (typically known by her initials, AKK) had sought to deflect attention from Germany's failure to meet its NATO defence-spending commitments by emphasising her commitment to 'A2A' or 'the ability to act'. This push to focus on capabilities is likely to be intensified by the fallout from COVID-19. Yet ongoing concerns about the operational readiness of many of the Bundeswehr's primary weapons systems mean that such a switch in focus is unlikely to help Germany's global military standing.

There have, though, been some positive developments on defence. In October 2019, France and Germany reached a legally binding agreement on arms-export controls. Joint work on a Future Combat Air System (FCAS) continued to progress, albeit slowly. In February 2020, the Bundestag approved €77 million (US$86m) for the development of the first prototypes under this transnational project. In April 2020, to some domestic disquiet, AKK confirmed her desire to replace some of Germany's ageing fleet of *Tornado* fighters with US-made F-18s, in order to guarantee that the country could continue to fulfil its nuclear-sharing obligations. The German destroyer *Hamburg* had been scheduled to set sail in June on a five-month training mission to Australia, suggesting belated German interest in cultivating closer political–military relations in the Asia-Pacific. Although COVID-19 forced the suspension of these plans, Germany has committed to sending another ship to the rescheduled Indian Ocean Naval Symposium that the *Hamburg* would have attended.

Looking forward, with AKK's shot at the chancellorship gone following her unexpected resignation as chair of the CDU in February 2020, and any temptation for her to self-censor in her push for the top job consequently diminished, Germany has a defence minister ready to fight her department's corner during the cross-government budgetary reviews that will follow. For example, a €130bn (US$145bn) fiscal-stimulus package unveiled by Merkel on 3 June 2020 included – with notably little fanfare or publicity but much careful curation behind the scenes – an extra €10bn (US$11bn) for the arms industry.

Nevertheless, old themes linger. In foreign and security policy, Germany's reality still rarely matches its rhetoric. For instance, AKK's call, in late 2019, for the creation of a 'safe zone' in northeastern Syria may have been an important statement of intent domestically, but it seemed naïve to the realities of the international situation. In July 2019, concerns about aligning too closely with its transatlantic ally prevented Germany from accepting the invitation to join a US-led naval-protection mission in the Strait of Hormuz. Germany argued for a European alternative, but few were surprised when no such alternative transpired.

The future face and focus of German leadership

The immediate trigger for AKK's resignation as CDU chair, and her renunciation of her claim to become the party's candidate for chancellor, was a local crisis in the state of Thuringia, where she failed to dissuade party colleagues from voting with the far-right Alternative for Germany (AfD). But her decision also reflected wider difficulties in settling internal frictions over the future direction of the party. Three candidates announced their ambitions to replace AKK as party chair, and likely therefore Merkel as chancellor following federal elections in autumn 2021. The election, however, originally scheduled for April, was postponed by the onset of the pandemic in Germany.

Long-standing Merkel critic Friedrich Merz would move the CDU back to the right, countering AfD popularity – including with more restrictive immigration policies. But his leadership bid is undermined by the potential difficulties he would face in forming a governing coalition. His more conservative economic instincts might also hold less appeal in the midst of a pandemic. By contrast, North Rhine Westphalia's minister president, Armin Laschet – who remains the front runner – would keep the party more on the centre ground, including on foreign policy, where he has warned against the 'demonisation' of Russia and 'anti-Putin populism'. Chair of the Bundestag Committee on Foreign

Affairs Norbert Röttgen is the outsider in the race. His instincts on key foreign-policy issues, including Russia and China, are notably more hardline than Merkel's.

All three candidates struggled to maintain momentum and profile during the pandemic. Should this continue, the CDU might opt to elect another candidate for chancellor. Merkel herself could unexpectedly reverse course and seek a fifth term, although she has repeatedly ruled this out. More likely, other contenders will come to the fore. For example, Markus Söder, Bavaria's state premier and leader of the CDU's sister party, the Christian Social Union (CSU), proved the rising star of the early months of the pandemic, registering record approval ratings among currently serving German politicians. Yet the historical precedents are not favourable for him: only two CSU candidates have ever run for chancellor for the CDU/CSU, and both lost.

Another name long billed as a 'chancellor-in-waiting' is Health Minister Jens Spahn. His credentials will depend heavily on Germany's management of the pandemic. One of the few potential contenders to have deliberately cultivated close relations with the circle around Trump, a Spahn government would be fundamentally Atlanticist in orientation, and sceptical of deepening EU integration.

Whoever emerges as Merkel's successor will inherit a political context transformed by the pandemic. Prior to the onset of COVID-19, tensions within Merkel's governing coalition appeared likely to prevent her serving a full fourth term. She would have departed office as the chancellor whose controversial handling of the migration crisis ended up fuelling the rise of the far-right AfD, dividing her own party and precipitating its plunge in popularity. But by mid-2020, Germany's success in responding to COVID-19 had helped restore Merkel's leadership credentials and, with them, the political fortunes of her own party. By June 2020, support for the CDU/CSU had reached nearly 40% in the polls, a jump of more than 10% from the onset of the crisis.

At the same time, support for the AfD fell by around one-third: in June 2020 it was polling around 10%. If the political threat it poses can be contained at that level, tensions within the CDU over how far it should move to the right are likely to ease. The pandemic might yet gift AKK's successor a less fractious, more united party in a more sombre country, focused on nursing the German economy back to health rather than flirting with extremism. But there is a risk that the prolonged economic pain of COVID-19 will fan the flames of that extremism, especially in deprived areas of eastern Germany where AfD support has been strong.

The boost to the CDU's political fortunes has also come at the expense of Germany's Greens, as voters prioritise more immediate economic concerns. Nevertheless, a CDU/CSU government, with the Greens as the junior coalition partner, continues to be the most likely outcome of the autumn 2021 elections. Significantly, though, the fall in support for the Greens means that the prospect of a centre-left coalition (of the Greens, the Social Democratic Party and the Left) coming to power in 2021 – with the radical shift in foreign and defence policy this would likely entail – has receded still further.

Conclusion

For several years, Germany has been increasingly capable and willing to exercise influence, notably at the European level. The events of 2020 raised the pressure to summon a still-more substantive response. The pandemic has forced Berlin to rethink what is required to demonstrate EU solidarity, while deteriorating US–China relations have fuelled a growing debate on the future management of Europe's relations with both powers.

COVID-19 also appears to have fundamentally changed the legacy that Merkel will bequeath her party and her successor. Yet the self-restraint and continuity that has framed German policy-making since reunification will not be easily broken. As the American conservative political commentator Walter Russell Mead noted in May 2020, if it really wants to exert the influence in international affairs commensurate with its economic status, 'Germany will have to break some of its deepest taboos, and begin to think and act like a great power again'.

China's Influence in the European Union

Why has the past year been so difficult?

The year to June 2020 was a difficult one for the relationship between China and the European Union. Chinese influence in Europe and – in competition with the United States – over Europe came under growing scrutiny, prompting a pushback by the EU and the United Kingdom. The EU adopted more defensive measures and became openly critical of China. By mid-2020 China's influence campaigns appeared to be struggling. Little progress had been made on planned cooperation agreements. These developments confounded the hopes expressed by the two sides a year earlier. This period may prove a turning point in their relationship.

The year that was to be

In early 2019, the EU began adopting key initiatives to push back against China's influence for the first time. Beijing had been dragging out or even stalling ongoing negotiations with the EU. Ahead of the April 2019 EU–China summit, the EU made its own moves. Firstly, it sought to commit China to signing technical agreements on civil aviation and geographical indications of origin by the end of the year.

Secondly, and more significantly, in a March 2019 communication titled 'EU–China – A Strategic Outlook', the EU Commission designated China as simultaneously a 'negotiating partner', 'economic competitor' and 'systemic rival'. It set out ten action points, many of them policy or rule changes aimed at countering China's leverage within Europe. These included tighter regulation of public procurement, guidelines on state subsidies, recommendations on 5G networks and 'swift' implementation of foreign direct investment (FDI) screening. Developing its own policies rather than relying on Chinese undertakings represented a major shift for the EU.

Overall, relations stayed on a positive note during this time. The April 2019 summit concluded with a pledge to achieve a Comprehensive Agreement on Investment (CAI) – negotiations on which had begun in 2014 – by the end of 2020, with significant progress to be made in 2019.

Both sides seemed to agree that the year that followed would be one of delivery for the EU–China relationship. China designated 2020 as a 'Year of Europe'. Germany's presidency of the European Council in the second half of 2020 was expected to create new momentum. An additional EU–China summit in Leipzig, involving all 27 EU member states, was planned for September 2020. The event would also serve to bring back into the fold the European states involved in the Cooperation of Central and Eastern Countries with China (CEEC) initiative – now the '17+1' since Greece joined in 2019. The lead-up to Leipzig would ensure the completion of the CAI. In Berlin, as in Brussels, there were concerns that the high stakes for the event might induce German Chancellor Angela Merkel to settle for less in the negotiations. But Europeans appeared to share some views with China that ran counter to those of the US administration of Donald Trump. These included Iran policy and rescue of the Iran nuclear deal; the preservation of multilateral institutions such as the World Trade Organization's Appellate Body; and the implementation of the climate-change-mitigation goals of the 2015 Paris Agreement.

Relations turn fractious

However, except for two aviation agreements for which the April 2019 summit had set a deadline, as of June 2020 there had been no delivery from the Chinese side. Even an agreement on geographical indications of origin, also under an end-of-2019 deadline, remained unsigned. Unnoticed by most Europeans, China increased its use of coal compared to 2018, and its post-COVID-19 recovery plan included a boom in new coal-fired thermal plants.

Rather than cooperating more closely with the EU, China's diplomacy towards several member states turned bitter in the second half of 2019, and its use of social media and cyber tactics sharpened considerably. In response, the EU increased its defensive actions. At the same time, it continued to negotiate on potential agreements, in particular signing a CAI by the end of 2020 that would entail significant structural changes in China's political economy. These included limiting state subsidies, granting reciprocal access on public procurement and the service sector, and protecting intellectual-property rights.

But by June 2020, agreement looked increasingly unrealistic. Chinese negotiators' attention had been focused on the US-initiated trade conflict since 2019. A transactional trade deal with the US (the Phase One trade agreement) concluded in January 2020 did little to change this. Sino-US relations entered an unprecedented crisis. The EU cannot compete with the US for leverage or influence. Furthermore, simmering problems in Europe – a protracted Brexit process, populist protests and a slow process for nomination of the new EU Commission, along with Trump's invective against the EU – led Feng Zhongping, Vice-President of the China Institutes of Contemporary International Relations, to conclude that Europe was 'powerless internationally in spite of its intentions'. Others used the milder expression 'pragmatic' – that is, amenable to compromises proposed by China.

Chinese diplomats also engaged in 'wolf-warrior diplomacy'. This term, taken from a popular nationalist novel, describes campaigns of influence, including in some cases insults or threats, towards several EU member states. Cases in point included attacks on Sweden (November 2019) over one of its citizens held in China; on the Czech Republic over its pushback on local Chinese influence and planned parliamentary visit to Taiwan (January 2020); and on Denmark over a media caricature (January 2020).

China also asserted itself against the largest European states. In December 2019, its ambassador to Berlin hinted at reprisals against the German automotive industry over the 5G and Huawei issue, while in June 2020 Beijing's envoy to London warned of possible cancellation of Chinese participation in nuclear plants and high-speed rail projects over the same issue. More broadly, China's social-media and hybrid cyber-warfare tools moved in 2019 from positive advocacy of China's governance and model to active criticism, slander and propagation of fake news targeting Europe's political and social systems. Chinese trolls and bots, which until recently were active mainly only in Taiwan, appeared in Europe. In the wake of these developments, in June 2019 the European Commission and External Action Service (EEAS)'s Disinformation Review, which initially covered Russia, began covering China as well.

China's tactics did not help its image. A December 2019 opinion poll by the Pew Research Center found that in nine out of 14 European countries polled, unfavourable views of China exceeded favourable views. The only exceptions were Greece and four Eastern European states, where more of those polled had a favourable view of China.

COVID-19 sharpens the edges

The advent of COVID-19 – and China's need to defend itself against accusations regarding its response during the first phase of the pandemic – sent its propaganda efforts into overdrive. Particularly notable was the notice on the website of the Chinese embassy in France in April 2020 that accused staff in retirement homes of 'leaving the pensioners to die from hunger and disease'. When the Netherlands altered the name of its representation in Taiwan to Netherlands Office Taipei, official Chinese media hinted that 'Chinese citizens' wanted Chinese companies to stop exporting medical supplies to the country. But in an extraordinary image reversal, China began using 'mask diplomacy' – shipping medical supplies to European countries – to showcase its goodwill and assistance. In several EU countries including Italy, government ministers welcomed Chinese cargo planes in person. While European medical donations to China were kept confidential at the behest of the Chinese government, China trumpeted its supplies to Europe: these had some impact on public opinion, even if in most cases they were sales rather than donations. In May, an opinion poll found that Germans now considered their country's relationship with China to be almost as important as its relationship with the US.

In June 2020, the EEAS accused China (and Russia) of targeted disinformation campaigns regarding European responses to the COVID-19 crisis. After a short virtual summit with China on 22 June, European Commission President Ursula von der Leyen pointed to China as the origin of cyber attacks on European hospitals and said that 'facts and figures' relating to disinformation had been presented to the Chinese leaders.

Europe's response

By mid-2020 Europe had been more successful in fostering internal policy cohesion and resilience to Chinese actions than in obtaining concessions in its relationship with China. Europe made progress with plans to enact an investment-screening directive by October 2020, and in introducing additional measures to avoid predatory takeovers at a time of sharp economic recession. In June 2020, the Commission published an ambitious White Paper covering foreign subsidies, with the aim of bridging the existing gap with intra-EU rules that did not apply to non-EU companies. Industrial policy, technology-export controls and mitigation of supply-chain risks entered the European lexicon.

5G became an urgent issue. With most 5G tenders due to play out in 2020, a pitched battle was waged, country by country, between the only widely available equipment suppliers, Ericsson, Nokia and Huawei. Debate ensued regarding security concerns about installing Huawei equipment and the interests of telecoms companies. These were concerned that they would bear the added cost of ripping out Huawei equipment, on top of the cost of purchasing 5G frequencies.

In January 2020, the EU Commission published a 'toolbox of risk-mitigation measures' for the cyber security of 5G networks. Sidestepping the issue of discriminating against a specific country or company, it nonetheless spelled out clear security criteria for the different layers of 5G equipment and was also consistent with the UK's National Cyber Security Centre's findings of the same month. Several countries, such as the Czech Republic, Denmark and Poland, opted for European lead suppliers. In France, while the government avoided an explicitly negative stance, the administration warned telecoms companies that authorisations for supplies by Huawei would not be extended beyond 2028. In Germany, while telecoms operators took advantage of a long silence by the government to choose Huawei, the public debate raged on.

Similar trends applied to decoupling – which the EU calls 'diversification of suppliers' – and to investment screening. Until 2018, many member states in need of capital remained sceptical of investment screening, which they feared would hamper their own efforts at rescuing their economies. Chinese money had come to the rescue of Portuguese utilities and banks after the 2008 global financial crisis. Other countries, such as the Netherlands and Sweden, feared screening could lead to retaliation against their firms in China. This trend has reversed, and the scope of investment screening has widened in several countries and at the EU level. With COVID-19, the pharmaceutical and medical supply industries acquired new relevance, along with ports, and aerospace and automobile value chains. Media are also increasingly considered vulnerable to foreign financial influence. Italy, previously open to Chinese FDI and takeovers, tightened its 'Golden Rule' that allows the government to stop acquisitions. In France, the home of proponents of European economic sovereignty and strategic autonomy, Chinese acquisitions declined. In the UK, emergency action blocked Chinese acquisition of a leading graphic chipset manufacturer, and a bipartisan group in Parliament was set up to counter Chinese influence.

China's crude diplomatic conduct did not help its economic relations. Hungary, Beijing's staunchest supporter at EU meetings, and Italy, the first G7 country to sign up to the Belt and Road Initiative (BRI), saw their exports to, and inward investment from, China decline in 2019. The CEEC 17+1 initiative began to lose its attractiveness after years of limited results on the investment front. Chinese investment flows mainly to Western European countries, while loans for infrastructure projects go to southeastern European EU accession candidates. In Montenegro and in Serbia, these loans create some path dependency, since they are either for major public works or bailouts that would not have been financially sustainable under market conditions. Here, as in Eastern Europe, the Chinese offer serves as leverage for local governments for more EU structural funding.

The economic narrative and elite capture

China's priority is to retain access to the European market while Europe complies with China's policy goals and – more importantly – practises 'non-interference' in its domestic affairs. China's main tool of influence is elite capture through perceived economic interest. The key word here is 'perceived'. Among EU states, only Germany's trade with China is significant: in 2019 it accounted for around half the total EU-27 exports to China. Yet even for Germany, China accounted for only 7.2% of its global exports in 2019. But these figures understate the wider impact of China's economy for the EU, which includes subcontractors to German and Swedish companies from southeastern Europe, finance and insurance interests largely operating via Hong Kong, and multinational companies with direct investment stakes in China. As for public decision-makers, there is no shortage of regional officials who are sensitive to the lure of real or potential Chinese investment.

China's public diplomacy is also more fine-grained than it might first appear. It uses myriad associations – run or controlled from behind by the Chinese Communist Party (CCP)'s United Front Department – to get across its messages. In May–June 2020 the aggressive posture of Chinese ambassadors began to recede. This is consistent with CCP techniques for domestic political campaigns, which alternate phases of struggle and unity: both words are now also used in an international context. For example, in May 2020 the Chinese ambassador to Berlin promised growth for the automobile, chemical, mechanical-engineering and pharmaceutical sectors in China. In the same month, his counterpart in Brussels urged Europeans to meet China 'halfway' on an investment treaty – despite no sign of concessions from China. After a virtual EU–China summit in June 2020, China's Ministry of Foreign Affairs quoted the EU high representative for foreign affairs and security policy as denying 'rivalry' with China, prompting a sharp rebuttal from the EU.

China's efforts to use more constructive rhetoric provides fodder for the advocacy of interest groups in Europe that support better relations – through friendship and professional associations, city-twinning schemes, chambers of commerce and young leaders' groups, for example. Through such groups, China seeks to manage the discourse of EU–China relations by insisting on the need to avoid polarisation or criticism. Invitations to China are frequent, but they epitomise the asymmetry in the relationship: while each side may in principle choose its members for these groups, in practice the CCP ultimately decides all Chinese participants. While

Chinese visitors usually enjoy free access once in Europe, visitors to China now face even more scrutiny than they did previously.

China also buys paid supplements in the European press at undisclosed prices and signs agreements to provide content to news outlets. A key association active in the EU Parliament is the EU–China Friendship Group. Its exact membership remains unclear, but it is managed from China. Similarly, the Chinese Academy of Social Sciences has established in Budapest a China–CEE Institute as an offshoot of the 17+1 initiative, with a legal European status that would enable it to create branches in any EU nation. In Germany, the China Bridge networking platform was created in January 2020 to foster elite ties. There are many similar clubs in the UK, where university alumni links also play a large role. France has several business and elite clubs that compete for attention, and several former prime ministers such as Jean-Pierre Raffarin and Dominique de Villepin are seeking a role in the relationship.

China is not the first country to use such methods. It likes to point out that US public diplomacy did so to develop networks of transatlantic links in the post-war period. The difference, though, lies in the asymmetry of openness between Europe, a community of pluralist societies, and China, where the CCP dominates all public life. To reinforce this control, a Foreign Non-Governmental Organisation (NGO) Law passed in 2017 strictly constrains the activities of foreign entities in China. China's 2015 National Security Law, its 2020 Hong Kong National Security Law and its 2017 National Intelligence Law put cooperation and contacts between Chinese and foreign individuals or entities under pressure.

Europe as a US–China battlefield

A larger struggle for influence over Europe's China policy has begun. In June 2020, the Trump administration launched a public-diplomacy campaign targeting European governments and public opinion. Meanwhile, several Chinese envoys to Europe, notably in France, Poland and Hungary, took to polemic with the US. As prospects of implementing the US–China Phase One trade deal faded away, battle lines were increasingly drawn in third countries.

The EU is grappling with the notion of 'strategic autonomy' or a 'sovereign Europe'. In recent months it has contended with multiple pressures from China. In April 2020, the leak of a draft EEAS letter appeared to show that reference to a China-led 'global disinformation campaign' around COVID-19 had been deleted from the final published version. The EEAS denied that it had responded to Chinese pressure to water the letter down. In May, China's official media censored an opinion piece signed by all EU ambassadors to Beijing. The EEAS publicly rebuked its own envoy over his acceptance of the cut.

More broadly, the EU maintained the three elements of its China policy: strategic cooperation, economic competition and systemic rivalry. In June, its report on COVID-19 disinformation denounced China as well as Russia. Von der Leyen and EU Council President Charles Michel adopted an unprecedentedly strong and public line of criticism towards China during the short virtual summit that began on 19 June 2020. Virtually no topic was left untouched, including cyber attacks on European hospitals, disinformation campaigns, Hong Kong and Xinjiang. The fate of a proposed 'Agenda 2025' cooperation framework would be in question if China did not sign a CAI by the end of 2020. Unprecedented emphasis was put on 'non-negotiable' human rights and issues of reciprocity during the summit.

China's influence campaigns in Europe thus appeared to be struggling at the highest levels. Yet it has not disowned its tactics nor clearly reacted to recent EU concerns. The EU–China Leipzig summit planned for September was postponed, ostensibly due to COVID-19. China may be tempted to wait until after the November US presidential election in order to understand its implications before reassessing its Europe policy.

The tide of Sino-European relations has turned in 2019–20. Europe began to adopt defensive measures against China's influence, asymmetric trade and investment practices, and its acquisition of critical technologies. Many, if not all, EU member states were more ready to criticise China's violations of human rights and threats to international law. As of June 2020, China remained unresponsive – though it had not made good on threats of economic reprisals. China must decide whether to persist with or modify its European policy. Europe will need to decide how far to shield itself from Chinese influence – and how to strengthen the transatlantic and global partnerships that will help it to manage and better engage with China.

European Strategic Autonomy
From symbolic to substantive?

Since 2016, 'strategic autonomy' has acquired the symbolic status both of an alliance-splitting concept and of an aspirational European defence strategy. Understanding the origins and development of this concept may offer some insight into the future prospects for European strategic autonomy, bearing in mind the withdrawal of the United Kingdom from the European Union, the economic impact of the COVID-19 pandemic, and external pressures from China, Russia and the United States.

Yet, on the face of it, there is little reason for the debate within the EU about the concept of strategic autonomy. Autonomy is a distinctly weaker notion than independence: in the political realm, autonomy is a status awarded to those unwilling or unable to be independent. *Las autonomias*, for example, are 17 regions and two cities in Spain, including Andalusia and Galicia, most of which do not aspire to actually become independent. Autonomy implies a degree of dependence.

In the strategic realm, one would not normally expect autonomy to be so hotly debated. After all, many of the sovereign nations that comprise the North Atlantic Alliance have on occasion demonstrated their willingness to act not only autonomously but with scant regard for NATO's shared strategic priorities, even for policies or operations resulting from collectively decided undertakings in the geographical ambit covered by the 1949 Washington Treaty. Half of NATO's members exercised strategic autonomy by not joining NATO's operations in Libya in 2011, for example. Greece did not participate in NATO's operations against Serbia in 1999, while Denmark studiously displayed strategic autonomy by refusing to endorse the deployment of US *Pershing*-II and land-based cruise missiles during the Euromissile Crisis. France withdrew from key NATO commands between 1966 and 2009. In order to reconquer the Falkland Islands after the Argentinian invasion of 1982, the UK withdrew with little advance warning the bulk of its vital naval and air contribution to the NATO Supreme Allied Commander Atlantic (SACLANT) array guarding the so-called Greenland–Iceland–United Kingdom (GIUK) Gap from the threat of a Soviet naval breakout. Some of these decisions were the source of much debate, but none were disputed solely or mainly on the grounds that the countries involved were invoking strategic autonomy per se.

Genesis of a concept

In semantic and political terms, strategic autonomy as a concept is rightly considered to be of French provenance. French diplomacy pushed strategic autonomy into the European realm at the time of the drafting in 2015–16 of what the EU presented as its new global strategy: 'Shared Vision, Common Action: A Stronger Europe: A Global Strategy for the European Union's Foreign And Security Policy'. The still-current document was published on 24 June 2016, the day after the UK referendum on membership of the EU. At the time the strategy was being drafted, neither Brexit nor the election of Donald Trump or Emmanuel Macron were seen as sufficiently likely prospects to weigh on the content of the document, to which the UK was a major contributor. In the third chapter, entitled 'The Priorities of our External Action', strategic autonomy is highlighted: 'An appropriate level of ambition and strategic autonomy is important for Europe's ability to promote peace and security within and beyond its borders.' The concept also features in the foreword, which states: 'The Strategy nurtures the ambition of strategic autonomy for the European Union.'

This language caused some diplomatic controversy at the time, but rather less so than that of the 1998 Franco-British Saint-Malo declaration, which affirmed the 'need to give the EU the capacity for autonomous decision-making and action … in order to respond to international crises when the Atlantic Alliance is not involved'. This drew an immediate riposte from then-US secretary of state Madeleine Albright, who roundly and very publicly denounced what became known as the 'three Ds': 'delinking' European defence from NATO, 'duplicating' existing efforts and 'discriminating' against non-EU allies. In the case of strategic autonomy as promoted in 2016, serious political trouble only began after Trump and Macron came to office.

Being of French origin, strategic autonomy was from the start suspected of being Gaullist wine being poured into a bigger European bottle. When the concept was put on the European stage in 2016, such thinking would have been a misperception. In the Gaullist lexicon, independence, not autonomy, was the watchword. France's 1972 Defence White Paper underscored 'the independence and strength of France' and the 'independence of the nation', and even the concept of 'non-alignment' appeared in reference to a hypothetical European defence strategy. In the subsequent post-Gaullist White Paper in 1994, after the collapse of the USSR, strategic autonomy appeared in the fourth chapter, with a close link to France's nuclear deterrent, which is simultaneously an enabler and a product of strategic autonomy. The term did not appear in the 2008 French White Paper on Defence and National Security but was replaced by a close proxy, 'autonomy of decision'. It then returned prominently in the 2013 White Paper, in which strategic autonomy was one of four guiding principles for determining defence capabilities. The perspective was clearly post-Gaullist: indeed, those were the years in which France returned to the full suite of NATO bodies, with the sole exception of the Nuclear Planning Group (NPG), which was created after France withdrew from key NATO commands in 1966. For the first time, the 2013 White Paper made the ability to participate as a joint-force corps-commanding first-entry partner in a major NATO operation the dimensioning factor of French defence capabilities.

It is also worth noting that in French doctrine documents, strategic autonomy is viewed as a public good of a generic nature: no specific foes or allies are mentioned. It is something that enables the achievement of aims in the field of deterrence and defence. In this context, strategic autonomy in France signalled a coming to terms with strategic reality, rather than an attempt to subvert it. The Gaullist yearning for independence was superseded by a more pragmatic process of adaptation. From this French perspective, the extension of strategic autonomy to the broader European context was a natural progression.

Enter Trump and Macron, exit Britannia

Trump's inauguration in January 2017, Macron's accession to the French presidency in May 2017 and the UK decision in March 2017 to trigger

Article 50 of the Treaty on European Union – leading to Brexit – created a new situation. Doubt was cast on the future of the United States' commitment to Alliance defence, prompting European efforts to hedge against the possibility of a sudden or gradual US opt-out and its strategic consequences, complicated by the planned withdrawal of the UK from EU councils. The possibility of a US withdrawal left many Europeans stunned and unprepared. This was notably the case in Germany, for which strategic life without a US-led NATO was politically impossible to seriously contemplate. However, shock and dismay did not provide guidance on averting the risk or dealing with its eventuation. Conversely, with strategic autonomy, the French had a concept at hand. Macron added his own semantic twist by developing the concept of 'European sovereignty' as the political companion of strategic autonomy.

On 26 September 2017, two days after Germany's federal election, Macron gave a major speech at the Sorbonne laying out his vision for the evolution of the EU, including its defence dimension. Notably, he proposed the European Intervention Initiative (EI2) to develop European military capabilities and a common strategic culture. Effectively launched in June 2018 with nine (now 14) European partners outside the EU framework, the EI2 has the strategic and military benefit of providing a common framework for the UK alongside several EU members, including Germany.

In parallel, France had played, with Germany, a leading role in putting together the Permanent Structured Cooperation (PESCO) framework, in which most of the EU's 27 members commit to EU-scale defence rationalisation and standardisation. Whether this will be conducive to strategic autonomy remains a moot point, since all EU members joined PESCO except Denmark (which has an opt-out vis-à-vis EU defence initiatives) and constitutionally neutral Malta. This inclusiveness, promoted by Germany, makes PESCO look rather like what already existed with limited effect in the framework of the EU's Common Security and Defence Policy (CSDP). The French would have preferred a more select and militarily more demanding avant-garde.

More promisingly, a European Defence Fund (EDF) was announced in 2017 to finance joint development and acquisition of defence materiel under the auspices of the EU Commission and the European Defence Agency. It was initially planned that the EDF would be able to leverage some €13.6 billion (US$15.2bn) between 2021 and 2027. This sum, some 7% of EU defence capital spending, would represent a credible sweetener for the joint development and procurement of defence equipment. This Europeans-only acquisition scheme was not to the liking of the Americans. For its part, the UK did not negotiate a working relationship with the EDF commensurate with the substantial cross-channel defence-industrial value chains with the EU countries, probably because the British government was distracted at the time by the domestic difficulties of developing a coherent strategy for Brexit.

The debate on European strategic autonomy inevitably has a nuclear dimension. Strategic autonomy in the French context is closely linked to France's policy of nuclear deterrence. France was a drafter and signatory of the 1974 NATO Ottawa Declaration, which underscored the contribution of French and British nuclear forces in strengthening the Alliance's overall deterrence. It is not generally understood, however, that France's nuclear forces extend deterrence of the sort the US provides, both directly and through NATO's nuclear missions, under the guidance of the NPG. Therefore, an as-yet low-key debate has begun, notably in Germany, posing several interconnected questions: will the US continue to extend deterrence through the basing of B61 nuclear weapons in Belgium, Germany, Italy, the Netherlands and Turkey for the execution of NATO nuclear missions? Will Germany continue to play its central role in implementing NATO nuclear missions, and what will happen if it does not? Could France step in and replace or supplement the US in this regard? Given the sensitivities and complexities involved, the protagonists tend to skirt around the issues rather than address them head-on. But they were clearly on Macron's mind when on 7 February 2020 he suggested holding talks with his European partners on nuclear issues and invited them to participate in the forthcoming annual French *Poker* nuclear exercises.

None of this was expected to meet with approval from a US president who had described the EU as a 'foe' on Twitter in 2017. By vocally promoting the creation of a European army and appearing to view the US as a technological threat, Macron inflamed matters when he seemed to be personally baiting his American counterpart on the eve of Trump's arrival in France in November 2018 for the centennial of the end of the First World War. Yet these tensions had few practical consequences. Despite German unease at the possibility of losing the Americans without a credible European 'plan B', and notwithstanding Polish attempts to forge a bespoke bilateral military deal with the US – centred on the funding by Poland of a US base to be called Fort Trump – the EU and its member states were not actively prevented from moving forward with their new policies.

The shadow of COVID-19

These organisational and material steps towards a degree of EU strategic autonomy were in some trouble even before the start of the COVID-19 pandemic. In the preparations for the EU's 2021–27 multi-year budget, the EDF was under strong pressure and unlikely to benefit from more than half of the expected funding. Defence-industry and procurement executives would take the previously proposed €13.6bn EDF budget seriously, since such a sum would generally be more than enough to cover the upfront costs and inherent difficulties of multi-country and multi-company programmes. Reducing that number substantially would put the scheme on the wrong side of the threshold of credibility.

An ambitious €6.5bn (US$7.2bn) European Action Plan on Military Mobility, adopted in March 2018, was in even greater trouble. Differing regulatory and normative regimes in Europe make it slow and difficult to move ground forces across the continent. In a strategic emergency, time would be of the essence. The Action Plan aimed to create the strategic equivalent of the single market by allowing unhindered force movement across the EU. This would be a substantial asset beneficial to all, including US forces based in Europe. The Action Plan may have been overpriced, but reducing the funding to a figure close to zero would be absurd.

The outcome of the traditional horse-trading between member states and within the EU institutions on the multi-year EU budget will give a sense of how strongly strategic autonomy weighs in the broader scheme of things. The EU's ability to be 'geopolitical', as the new EU Commission head Ursula von der Leyen put it, will depend substantially on the outcome of this process.

The sustained growth of defence spending in the EU, to which most of its members have contributed from 2015 onwards – reflecting the impact of Russia's aggression against Ukraine – is not in doubt. This led to NATO's unanimous commitment to raise members' defence spending to at least 2% of GDP by 2025. The subsequent increase in defence spending by NATO's European members, 20 of which currently belong to the EU, may simply have restored defence budgets to what they were before the 2008 financial crisis, and its meaning should therefore not be exaggerated. Yet in absolute terms, the cumulative increase in European defence spending in 2015–19 was around US$100bn, with further increases expected in the following five years in order to meet the NATO commitment.

COVID-19 and the subsequent economic hardship put all these plans up in the air. Strategic autonomy requires money. Before the pandemic, that should not have been a show-stopper. According to a recent IISS study, it would cost European NATO member states some US$47bn a year over a ten-year period to make up for key US military capabilities in case of a US force withdrawal from Europe, notably in terms of sea lines of communication and the ability to deal militarily with a state-level conflict between Russia and one of its NATO neighbours. This looks like a large number, but it could fit into NATO's 2% commitment.

The quest for strategic autonomy will now likely recede in the face of new realities, with the health and economic consequences of the pandemic sidelining defence as a major concern on the European agenda. In the competition for political and budgetary attention, the prospects for EU strategic autonomy do not look encouraging. Yet the European zeitgeist – particularly after the onslaught of the pandemic, which highlighted the usefulness of borders and the shoring of key economic value chains – is that of *une Europe qui protège* (a protective Europe). Macron did not invent this theme, but sensing its political potential, he had already begun to develop it during his 2017 election campaign. The pandemic has put it firmly in the EU political mainstream. Protecting the EU and its external borders implies an enhancement of European sovereignty and therefore of strategic autonomy. However, autonomy in the service of Eurocentric protection would likely signal a retreat of extra-regional ambitions relying on force projection.

Pressures on defence spending as a whole as a result of the economic impact of COVID-19 will probably outweigh the political momentum in favour of *l'Europe qui protège*. However, the net effect of these trends will also depend on how the economic impact of the pandemic might weigh quantitatively and qualitatively on Chinese and Russian military expenditure.

External pressures: China, Russia and the US

The prospects for EU strategic autonomy also depend on the actions of China, Russia and the US. A second Trump administration may bring to fruition the Trumpian vision of a US unencumbered by alliances. A Biden presidency would be no less focused strategically on China than the US currently is, forcing hard choices on Europe. And the Russia of Vladimir Putin and the *siloviki* ('strongmen') has not abandoned its revisionist agenda. Strategic autonomy may yet be thrust upon the Europeans.

Although it remains to be seen whether such pressures will have a centrifugal rather than a centripetal impact on the EU, initial signs point towards a greater degree of autonomy. In the pandemic's wake, Russia's means to divide the Europeans remain in place, but their virulence is limited by the combined effect of reduced Russian export revenues and Western sanctions, not to mention European and NATO efforts at deterrence and reassurance. 'Londongrad' as a symbol of a Russian-funded ecosystem in Europe is not what it was, and it is now on the 'wrong' side of the EU border. Russian influence operations may have had a material effect on the last US presidential elections and the UK referendum in 2016. Conversely, they failed within the EU-27 at the time, and they continue to have little obvious impact.

The rise of China initially had a divisive effect on the EU. From 2014 onwards, as part of the new Silk Road, China created the 16+1 forum at prime-ministerial level with Central and Eastern European former communist states (with Greece joining as a 17th member in 2019), including ten EU (and NATO) members. By creating a network of interests, this initiative has already stymied efforts at strategic autonomy on issues such as the establishment of EU-wide control of foreign investment and technology transfers. By late 2019, a progressively more ambitious and sometimes cocksure China came close to splitting the EU on the issue of prospective 5G systems controlling the online economy. US opposition served to enhance European divisions. Since then, China's perceived mishandling of the pandemic compounded by the unleashing of 'wolf warrior' Chinese diplomats in key European capitals has set back, at least for a time, Beijing's and Huawei's attempts to capture key positions in 5G. Here again, COVID-19 has attenuated rather than exacerbated divisions within the EU.

The influence of the US on European strategic autonomy will be highly dependent on the outcome of the November 2020 elections. But even if a better organised and more constructive administration came into office, the 'China consensus' would presumably remain as the one big strategic issue uniting the American body politic. The prospects for EU strategic autonomy will largely depend on Europe's ability to handle the increasingly conflicting pressures between the US and China in a united manner, whether in a NATO format or in a direct EU–US framework.

CHAPTER 8

Russia and Eurasia

RUSSIA AND EURASIA

Drivers of Strategic Change

REGIONAL SHARE OF GLOBAL POPULATION, GDP AND DEFENCE BUDGET

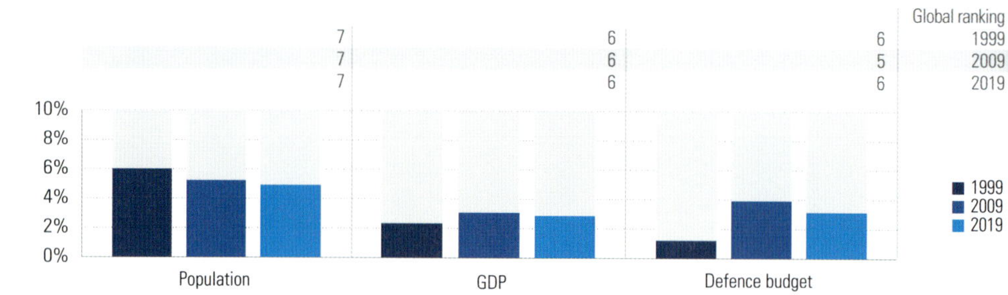

POPULATION

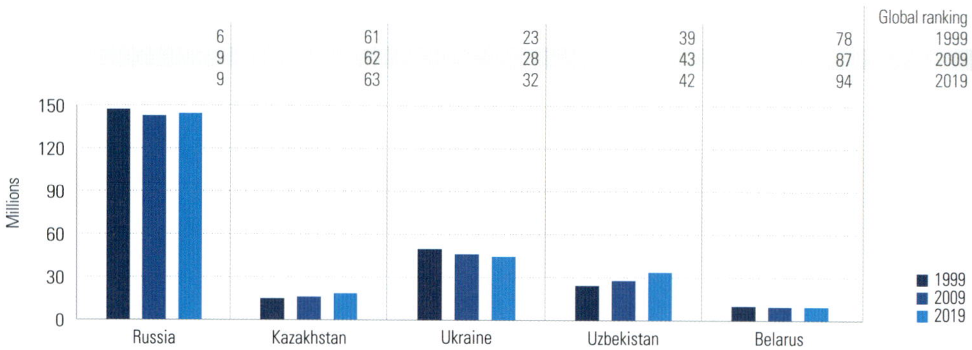

AGE STRUCTURE

(Percentage of national population)

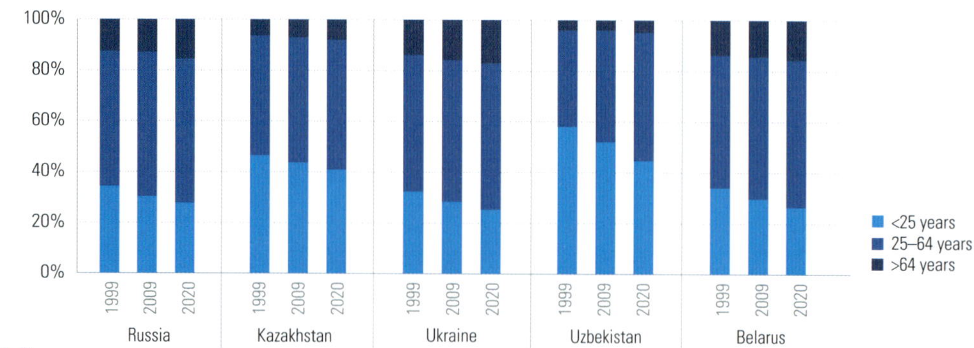

Median age	Russia	Kazakhstan	Ukraine	Uzbekistan	Belarus
2000	36.5	27.8	37.8	21.0	36.5
2010	38.0	28.9	39.5	24.5	38.9
2020	39.6	30.7	41.2	28.2	40.3

GDP
(Constant 2010 US dollars)

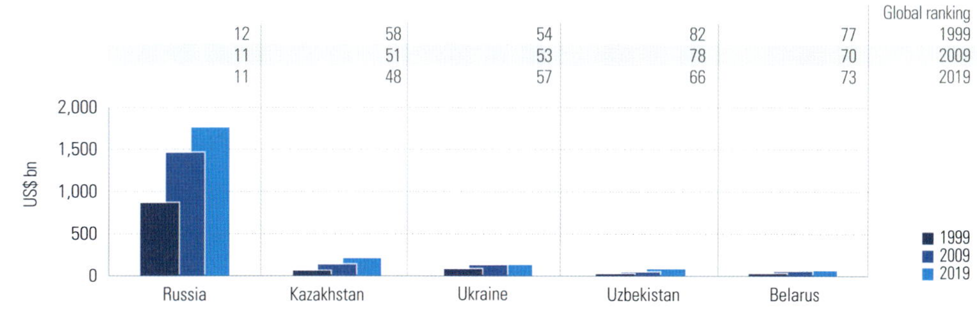

GDP PER CAPITA
(Constant 2010 US dollars)

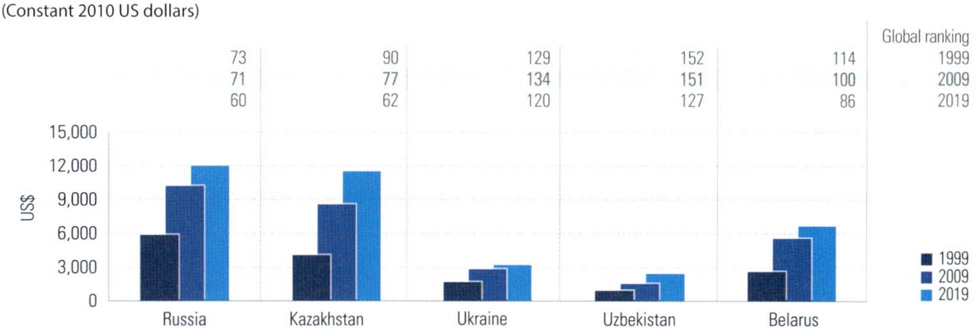

DEFENCE BUDGET
(Constant 2015 US dollars)

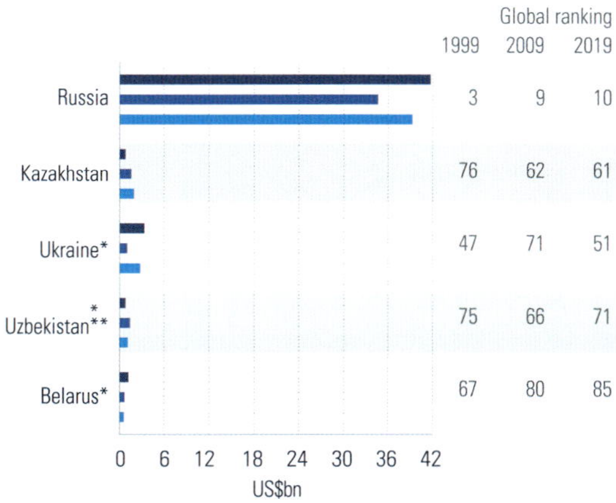

ACTIVE MILITARY PERSONNEL

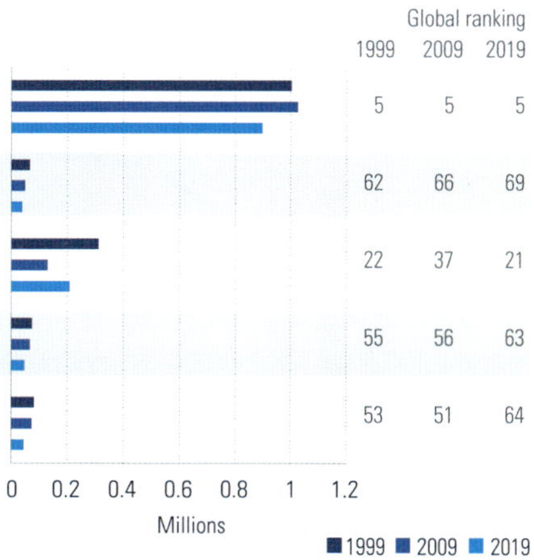

* 1999 defence budget values for Belarus, Ukraine and Uzbekistan are estimates, and may be distorted by high inflation rates in these countries
** No 2009 data available for Uzbekistan. 2010 data is therefore used in place of 2009 data

For explanation of drivers and sources, see page 7

HUMAN DEVELOPMENT INDEX (HDI)

(Score between 0 and 1, where 0 denotes a low level of development and 1 a high level of development)

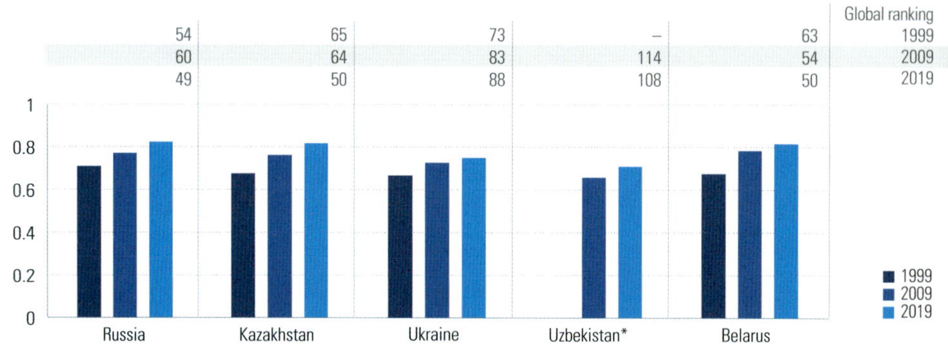

* No 1999 data available for Uzbekistan

POLITICAL SYSTEM

(Score between 0 and 100, where 0 denotes no political freedom and 100 fully free)

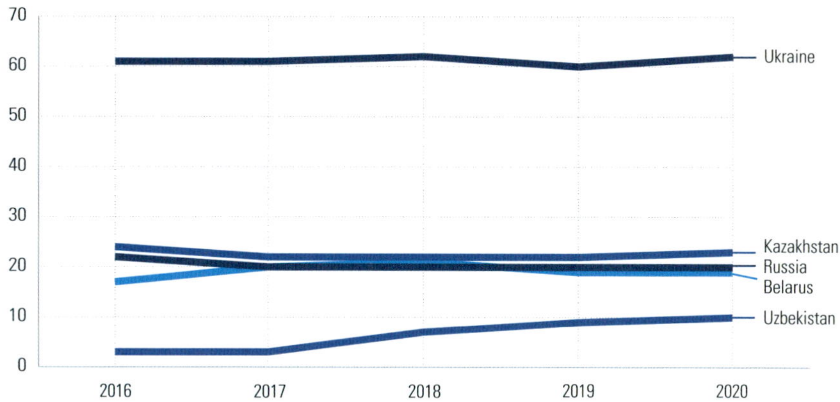

DOMESTIC APPROVAL RATING FOR PRESIDENT VLADIMIR PUTIN, AND RUSSIA'S GDP GROWTH RATE

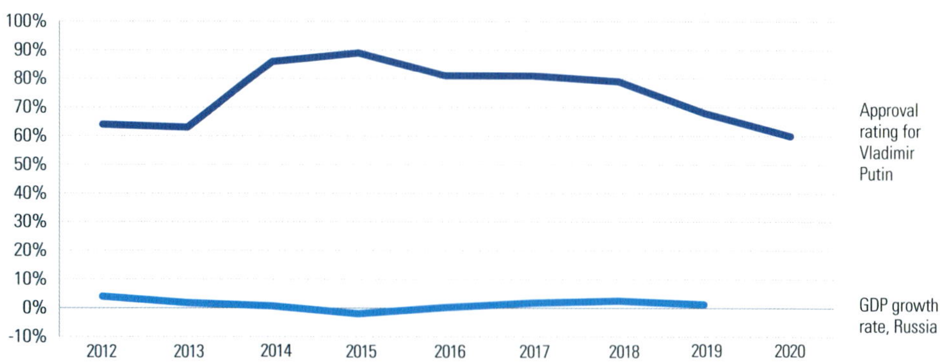

For explanation of drivers and sources, see page 7

2019–20 Review

Russia tightened its repression of political opposition and passed major constitutional reforms during the year to mid-2020. It pushed for, and other post-Soviet states resisted, further regional integration. Relations with the West remained poor, while those with China continued to deepen. Russia was especially active in the Middle East, extending its influence in Syria and escalating its involvement in Libya.

Ukraine's new president, Volodymyr Zelensky, deftly managed a set of delicate international challenges. But reform momentum at home began to wane. There was a wide range of policy responses to the COVID-19 pandemic across post-Soviet states. Belarus's poor performance deepened popular disillusion with President Alexander Lukashenko as he prepared to run for a sixth term.

Russia: repression and reform

Russia held gubernatorial or local elections in 18 regions on 8 September 2019. The authorities refused to register most opposition candidates, leaving only members of the 'systemic opposition' – parties, like the Liberal Democratic Party of Russia (LDPR) and Communist Party, tolerated as unthreatening competitors – to stand against the ruling United Russia (UR) party. Tens of thousands of people protested in the biggest street demonstrations since 2011. Over 3,000 were arrested, many were beaten and some were sentenced to up to five years in prison.

All 16 gubernatorial elections were won by UR or pro-Kremlin incumbents – though there were reports of significant electoral violations in St Petersburg. But the picture was less favourable to the Kremlin in Moscow. Leading opposition figure Alexei Navalny developed a 'smart voting' strategy to encourage people to support the candidate most likely to defeat UR. As a result, UR's majority on the city council fell by one-third. Overall, the election showed the declining popularity of UR and the potential of smart voting as an opposition campaign strategy.

Immediately after the election the authorities stepped up their pressure on Navalny. Offices of his Anti-Corruption Foundation (FBK), and hundreds of homes of activists and their relatives, were raided. Further raids, and confiscation of equipment, took place in October and December. Homes of activists of the independent election-monitoring organisation Golos were also raided. The Ministry of Justice designated the FBK a 'foreign agent' in October, subjecting it to closer scrutiny and possible suspension.

On 15 January 2020, President Vladimir Putin gave his annual state-of-the-nation address to parliament. He proposed a series of major constitutional changes covering a range of political, social and cultural issues. The most significant concerned the institutions of power. At first sight these appeared to grant greater powers to the Duma (the lower house of parliament) and to give the State Council, a hitherto merely consultative body, a major new role. Many observers speculated that Putin, ineligible to stand again for the presidency in 2024, would seek an alternative power base there.

The government resigned after Putin's address. Putin replaced Dmitry Medvedev, prime minister since 2012, with Mikhail Mishustin, a low-profile technocrat who, as head of the Federal Tax Service, had been credited with modernisation of the tax administration. Medvedev became deputy chairman of the Security Council under Putin.

In March, parliament approved the constitutional amendments. In a dramatic but clearly choreographed intervention, Valentina Tereshkova, a Duma deputy who in 1963 had become the first female astronaut, proposed that the limit on serving two consecutive presidential terms be 'reset to zero', thus allowing Putin to run again in 2024 and 2030. This was passed with no votes against.

Russia's initial response to the COVID-19 pandemic was to close its border with China at the end of January, before Russia had recorded its first case. The authorities subsequently took few measures. As late as March, Putin claimed that COVID-19 was 'on the whole under control'. On 6 March, Russia rejected oil-production cuts that other OPEC+ members had agreed to manage the expected downturn in global demand. This led to a price war with Saudi Arabia and severe market turbulence until a new deal was reached on 9 April, requiring Russia to accept disproportionate production cuts. This episode exacerbated the economic impact of the pandemic on Russia.

By late March, the Russian authorities realised the seriousness of the pandemic. On 25 March, Putin addressed the country to warn that the impact of COVID-19 could soon become as severe as in Europe. He gave regional governors the lead role in tackling it, despite their limited powers in the highly centralised system of power he had built. There was no nationwide lockdown.

Russia quickly scaled up a mass COVID-19 testing capacity. The daily infection rate peaked in mid-May. By 30 June, Russia had recorded 650,000 cases. However, the case fatality rate remained below 2%, one of the lowest in the world. This led some observers, including medical personnel, to cast doubt on the reliability of this statistic.

The pandemic compelled the Kremlin to postpone a lavish 75th-anniversary celebration of the end of the Second World War, which several world leaders had planned to attend. It also postponed the national vote to complete the adoption of the constitutional amendments, from 22 April to 25 June–1 July. According to the Central Election Commission, 77% voted for the amendments on a 68% turnout. But reports of the use of 'administrative resources' – state power to induce, pressure and persuade voters to support the desired outcome – and falsification of results cast doubt on the official result.

Difficult relations with the West, better ones with China

Russia's relations with the West remained difficult. In December 2019, the US imposed sanctions on companies involved in the Nord Stream 2 pipeline being built to transport Russian gas to Germany via the Baltic Sea. In March 2020, the US imposed sanctions on the trading arm of Rosneft, Russia's biggest oil company, for its role in transporting Venezuelan crude.

Putin kept up his rhetorical emphasis, notable in recent years, on Russia's strategic military power. In December 2019, he announced that Russia was the first country in the world to deploy hypersonic missiles. In June 2020, Russia published its strategic nuclear-deterrence policy for the first time. This set out the circumstances in which Russia might use nuclear weapons, including in response to non-nuclear attacks or the imminent threat of nuclear attack. In the same month, Russia and the US began talks, proposed by Russia in November 2019, on extending the 2010 New START treaty that limits strategic nuclear weapons and is due to expire in February 2021.

There were hopes that Russia's relations with Europe would improve following the decision by the Parliamentary Assembly of the Council of Europe in June 2019 to restore Russia's voting rights, which had been suspended in 2014 after its annexation of Crimea. But new problems arose. A former Chechen rebel commander, Zelimkhan Khangoshvili, was shot dead in Berlin in August 2019. A Russian national was charged with his murder. Germany accused the Russian state of ordering the killing. In May 2020, German Chancellor Angela Merkel revealed that the German authorities attributed the 2015 cyber attack on the German parliament to Russian military intelligence.

More broadly, the politics of memory clouded the relationship between Russia and the European Union. In September 2019, the European Parliament passed a resolution condemning the 1939 Nazi–Soviet pact. This led to a war of words between Russia and the EU. Putin raised the issue several times, even writing a 9,000-word article on it. In December, Russia protested the passage of a Czech law establishing a national day to commemorate the 1968 Soviet invasion of Czechoslovakia. In response, President Milos Zeman, who had sought to maintain a good relationship with Russia, criticised this 'absolute insolence'. Meanwhile, the EU and Russia renewed their respective sanctions, in force since 2014, against one another.

Russia used the pandemic in its foreign policy. In March, it sent eight mobile brigades of military virologists from its radiation, chemical and biological defence forces to Italy to provide assistance. In April it flew a consignment of equipment including ventilators and personal protective equipment – apparently sold, not donated – to New York. These incidents sowed confusion in the EU and the US.

Russia's ambitions to remain a dominant gas supplier bore fruit. Two major pipelines were inaugurated: Power of Siberia to China on

2 December 2019, and TurkStream to Turkey on 8 January 2020. These will provide up to 38 billion cubic metres (bcm) and 31.5 bcm of gas per annum respectively.

Russia's relationship with China continued to develop following their announcement of a 'comprehensive strategic partnership' in June 2019. In July, they conducted their first long-range joint air patrol in the Asia-Pacific region. South Korean jets fired warning shots at Russian aircraft that violated South Korean airspace during the exercise. In September, Russia hosted the *Tsentr*-2019 exercises in which China, India, Pakistan and some Central Asian states participated. In all, 128,000 personnel were involved. In October, Putin announced that Russia was helping China to create a missile-defence system. The two countries also cooperated to reduce their dependence on the US dollar. By mid-2020 more than half of their trade was conducted in other currencies. But the arrest in June 2020 of a leading Arctic researcher on charges of spying for China hinted at tensions.

Deepening influence in Syria and Libya

Russia's most significant diplomatic and military theatre was the Middle East and North Africa. In October 2019, following the US withdrawal from northeastern Syria and the subsequent Turkish offensive against Kurdish forces, Russia entered the area with Syrian government forces and took over the abandoned US base at Manbij. These developments marked a further extension of Russia's presence in Syria at US expense.

From November, Russia escalated its air support for the Syrian government's offensive against Idlib, the last major rebel-held part of the country. This led to the deaths of 34 Turkish soldiers in airstrikes in February 2020 and a Turkish intervention soon after. Russia and Turkey scrupulously avoided direct contact and agreed a ceasefire that they forced on their local allies. In June 2020, Russia resumed airstrikes to erode rebel positions ahead of a future campaign but also to pressure Turkey. As before, these included indiscriminate attacks on civilian facilities. Overall, Russia cemented its position as the major power broker in Syria as a result of its military ruthlessness and its effectiveness in capitalising on US policy inconsistencies.

Russia began to deepen its involvement in the Libyan civil war from summer 2019. It escalated support for Khalifa Haftar's Libyan National Army (LNA) – fighting against the UN-recognised Government of National Accord (GNA) – through forces of the Russian Wagner Group 'private military company' (in reality, state-linked mercenaries) with air support from Russian fighters and frequent supply flights from Russia's air base in Hmeimim, Syria. A UN report in May 2020 assessed that up to 1,200 Russian mercenaries were fighting in Libya. Russia also recruited and deployed as many as 2,000 Syrian militiamen on the Libyan battlefield.

In July 2019, Russia began to send elements of the S-400 air-defence system to Turkey, a move the United States strongly opposed. In October 2019 Putin hosted the first Russia–Africa Summit in Sochi, which 43 heads of state or government attended. The event focused on economic and military cooperation.

In September 2019 Russia joined the 2015 Paris Agreement on climate change. The 1990 baseline for calculating emissions reduction is undemanding for Russia. Nonetheless, the government scrapped plans to introduce company-specific targets following business lobbying. Meanwhile, Arctic temperatures continued to rise at a faster rate than the global average. Unusually large fires affected Siberia and the Russian Far East in 2019 and 2020.

Regional integration: pressure and resistance

Russia sought to deepen its regional influence by proposing that the Eurasian Economic Union (EAEU), which it dominates, expand into new policy areas. In May 2020, Kazakhstan's President Kassym-Jomart Tokayev led objections to an ambitions EAEU development plan. Meanwhile, Uzbekistan gained observer status at the EAEU in April. President Shavkat Mirziyoyev said his country would not become a full member, as 'nobody will ever give away our independence to anyone'.

There was little progress in ending the Russia–Ukraine conflict. In September and December 2019, and in April 2020, the two countries agreed prisoner exchanges. The first of these controversially included the return to Russia of Vladimir Tsemakh, who was wanted for questioning by the Dutch team investigating the shooting down of Malaysian Airlines flight MH17 over Ukraine in 2014.

Zelensky met Putin for the first time in Paris on 9 December in an effort to relaunch talks. A further meeting of the two leaders on the same day with Merkel and French President Emmanuel Macron marked the first meeting since 2016 of this Normandy Format for resolving the conflict in Ukraine. Despite renewed commitment to a ceasefire, Ukrainian forces and civilians continued to suffer casualties. In late December, Ukraine and Russia agreed a five-year gas-transit deal.

Deputy head of the presidential administration Dmitry Kozak replaced Vladislav Surkov as chief Russian negotiator with Ukraine in February 2020. Kozak had long experience in managing policy towards the Russia-dependent Transnistria region in Moldova. Meanwhile, Russia continued issuing Russian passports to Ukrainian residents in the eastern Ukrainian region of Donbas, a policy it had begun just after Zelensky's election in April. By June 2020 it had created 200,000 Russian citizens in this way, building a potential pretext for maintaining longer-term involvement in Ukraine.

Russia was not Ukraine's only foreign-policy challenge. In July 2019, US President Donald Trump asked Zelensky to investigate US presidential candidate Joe Biden and his son Hunter in return for an invitation to the White House. Although this led to Trump's impeachment, Zelensky managed to avoid offending the White House and getting pulled into US domestic partisan conflict. Despite his lack of political experience, Zelensky's handling of relations with Russia and the US alike was adroit.

However, his attempts to tackle domestic corruption were more mixed. A strong reform team under Oleksii Honcharuk, Zelensky's first prime minister, made good initial progress against oligarch resistance. But in March 2020, Zelensky replaced Honcharuk and more than half the cabinet. The new team completed negotiations with the IMF for a US$5bn loan, but reform impetus slowed. Soon after, in a worrying development, several investigations were opened against Zelensky's predecessor, Petro Poroshenko. The US and EU expressed concern that this could amount to political persecution.

Russia stepped up pressure on Belarus to accept closer integration under the 1999 Union State treaty, a largely paper agreement. At the end of 2019, Russia briefly cut oil supplies. Even after these were restored, Belarus sought alternative sources, including from the US. More generally, Lukashenko sought to improve relations with the US. John Bolton, then US national security advisor, visited Minsk in August 2019, and Secretary of State Mike Pompeo visited in February 2020. In April, the US appointed an ambassador-designate to Belarus for the first time since the previous ambassador was forced out in 2008. She was due to take up her post later in the year.

In Kazakhstan, Tokayev, who had succeeded Nursultan Nazarbayev as president in March 2019, settled into his new role. But in October he unexpectedly issued a decree requiring him to consult Nazarbayev when making key government and regional appointments. In May 2020, Dariga Nazarbayeva, daughter of the former president, lost her position as speaker of the Senate. These moves hinted at ongoing tension in the presidential transition.

Drumbeat for the Belarusian election

Beyond Russia, responses to COVID-19 across the post-Soviet space broadly correlated with regime type. The more democratic the country, the earlier it tended to respond to the pandemic. Georgia was especially effective: by mid-2020, it had recorded fewer than 1,000 cases. At the authoritarian end of the spectrum, Turkmenistan implausibly did not record a single case, amid reports that evidence of the virus was being suppressed. There were exceptions to this pattern: though not democratic, Kazakhstan dealt with the first months of the pandemic well, enforcing rigorous lockdowns and providing targeted financial support. But across the region there were concerns that a second wave could hit later in the year.

Belarus adopted perhaps the worst, and politically most consequential policy. Lukashenko dismissed the pandemic as 'corona-psychosis', claimed it could be cured by drinking vodka and driving tractors, and derided its victims. He insisted on holding the annual 9 May Victory Day parade, which was postponed or cancelled elsewhere (except in Turkmenistan, which held it for the first time). Civil society filled this leadership vacuum with grassroots activity, such as crowdfunding support for medical staff and making needed medical equipment.

Against the background of an increasingly disillusioned yet emboldened population, Belarus's presidential election campaign got under way in May 2020. Lukashenko sought re-election for the sixth time to continue his 26 years of rule. Despite the highly authoritarian system, three significant opposition candidates emerged, only to face persecution. On 29 May, prominent activist and video blogger Sergei Tikhanovsky was arrested. On 18 June, Viktor Babariko, chairman of a major bank, Belgazprombank, was also arrested and charged with embezzlement and bribery. A week later, Valery Tsepkalo, former head of the successful Minsk IT-Park, fled abroad for his safety after being barred from running.

The only candidate of note was Tikhanovsky's wife, Svetlana Tikhanovskaya, who stood in place of her husband. Lacking political experience, she was deemed unthreatening and permitted to stand. But the fact that important figures with elite ties had sought to do so, together with the unmistakeable growth of popular disaffection, suggested that Lukashenko might have more of a fight on his hands than he realised.

Russia: The Road to 2024
What challenges face Putin and his system?

The last day of 2019 marked 20 years of Vladimir Putin's rule. He had become acting president on Boris Yeltsin's resignation on 31 December 1999 and cemented this role by winning the presidential election in March 2000. His position quickly became unassailable as he brought about far-reaching change to the way Russia was governed.

But at the start of 2020, Putin faced two problems. Firstly, according to the constitution, he would have to leave office in 2024. Secondly, he faced a more serious combination of domestic and external policy challenges than at any other time in his presidency. These had begun to strain the performance and cohesion of the system of power he had built over the previous two decades.

These two issues – the future of Putin's presidency, and the future of his system – will dominate Russia over the next four years. How has Putin begun to tackle them, what do his choices tell us about his priorities, and what are the implications for Russia?

Solving the 2024 puzzle

In 2024 Putin will turn 72. He will have ruled Russia for 24 years, one-third of his life. In the past 200 years only Tsar Nicholas I and Joseph Stalin have done so for longer. But while Putin's political dominance, longevity and many of his methods of control invite comparison with tsars or Communist Party leaders, there are two formal constraints on his leadership with no precedent in previous regimes. The first is the requirement to win periodic re-election. The second, underlying this, is the set of rules governing the presidential office itself.

Putin has not found it difficult to win presidential elections. He has done so four times: in 2000, 2004, 2012 and 2018. The first was the freest and closest: Putin won only 53.4% of the vote. In his most recent victory in 2018 he received 76.7%, his highest share of the vote. The Kremlin has refined the use of 'administrative resources' – electoral conditions, methods and pressures – as well as falsification of results to achieve its desired outcome.

The rules about the presidency are more problematic. The 1993 constitution that Putin inherited stipulated that no president could serve more than two consecutive terms. Since Putin was elected in 2012 and 2018, as things stood at the start of 2020, he was prohibited from standing again in 2024. In principle, four options were available to him.

Firstly, he could repeat what he had done when this issue first arose in 2008. Then, he had become prime minister for four years, swapping places with Dmitry Medvedev, who became president. They swapped offices, but not effective power. Putin's dominance moved with him to his new role – a vivid demonstration of the primacy of networks and informal understandings over constitutional machinery. This Putin–Medvedev 'tandem' worked reasonably well, with only occasional friction, for the next four years. But when the question of who would stand as president in 2012 arose, Putin's casual public remark that he and Medvedev had agreed to return to their earlier roles sparked the biggest protests of the Putin era. This chastening experience may have deterred Putin from repeating this manoeuvre, even assuming he could find another trusted and unthreatening figure like Medvedev to ride the tandem again.

Putin's second option was to leave the presidency more permanently but to exert control, or at least decisive influence, in a different capacity. This would require him to find a different long-term role in Russia's hyper-presidential system. Some speculated that Putin could move into a guiding role behind the scenes, possibly as Lee Kuan Yew had done as 'Minister Mentor' in Singapore. Neighbouring Kazakhstan offered a local precedent. President Nursultan Nazarbayev began such a transition in March 2019, resigning in favour of Kassym-Jomart Tokayev while retaining a series of other posts, including head of the Security Council. But signs soon emerged that this arrangement had not stabilised: in October 2019, Tokayev agreed to consult his predecessor on key appointments, and in May 2020 Nazarbayev's daughter Dariga Nazarbayeva was removed as Senate speaker – a position that would have made her acting president in the event of Tokayev's removal or incapacitation. If Nazarbayev, widely popular and respected as 'leader of the nation', struggled to manage a transition of this kind, attempting something similar in Russia would be more perilous for Putin. His own example was testimony: his power and priorities as leader evolved differently from the way that Yeltsin had expected when choosing him as successor.

Putin's third option was to find a way to lift the constitutional constraint. One much-discussed way would be to create a 'union state' with Belarus that Putin could then lead. But this would require the coercion or consent of Belarus, a sovereign country – a difficult goal with major foreign-policy implications. A more feasible approach would be to amend the current constitution to allow him to serve more than two consecutive terms. Putin had rejected this in 2008 in favour of the tandem, emphasising the importance of constitutional stability. While a new law the same year extended the presidential term of office from four to six years, the constitution itself remained largely untouched.

Finally, rather than remain in power, either as president or in another capacity, Putin could leave public life and enjoy retirement as a private citizen. This was in principle the simplest option: it only required Putin to step down. It may even be attractive to him: his growing distraction and boredom in recent years have been much noted. But retirement was probably the least likely option. With the exception of Putin's immediate predecessor, Yeltsin, there is no example in modern Russian or Soviet history of a leader giving up power rather than having it taken from them by death, overthrow, revolution or, in the case of Mikhail Gorbachev, the dissolution of his country. Recent practice in other post-Soviet states is no more reassuring. In authoritarian ones, death and overthrow have been the normal modes of leadership succession. In more democratic ones, leaders who have stepped down (usually after electoral defeat) have sometimes then faced prosecution. From early 2019 to mid-2020 alone, this fate befell four of them: Robert Kocharian and Serzh Sargsyan in Armenia, Almazbek Atambayev in Kyrgyzstan and Petro Poroshenko in Ukraine. Such cases are likely to have amplified Putin's own concern that it was unsafe for him to leave office.

This, then, was the problem that Putin faced on his 20th anniversary as president. He almost certainly felt obliged, for his own safety, to stay in power. But the constitution, which he had repeatedly said he would not amend, prohibited him from doing so. He soon began to solve this problem. In his annual state-of-the-nation address on 15 January 2020 he proposed major constitutional changes. These encompassed three main areas. The first concerned political institutions, in particular granting greater powers to the Duma (the lower house of parliament) to appoint the government, and raising the status of the State Council, a hitherto merely consultative body. Putin also proposed that the president now be limited to no more than two terms *in total*, not just consecutively, thus further tightening the restrictions.

The second area of proposed constitutional change was economic. It enshrined pension and wage guarantees that were bound to be popular. The third area comprised conservative social and cultural measures. These included specifying marriage as a union between a man and a woman and protecting the 'historical truth' of the Second World War.

The first group of changes attracted most attention. They were scrutinised for clues as to how Putin would handle the '2024 question'. Many observers speculated that Putin was moving towards the second option mentioned earlier: staying in power, but in a non-presidential role, either as chairman of the newly enhanced State Council, or perhaps as a stronger prime minister. But ambiguity remained: a few days after his speech, Putin emphasised the need to maintain a 'strong presidential republic' and explicitly rejected the Singapore Minister Mentor scenario. The only clear conclusion was that Putin's state-of-the-nation speech was not the end of the story.

Developments took a dramatic turn in March. As the Duma was approving the constitutional changes, an amendment was added to 'reset' the number of Putin's presidential terms to zero. At a stroke, the effect of the changes on Putin's tenure in office became the opposite to the impression his January announcement had created. Then, he had appeared to impose new limits on presidential tenure. Now, he would be able to remain in office until 2036. Of the available options set out earlier, he had chosen the third – to amend the constitution to stay on as president – not the second – to find another, non-presidential way to stay in power. But by revealing his intention only later, he had kept elites guessing.

Putin also insisted that the constitutional changes be put to a national vote. This was originally set for 22 April, but reluctantly rescheduled to 25 June–1 July due to the COVID-19 pandemic. By the time it took place, official enthusiasm had waned and the vote felt more like something to get through than to celebrate. But the authorities still had to mobilise support for it and did so mainly by emphasising the changes relating to welfare measures rather than Putin's potential continuation in office.

The outcome was never in doubt: copies of the amended constitution were on sale even before voting had ended. Despite calls for a boycott from leading opposition figure Alexei Navalny, the Communist Party and others, the official 'yes' vote was 77% on a turnout of 68%. Both figures were almost identical to Putin's 2018 election victory. Independent electoral expert Sergei Shpilkin estimated, on the basis of statistical anomalies, that the true result was closer to 65% on a 43% turnout, figures consistent with earlier opinion polls. But Russia moved on and there were no significant protests.

The national vote was not legally required to pass the amendments. Putin's purpose in holding it was not entirely clear. A plausible explanation is that he did so to show elites that he was still able to achieve a decisive popular victory. As with the manner of revealing his intentions, so with his method of implementing them: the major audience was the elites around him, rather than the wider population. Similarly, Putin refused to confirm that he would stand again in 2024, saying only that he had not yet decided. Once again, this appeared to be a way of keeping those around him off balance, making it harder for them to plan confidently.

While the future of Putin's presidency attracted most interest, the other constitutional amendments – more than 200 of them – are also important. Those concerning political institutions risk creating confusion. The new role of the State Council, and its future relationship with the government, are especially unclear. In such uncertainty, the hidden software of informal networks will become even more important.

Another amendment grants former presidents immunity from prosecution. This would protect Putin if he chose to leave office. But having gone to the trouble of making it possible to run again, it seems likely that he will do so. At first sight, re-election should not be difficult. The national vote confirmed the Kremlin's ability to deliver electoral outcomes. Only the tandem arrangement in 2012 provoked major protests around a presidential election. Nonetheless, the next four years will pose a combination of challenges Putin has not faced before.

Declining performance, growing repression

Economic performance, a key determinant of popular support, is likely to remain stagnant. Three factors are at work. Firstly, the hydrocarbon revenue-growth model that doubled GDP in Putin's first two terms is exhausted, beset by a large and growing state, corruption, insecure property rights and other structural problems. Tackling these dysfunctions requires major reforms that would challenge the powerful elite interests that benefit from the status quo. The Kremlin has no appetite for this. There has been no significant economic reform since 2003.

Secondly, even though oil revenues no longer fuel strong growth, Russia remains heavily dependent on them. But oil prices have been persistently low since late 2014 and are unlikely to rise in the near term. The onset of a post-carbon age, hastened by COVID-19, calls into question whether they will ever do so. Forecasts by BP suggest that global demand for oil could peak as early as the mid-2020s. Finally, US and European sanctions, imposed in 2014 in response to Russian intervention in Ukraine, remain in force and are unlikely to be lifted. If anything, a further foreign-policy crisis could lead to their escalation.

This combination of a sclerotic political economy, lower oil prices and an adverse external environment offers a bleak economic outlook for Russia. When stagnation first set in and mass protests hit Russia in 2012, the Kremlin's response was to build up nationalism as a pillar of popular legitimacy. This took on a sharper anti-Western edge than in the past, notably in vilification of the European Union as a source of degenerate values that were alien to Russian culture and spirituality. Russia's annexation of Crimea in 2014 stoked up this fervour, pushing Putin's popularity up to record levels.

But this was only a temporary solution to the problem of political legitimacy. By 2018 the effect had begun to wane. Polls by the respected independent Levada Center showed that Putin's approval ratings had fallen to a record low of 59% in April 2020. Nor does it seem likely that Putin could rekindle nationalist support with another military adventure – though this may not prevent him from trying. Crimea was a unique case: its large Russian population and Russian military base made it relatively easy to annex before the West could respond. Nor is there a popular appetite now to confront the West: Levada Center polls showed that the number of Russians with a favourable attitude to the EU more than doubled (from 20% to 49%) and to the US more than trebled (from 12% to 42%) between January 2015 and January 2020.

As 2024 approaches, Putin will have neither a vibrant economy nor nationalist fervour to draw upon for popular support. It is unclear what he will appeal to, beyond the cultural conservatism embodied in some of the recent constitutional changes. But does this matter, given the Kremlin's proven record of electoral success? 'Administrative resources' are not foolproof. They have been unable to stem decline in support for the ruling United Russia (UR) party, which Navalny has labelled 'the party of crooks and thieves'. Navalny's new 'smart voting' strategy, which directs voters to support the candidate most likely to defeat UR, embarrassed the party in the September 2019 Moscow city-council elections. So toxic had the brand become in Russia's capital city that UR members seeking re-election stood as independent candidates rather than remain associated with their own party. This did not save several of them, including the local UR head, from defeat. Navalny has since developed plans to roll out smart voting to other regions.

The Kremlin soon showed how seriously it took this threat by mounting police raids on Navalny's Anti-Corruption Foundation offices across the country. It sees even limited and local opposition gains as a threat, and Navalny as a particular danger. His ingenious exposés of high-level corruption have angered senior figures. Unwilling to give him the national platform of an election campaign, the authorities barred him from standing in the 2018 presidential election, even though he had no prospect of winning. This illustrates Russia's larger direction of travel: as popular support declines, restrictions on civil society grow tighter. The state not only has formidable repressive capacity but is investing in more. A case in point is the Russian National Guard (Rosgvardia). Created in 2016 and directly subordinate to Putin, the key purpose of this new law-enforcement body may be to avert a 'colour revolution' – a mass movement to bring about political liberalisation – of the kind that has erupted in several post-Soviet states over the past two decades. In 2018, Rosgvardia's head, Viktor Zolotov – who is Putin's former bodyguard – challenged Navalny to a duel and threatened to turn him into a 'juicy beefsteak'.

There are still limits to repression when public opinion mobilises against this. Two students, Yegor Zhukov and Pavel Ustinov, were unexpectedly released in December 2019 following outcries against the prison sentences they faced for taking part in demonstrations. In January 2020, five police officers who had framed investigative journalist Ivan Golunov on drugs charges were arrested. But such concessions are rare. Despite their many differences, Russia in the 2020s may come to loosely resemble the Soviet Union in the early 1980s. The latter system was stagnant, corruption was serious and disillusionment was deepening. Yet it retained the coercive means to prevent discontent from posing an organised threat to the regime – until a reforming leader, Gorbachev, came to power.

Putin's management of elites

Declining legitimacy matters, but its political impact is likely to remain limited in the near term. The major challenge Putin faces on the path to 2024 is not controlling civil society but managing elites. Across post-Soviet space and beyond, major change in authoritarian systems usually takes place only when a split emerges within the regime. If elites remain united in defending the status quo, no amount of 'people power' from below can challenge it. But if a critical mass defects from the regime and sides with demands for change, an apparently strong system can weaken quickly.

How likely is this in Russia? There are several reasons why some Russian elites might contemplate the possibility of a different future. They include the weak rule of law; intense infighting over assets, cash flows and influence in a stagnant economy; predation by *siloviki* (officials from the security and defence sectors); and personal insecurity. Ageing oligarchs want to bequeath their fortunes safely to their children. Western sanctions that might target these are a further worry. But it will be difficult, even for the wealthiest figures, or for technocrats whose expertise is useful to the Kremlin, to exert enough influence to shift policy onto a more moderate course. The logic of growing repression of civil society means that *siloviki* will grow stronger at the expense of more pragmatic or liberal voices.

But it is clear that relations among elites will become more fractious. The conservative ideology that Putin has developed does little to bind them. Elites cannot merely be repressed – they must be managed. Putin's priority is to keep them loyal, dependent, afraid or uncertain – hence his determination, in managing constitutional reform, to retain the initiative and keep those around him guessing. If even Putin feels insecure enough to stay on in office, how much less secure is the system for others.

An uphill path to 2024 – and beyond

Putin's personal position remains strong. He faces no direct challenge from below or within. He has shown he can change the constitution at will and win elections. But for reasons of personal security, he probably feels obliged to remain president. To this extent, he is ironically a prisoner of the system that he has built and dominates. This system is performing poorly for domestic reasons that Putin will not, and external reasons he cannot, change. Growing repression testifies to the Kremlin's sensitivity about the resulting loss of support. Elite manoeuvring and infighting testify to the growing anxieties even of those who have benefited most from the regime.

But beyond this turbulence and unease, a deeper force may be driving change. Despite the problems Russia faces, it is an upper-middle-income country with a sophisticated, highly educated and globally connected population. The more rigid and repressive the system that rules it, the more incongruous, even incompatible, the relationship between state and society appears. For this reason too, the path to 2024, and beyond, will become harder.

Zelensky's First Year as Ukrainian President
Is he fulfilling his promises?

When Volodymyr Zelensky took office on 20 May 2019, he became the sixth president of post-Soviet Ukraine and the third to promise a radical break from the country's system of governance. Instead, he is both adapting to that system and bending its rules.

Zelensky inherited a challenge that has defined and constrained the ambitions of every reformist president of Ukraine. This challenge is to carry out, by lawful and democratic means, far-reaching reform of a system whose core institutions are compromised, and in some cases captured, by entrenched corporate ('oligarchic') interests operating in the shadows of legal authority.

But Zelensky also inherited uniquely favourable political circumstances. He was elected with an unprecedented 73% of the vote, securing a majority in all but one of Ukraine's 24 oblasts (regions). Equally unprecedented was the absolute majority his party, Servant of the People, won in the 450-member Verkhovna Rada (the Ukrainian parliament) in the subsequent parliamentary elections of 21 July 2019. However, even these political fortunes have not enabled Zelensky to surmount the reform challenge.

Zelensky set himself an equally ambitious foreign-policy goal: to end the conflict with Russia in the eastern Ukrainian region of Donbas. This gave rise to a second challenge for the new president: how to reconcile Russia's interests, supported by its formidable power, with the preservation of Ukraine's sovereignty and independence. Since the Russian Federation recognised Ukraine's independence *de jure* in 1991, it has sought to limit it de facto. For more than 20 years it pursued this aim with a combination of underhand and heavy-handed, but peaceful, means. But since early 2014, it has done so largely by means of a war that it denies waging.

Breaking the system

When Zelensky announced his candidacy for the presidency in December 2018, he was an actor and comedian with no political experience. For those who wanted a clean break with the past, this was one of his principal attractions. Yet he was not without systemic connections. The '1+1' channel that broadcast his celebrated television series, *Sluha Narodu* (Servant of the People), is owned by one of Ukraine's most powerful and aggrieved oligarchs, Ihor Kolomoyskyi, whose principal asset, Ukraine's largest commercial bank, PrivatBank, had been nationalised by Zelensky's predecessor, Petro Poroshenko, in 2016 with Western backing after US$5.5 billion disappeared from its accounts. Moreover, some of Zelensky's inner circle had questionable connections with the administration of former president Viktor Yanukovych. It was Yanukovych's abandonment, under heavy Russian pressure, of negotiation of an Association Agreement with the European Union that led to the Euromaidan mass demonstrations in Kyiv in 2013–14. Following Yanukovych's flight to Russia in February 2014, Russia began to intervene in Donbas, and in March annexed Crimea. Within weeks of Zelensky's election, several Yanukovych allies returned (many from Russia) to Ukraine. Since returning, Andrii Portnov, the former deputy head of Yanukovych's administration, has laid much of the groundwork for the prosecution of Poroshenko. Valerii Khoroshkovskyi, the former first deputy prime minister and head of the Ukrainian security service under Yanukovych, was, according to Zelensky, offered the post of head of the state customs service, but declined it.

Despite these connections, Zelensky wasted no time signalling that serious changes were in store. One day after his election, he dissolved parliament and brought parliamentary elections forward from October to July 2019. His first cabinet was almost a dream team. Not only were most of the key ministers committed reformers, many were people of proven ability. In six months, the cabinet of Zelensky's first prime minister, Oleksii Honcharuk, methodically began to unstitch the bonds of oligarchic control. Much of this work was undertaken by Ruslan Riaboshapka, a figure of enormous experience and single-mindedness, who replaced the compromised Yurii Lutsenko as prosecutor general in August 2019.

Several reform-minded officials of the Poroshenko era were also appointed: the apolitical minister of finance, Oksana Makarova; her more high-profile predecessor, Oleksandr Danyliuk, appointed secretary of the National Security and Defence Council; and former head of the Defence Reforms Project Office Andriy Zahorodniuk, who became defence minister. With fewer plaudits, Zelensky also reappointed Poroshenko's interior minister and most potent rival, Arsen Avakov, whose ministry and force structures have 300,000 employees and a budget of US$3.5bn at their disposal.

From the outset, momentum was impeded by structural obstacles as well as the president's own priorities. In contrast to the cabinet, the team that Zelensky brought into the newly renamed Office of the President largely came from his television company, Kvartal 95. The highest profile was Andrii Bohdan, Kolomoyskyi's personal lawyer, whom Zelensky appointed as head of the presidential administration in May 2019. These appointees, who did not require parliamentary confirmation, swiftly emerged as the locus of decision-making and, for some cabinet ministers, a chronic source of interference. Danyliuk, in particular, found the 'secretive games' and obstruction from Bohdan so destructive that he resigned in September 2019. Bohdan's own dismissal in February 2020 did little to alter this dynamic.

During Honcharuk's premiership, policy reform was dramatic. Tax revenues were increased, the burdens on small business lightened, utility prices cut, state gas company Naftohaz unbundled and several corrupt officials dismissed. But institutional reform was erratic and half-hearted. Systemic transformation is inherently disruptive, and the reward for institutional success can be loss of popularity and even power. This was not a risk Zelensky appeared willing to take. Initially, he accepted Honcharuk's argument that administrative reform could not succeed without competitive salaries. But when these salaries provoked public outcry, he reversed his decision. He was deaf to entreaties to retain the accomplished but controversial Uliana Suprun as acting minister of health. In the brief time since her resignation in August 2019, she has had three successors.

Thirty years after independence, Ukraine's Ministry of Defence is still dominated by those who define 'defence reform' as the addition of more resources to an unchanged structure. The ministry has few civilians in senior roles, and there is little proper civilian oversight. It also remains a commercial as well as a military establishment, with much state property under its management. The promotion to defence minister of Zahorodniuk, who earned plaudits as head of the Defence Reforms Projects Office, thus promised change.

In contrast to the Ministry of Defence, which had no analogue in Soviet Ukraine, the Security Service of Ukraine (SBU) is the lineal descendant of the republican-level KGB. At least until 2014, it never fully overcame its post-Soviet hangover, namely, informal ties with the networks of the successor security services of the Russian Federation. Moreover, the SBU's operatives occupy a critical space at the intersection of the legal and criminal worlds that is not easily supervised or monitored. Even its most effective heads have lacked the time, authority or tenacity to root out these ills. In his first six months, Ivan Bakanov, a childhood friend of Zelensky's with no security background, racked up several headline successes against corruption. But his will and capacity to change the culture of this service have yet to be demonstrated, let alone tested.

Thus far, the Zelensky presidency has largely neglected the most important reform of all: the establishment of a judiciary accountable to law rather than power. Zelensky's flagship November 2019 law on judicial reform introduced several worthy, but secondary, reforms. It failed to address reform of the High Council of Justice, the body that appoints, disciplines and removes judges. Until this body is rejuvenated, judicial reform will be stymied. No steps have been taken to loosen Avakov's hold on the investigation of officers of the National Police accused of abusing their authority. Pressures to break the autonomy of the National Anti-Corruption Bureau continue to impede its authority and work.

One of the most disturbing developments of the first year of the Zelensky presidency was the decision in June 2020 to open criminal investigations against Poroshenko. The multiplicity of charges brought against him ranged from corruption to treason. Such charges bring to mind the imprisonment of former prime minister and presidential candidate Yulia Tymoshenko in 2011, which deeply damaged Ukraine's standing in Europe. On 1 July, the co-chairs of the European Parliament's former Special Commission, convened in 2012 to address that issue, released a joint statement of concern suggesting that a 'worrying slippage in standards is beginning to emerge, a step backwards to what risks evolving into a concerning post-Maidan revisionism'. In the face of such concern, it was unclear why Zelensky appeared determined for the trial to proceed.

Return of the system

On 4–5 March 2020, Zelensky dismissed ten of his 16 ministers, most notably Prime Minister Honcharuk, Prosecutor General Riaboshapka, Minister of Defence Zahorodniuk and Minister of Finance Makarova. This dramatic development unsettled foreign investors and depressed the value of government bonds. In November 2019, Zelensky had warned that ministers would be replaced if they did not produce results in three months. From the standpoint of reform, this benchmark was misconceived. But from the standpoint of popularity, Zelensky's deadline made sense. Between November 2019 and February 2020 his approval ratings fell from 80% to 50%, and those of the government to 30%. Moreover, his ministers remained independently minded and even defiant. Riaboshapka adamantly refused to endorse the charges against Poroshenko, calling them 'legal trash'.

But the underlying cause of the dismissal of reformers was the reassertion of power by major oligarchic interests. In its short tenure, the Honcharuk cabinet had managed to unsettle these interests but reaped no immediate dividends in terms of economic growth or public support. Confronting a near-perfect storm, Zelensky was of no mind to wage a struggle with the 13 financial–industrial groups that dominate the country. Instead, he threw the crew off the ship.

The ministers were replaced by an assortment of the managerially able, the lacklustre, the retrograde and the unknown. Their principal commonality is subservience. Prime Minister Denys Shmyhal was a competent manager of an enterprise owned by oligarch Rinat Akhmetov, the richest person in Ukraine, before becoming a well-regarded governor of

Ivano-Frankivsk oblast. Zahorodniuk's successor, Lt-Gen. (retd) Andrii Taran, prioritises the military capability of the Ukrainian Armed Forces over democratic oversight and anti-corruption efforts. Igor Umanskyi, the replacement finance minister, has consistently opposed Ukraine's 'subservience' to the IMF, in contrast to his predecessor, Makarova, who enjoyed a close working relationship with it. As acting head of the State Investigations Bureau, Riaboshapka's successor, Iryna Venediktova, had appointed Yanukovych's defence attorney as her first deputy. Her new appointees as prosecutor general had, before Zelensky took office, blocked corruption prosecutions as well as investigation into murders during the 2013–14 Euromaidan protests.

Amid these changes, the COVID-19 crisis appeared to be a clarifying moment. Its economic impact swiftly persuaded Zelensky that Ukraine could not dispense with the IMF, whose US$5bn Stand-by Arrangement for Ukraine was expected to unlock a further US$5bn from other donors. On 30 March 2020, Umanskyi was replaced by Serhii Marchenko, who supported the IMF's conditions. Zelensky's success in securing the IMF package on 9 June over tenacious parliamentary opposition, including the Kolomoyskyi faction of his own party, redounded to his credit. But appearances were deceptive. On 1 July, shortly after the first IMF tranche had been disbursed, Governor of the National Bank of Ukraine (NBU) Yakiv Smolii resigned, citing 'systematic political pressure'. Since the tenure of governor (later president) Viktor Yushchenko in 1993–99, the NBU has earned respect as an anchor of Ukraine's macroeconomic stability. Smolii's resignation prompted expressions of concern from the IMF, the G7 and the business community.

Honcharuk said in June 2020 that 'the Zelensky that I started working with and Zelensky today are two different people'. Danyliuk believes that Zelensky, like the rest of Ukraine's establishment, was profoundly unnerved by US President Donald Trump's evident disdain for Ukraine and by the controversy around the impeachment process against Trump. In September 2019, reports emerged that Trump had pressured Zelensky to investigate Ukrainian gas company Burisma Holdings' executive Hunter Biden, the son of US Democrat presidential candidate Joe Biden.

For almost 30 years, US support for Ukraine's independence and its Euro-Atlantic aspirations underpinned Kyiv's confidence and its policy. The damaging Trump–Biden affair strengthened those who believe that the West regards Ukraine as a bargaining chip with Russia and a resource to be exploited by internal forces in domestic battles unrelated to its well-being and security. The result, in Honcharuk's view, is a new susceptibility on Zelensky's part to fake news and anti-Western myths.

The Russia conundrum

Zelensky inherited the fruits of the two Minsk agreements to end the conflict in Donbas, which were concluded respectively in September 2014 and February 2015 with the participation of France, Germany, Russia and the Organization for Security and Co-operation in Europe (OSCE).

The Package of Measures for Implementation of the Minsk Agreement of 12 February 2015 (Minsk II) committed the parties to a set of what soon came to be called security and political provisions. The former included an immediate and comprehensive ceasefire; withdrawal of heavy weapons, foreign forces, equipment and 'mercenaries'; the reinstatement of Ukraine's full control of the state border the day following local elections; and monitoring and verification of all measures by the OSCE. The key political provisions included the holding of local elections in accordance with OSCE standards; the creation of a regime of 'special status' in the former conflict zone; Ukrainian constitutional reform; and the entry into force of a new constitution enshrining these provisions. At the same time, these provisions were to be implemented in accordance with existing Ukrainian legislation.

Disputes regarding the sequencing of these provisions and their precise interpretation quickly produced discord and deadlock. There is no consensus on who is subject to the agreement and who must answer for violations. Russia denies being a party to the conflict, and Ukraine refuses to grant official recognition to the representatives of the so-called Donetsk and Luhansk 'people's republics'. Russia insists that the political provisions be implemented without delay. Ukraine insists that the security provisions must come first. None of the 30 ceasefires established so far have endured. More Ukrainian military personnel have died in the fighting since the Minsk agreements came into effect than in the months before they were concluded.

Since taking office, Zelensky has changed the dynamics of the conflict in three ways. Firstly, as a Russian speaker and an eastern Ukrainian of Jewish origin, he has done much to discredit the Russian charge that post-Maidan Ukraine is an ethno-nationalist, quasi-fascist state. Alongside his rehabilitation of the Russian language, his emphasis on the humanitarian, as opposed to military, dimension of the conflict has provided Ukraine with an element of soft power that it hitherto lacked despite the fact that the majority of the troops that nearly defeated the separatist 'militias' in 2014 were Russian-speaking eastern Ukrainians.

Secondly, he has accepted the 2016 formula of former German foreign minister Frank-Walther Steinmeier, initially unwritten and soon forgotten until Russia made its acceptance a condition for holding a further Minsk summit. The formula accepts Moscow's core demands: the holding of elections in the occupied zones of Donetsk and Luhansk (the so-called 'ORDLO') followed by the granting of special status. Yet it also stipulates that the elections be 'free and fair', conducted under OSCE supervision and according to OSCE standards. When Zelensky accepted the formula on 1 October 2019, he maintained that these standards would require the withdrawal of troops, the disarming of the insurgents, full media access and the deployment of a UN–OSCE peacekeeping force (which Poroshenko had also demanded). In essence, Zelensky reintroduced the security provisions through the back door. Although Moscow rejected this interpretation, it withdrew its objections to the 'Normandy Four' (France, Germany, Russia and Ukraine) summit that took place in Paris on 9 December 2019.

Zelensky's third innovation, unveiled at the summit, was the introduction of several new principles: reinstatement of Ukraine's full control of the state border before the elections; representation of the 1.5 million Ukrainian internally displaced persons (IDPs) in devising the framework of the elections; and the right of the IDPs to vote in the elections themselves. Whereas the border provision contradicts the

Minsk II agreement, the IDP provision is consistent with it, although not explicitly specified. At the closing press conference, he also reiterated three existing 'red lines': no 'federalisation', no foreign influence over Ukraine's 'vector of development' and no relinquishing of territory, including Crimea. Even Zelensky's domestic critics acknowledged that his performance at the summit was a tour de force.

With respect to all these changes, Zelensky finds himself in the cross hairs of public opinion. Although a solid majority supports his goal of a swift end to the war, an even stronger majority opposes any changes to the occupied territories' pre-war constitutional status. Zelensky's belief that Russian President Vladimir Putin would like to find a way out of the conflict, and that disengagement zones, prisoner exchanges and other small steps will enable this, has unsettled many in Ukraine. Putin has also disputed this, insisting that there is only one way to end the conflict: the implementation of the Minsk accords 'as written'.

In early February 2020, Putin replaced Russia's long-standing plenipotentiary for conflict settlement, Vladislav Surkov, with a seasoned negotiator and former deputy prime minister, Dmitry Kozak. Kozak wasted no time securing the agreement of Zelensky's new Chief of Staff Andrii Yermak to the establishment of a Consultative Council granting official recognition to representatives of the people's republics. A furious reaction from Ukrainian civil society (not least the Public Council of the Ministry of Foreign Affairs) swiftly unravelled this agreement. In response, Ukraine has now raised its game, assigning a deputy prime minister to the working-level Trilateral Contact Group of Ukraine, Russia and the OSCE, and adding IDP representatives to its delegation.

A flawed but unfinished project

Zelensky has learned from his year in office and been changed by it. But some initial shortcomings are still evident: an absence of state administrative culture, a dearth of economic and military knowledge, and a shallow understanding of Russia's strategy and goals. His warmth of personality is compromised by his need to be liked, and his ability to think creatively is challenged by a lack of experience in thinking systematically or strategically. His marginalisation of professionals and his reliance on a coterie of loyalists have not served him well.

At the same time, he has been tested to the limit by Trump's belief that Ukraine is 'Europe's problem'. Trump's jarring comment in September 2019 that 'I really hope that you and President Putin get together and can solve your problem' would have unsettled a leader with far more steeliness and experience than he possesses. Yet he has not buckled. While he lacks the resolution of his predecessor, he has demonstrated ingenuity and a degree of audacity in meeting the challenge that Russia poses. What he has yet to demonstrate, with regard to both Russia and the challenge of internal reform, is tenacity and steadiness. Too often he has vacillated when predictability is needed.

Zelensky's principal goal, the transformation of Ukraine, has been undermined by his own misjudgements and the aggressiveness of those determined to preserve the status quo. But it is too early to conclude that he is in thrall to the system he was elected to dismantle. There is every possibility that Ukraine will remain a country run from the shadows with the complicity of its politicians and officials. But a president as attuned to public opinion as Zelensky knows that he accepts this fate at his peril. His vitality is incontestable, and his potential is far from exhausted. He has been in office for a relatively short time, and it would be premature to pronounce his project a failure.

Central Asia: Connectivity, COVID-19 and Geopolitics

What next for the Eurasian heartland's international ties?

Since gaining independence in 1991, the five post-Soviet countries of Central Asia have striven to emerge from political and economic isolation. They are disadvantaged by geography: Kazakhstan, Kyrgyzstan, Tajikistan and Turkmenistan are landlocked, and Uzbekistan is double-landlocked. They also inherited a Soviet transport infrastructure that linked them to Russia rather than to the outside world.

Against an unpromising backdrop, these states sought to develop new routes to connect them to neighbouring and global markets. But their lack of financial and institutional resources made it difficult both to develop the necessary hard infrastructure of roads, railway links, ports, pipelines, electricity grids and fibre-optic cables; and to create and run the soft infrastructure of cooperation mechanisms, governance regimes and regulations for managing transnational flows with other states.

Fortunately, a variety of international actors have proved ready to invest in Central Asian connectivity. International development banks such as the World Bank, the European Bank for Reconstruction and Development (EBRD), the Asian Development Bank and the Asian Infrastructure Investment Bank have all done so to carry out their institutional agendas. Public and private actors from the European Union, China, Japan, Russia, South Korea and the United States have also pursued geopolitical goals and economic interests in Central Asia.

The year to June 2020 brought successes and failures of national and international efforts to foster connectivity in Central Asia. The normal course of events was interrupted by the outbreak of the COVID-19 pandemic. Quarantines, border closures, disruptions in supply chains and volatile commodity prices posed immediate challenges. Looming economic crises also reinforced ongoing geopolitical shifts, creating uncertainty about the future.

Whether the pandemic causes merely temporary disruption, or exerts a more profound and long-term impact, is of vital importance to the region. Will international support for Central Asian connectivity, first and foremost China's Belt and Road Initiative (BRI) but also the EU's efforts, decline because of the crisis? Will Central Asian countries be affected in dissimilar ways, given the considerable variation in their circumstances, wealth and institutional capacity? While it is too early to offer definitive answers to these questions, developments in the year to June 2020, both before and after the outbreak of COVID-19, offer important insights.

China's Belt and Road Initiative

Over the past decade, the main hopes for fostering connectivity in Central Asia have been linked to China and its expansion into Eurasia. These hopes were boosted in September 2013 by Chinese President Xi Jinping's launch of the Silk Road Economic Belt (which became part of the BRI in 2015) in Kazakhstan's capital Astana (now known as Nur-Sultan). Xi outlined its five priorities: policy communication and coordination; road connectivity (cross-border transportation infrastructure) between East Asia, West Asia and South Asia; unimpeded trade; monetary circulation through local currency convertibility and settlement under current and capital accounts; and people-to-people exchanges.

The scope of China's initiative expanded over the next two years. In the 'Vision and Actions on Jointly Building Silk Road Economic Belt and 21st Century Maritime Silk Road' issued in March 2015, 'road connectivity' was upgraded to 'facilities connectivity', encompassing all types of infrastructure including aviation, energy infrastructure, and optical cables and other communication trunk-line networks. The section on unimpeded trade focused on investment facilitation and the expansion of mutual investment areas, as well as division of labour and distribution of industrial chains.

Central Asian governments joined the initiative enthusiastically, finding great complementarities between their national development strategies and the BRI. In 2016, Kazakhstan, the regional economic powerhouse whose cooperation with China was most advanced, linked its new economic policy *Nurly Zhol* ('Bright Path') to the BRI, focusing on three priorities: transport infrastructure, energy and manufacturing. Uzbekistan, the second-largest economy of the region, has been catching up fast with Kazakhstan since Shavkat Mirziyoyev became president in 2016. In 2017, Beijing and Tashkent agreed to strengthen cooperation in the context of the BRI in trade, investment, finance, transport and communication, agriculture, industrial parks and other areas. Tajikistan's National Development Strategy 2030 (adopted in 2016) focused on energy security, transit potential and agriculture, areas in which China plays the role of the country's main partner and investor.

International development banks recognised the significant potential benefit of the region's engagement with China. The EBRD's 2018–19 Transition Report predicted that, in some Central Asian countries, BRI investment in infrastructure would raise real GDP per capita by 4–6%. In 2019, the World Bank Group published a working paper on the BRI's role in reshaping economic geography in Central Asia. Its authors calculated that aggregate real income gains from the initiative would be 2–3% of regional income.

However, despite Central Asian governments' enthusiasm for pursuing greater connectivity and prosperity, progress in transport and energy had proved limited even before the COVID-19 pandemic. Indeed, the biggest China–Central Asia connectivity projects preceded the BRI. These include the Kazakhstan–China oil pipeline; the Central Asia–China gas pipeline (Lines A, B and C) connecting Turkmen gas fields with China via Uzbekistan and Kazakhstan; the Khorgos-Eastern Gate dry port on the Kazakhstan–China border; Kazakhstan's portion of the Western Europe–Western China highway; and the China–Kazakhstan–Turkmenistan–Iran railway – all of which were completed before 2013. The plans to construct a China–Kyrgyzstan–Uzbekistan railway, and Line D of the Turkmenistan–China gas pipeline crossing Uzbekistan, Tajikistan and Kyrgyzstan, also announced before the BRI, were placed under the BRI umbrella but have made little progress since then.

Uzbekistan is the only Central Asian country that can boast major successes in improving connectivity in recent years. In April 2019, after reaching an agreement on complex tariff rates, it joined the China–

Kazakhstan–Turkmenistan–Iran railway corridor. In June 2020, the China–Kyrgyzstan–Uzbekistan multi-modal transport link was launched. The first train on this route left China's Gansu province and arrived at the Chinese–Kyrgyz border. There, containers were put onto trucks and transported across southern Kyrgyzstan to Osh, where they were then moved by train to Tashkent, the capital of Uzbekistan. This route, albeit less convenient than the planned railway, has created a shorter and more efficient transport connection between China and Uzbekistan.

The launch of the multi-modal route made the Kyrgyz government revisit plans to build a railway link with China, realising that the project was too challenging under the circumstances. In the meantime, the country also lost a BRI project that had been under development. In February 2020, the Kyrgyz government cancelled a US$275-million Chinese investment project to build a logistics centre in the Naryn Free Economic Zone at At-Bashy in central Kyrgyzstan. The decision followed a series of protests by local residents against the 49-year lease of 200 hectares of land to the joint Kyrgyz–Chinese venture that was to build the logistics centre.

A scandal around the Bishkek thermal power plant – upgraded by a Chinese company with a Chinese loan – may have aggravated the anxieties of the Kyrgyz government. In 2018, the plant broke down, leaving residents without heat during a harsh winter. Two former prime ministers of Kyrgyzstan were found guilty of lobbying for the interests of the Chinese company and were sentenced to long jail terms. In general, China and its activities in Kyrgyzstan tend to be viewed with suspicion by the Kyrgyz public due to fears of China's domination and loss of land to Beijing's potential territorial claims, as well as distrust of the integrity of the Kyrgyz authorities.

The country's prospects of developing connectivity under the BRI umbrella grew even dimmer with the outbreak of COVID-19. Previously, Kyrgyzstan took out loans from the Export–Import Bank of China for infrastructure projects such as road construction and repair and modernisation and construction of transmission lines. In 2019, Bishkek's public debt to China stood at US$1.7 billion (46% of total public debt), which already put it in a group of countries designated by the IMF as having 'moderate risk of debt distress'. With the pandemic crisis, the country's debt solvency has weakened.

Tajikistan is even more heavily indebted to China. In 2019, its public debt to China stood at US$1.5bn (or 52% of total public debt). The government pleaded and was able to postpone a payment of US$40m to the Export–Import Bank of China until 2021. Nevertheless, despite Tajikistan's worsening financial situation, Beijing is likely to continue supporting infrastructure projects in Tajikistan. In June 2020, Xi, during a phone conversation with his Tajik counterpart Emomali Rahmon, said the two countries should start planning for cooperation in the 'post-COVID era' in key areas including energy and interconnectivity. These projects are likely to foster domestic, rather than international, connectivity. If Line D of the gas pipeline to China is completed (at present, China's demand for Central Asian gas is falling), it will benefit Tajikistan only in the form of transit fees.

The Central Asian country most invested in Eurasian connectivity is Kazakhstan. Since early independence, it has poured significant resources from oil revenues into developing its transport and logistics systems. Flagship projects include Kazakhstan's portion of the Western Europe–Western China highway; the Khorgos-Eastern Gate dry port, conceived as a world-class hub for cargo flows along the new Silk Road; and the Aktau and Kuryk seaports on the Caspian shore. All but the first of these were developed by Kazakhstan Temir Zholy (KTZ), the national railway company transformed in 2011 into a transport–logistical holding tasked with developing 'Kazakhstan's transit capacity and global infrastructure integration'.

While these projects are largely completed, and the highway and ports operational, Kazakhstan's transit potential remains underdeveloped. Between 2010 and 2018, the country's Logistics Performance Index (LPI) score, calculated by the World Bank based on six core performance components (efficiency of customs and border clearance; quality of trade and transport-related infrastructure; ease of arranging competitively priced shipments; competence and quality of logistics services; ability to track and trace consignments; and timeliness of shipments), decreased from 2.83 to 2.81. For comparison, between 2010 and 2018, China's score increased from 3.49 to 3.61, and global top performer Germany's score from 4.11 to 4.20. The indices indicate that Kazakhstan's logistics and transport improvements are not enough, while other countries are growing more competitive.

Corruption is one of the problems hindering better performance in Kazakhstan. Systemic corruption has been a prominent feature of Kazakhstan's political economy in general, and of its transport-infrastructure development in particular. This extends from the top of the decision-making pyramid, where major projects require the payment of large kickbacks, to small-scale bribes solicited by customs officials at the border. In May 2019, a minor (by Kazakhstan's standards) scandal broke out when several customs officials were arrested for taking bribes in exchange for letting Chinese goods transit from Kazakhstan to Uzbekistan. In October 2019, the state anti-corruption agency accused a state-owned company, Astana LRT, of massive embezzlement in connection with its building of the Light Railway Transit (LRT) system, a high-profile BRI project financed by the China Development Bank (CDB) in Nur-Sultan. Construction had been suspended in April 2019 when the city administration and the CDB failed to reach an agreement on the terms of the next loan tranche. As in Kyrgyzstan, such scandals have fuelled public suspicions about China's activities and projects.

As a result of the pandemic, government resources will shrink. The IMF Regional Economic Outlook, published in April 2020, projected a 2.5% decline in GDP for Kazakhstan in 2020. Less funding will be available for ambitious infrastructure projects, and projects that go ahead will be under greater public scrutiny, resulting in more criticism, scandals and possibly protests. While the crisis will probably not undermine the strategic importance that Kazakhstan's government assigns to transcontinental connectivity, the capacity to implement it will be severely constrained.

There are also many uncertainties in the external environment, such as the impact of COVID-19 on China's BRI in Eurasia. The disruption of sea and air cargo shipments increased demand for the China–Europe rail land bridge. The two-way volume of trade along this route in the first quarter of 2020 almost doubled compared to the same period in 2019. However, this may be a temporary trend. When the situation normalises, the attractiveness of the railway option vis-à-vis air and sea will depend on its reliability and cost. It has been competitive only thanks to subsidies provided by the Chinese government to Chinese companies to support China–Europe railway cargo.

A February 2020 study of the development potential of Eurasian corridors, carried out by the International Union of Railways, projected that the reduction of Chinese subsidies from 50% to 20% would likely have a drastic impact on trade flows and lead to a reallocation of them across the routes. Such a reduction would first impact southern routes via Kazakhstan and reduce their volumes. The study also envisaged a possible reshaping of the market for transit services due to new opportunities for market players on the northern routes and fewer opportunities on the southern routes (via Kazakhstan and the Trans-Siberian and South Asia–Central Asia–South Caucasus Middle Corridor). Given Beijing's plans to phase out subsidies for Eurasian rail-freight traffic, such a scenario can be deemed likely; if it fully materialises, it will be a significant blow to Kazakhstan's plans and Central Asian connectivity.

The EU's Eurasian connectivity strategy

The EU has been providing assistance to foster Eurasian connectivity since the early 1990s. It launched the Europe–Caucasus–Asia Transport Corridor (TRACECA) programme in 1993 and Interstate Oil and Gas Transportation to Europe (INOGATE) programme in 1996 to help its eastern neighbours, including those in Central Asia, to access European and world markets. The Aktau and Turkmenbashi seaports were among the projects supported through these programmes. TRACECA evolved from a programme into an organisation composed of 13 states along the corridor.

China's expansion into Eurasia through its BRI rekindled the EU's interest in transcontinental connectivity. In October 2018, the European Council adopted 'Connecting Europe and Asia: Building Blocks for an EU Strategy'. This strategy outlined the 'European approach to connectivity' as 'sustainable, comprehensive and rules-based'. It argued that connectivity projects should be economically efficient and fiscally viable, should comply with high environmental standards, and should have high standards of transparency and governance. The strategy implied that China's approach to fostering connectivity lacked these features.

The EU included Central Asia among the regions where it planned to foster connectivity. In June 2019, the EU Council endorsed a new Strategy for Central Asia that prioritised the creation of partnerships on sustainable connectivity. The latter expressed their support and readiness to work with the EU. They were interested in attracting European investments in their infrastructure and in technical assistance. They also hoped that EU engagement could help them balance their growing dependence on China.

The outbreak of the COVID-19 pandemic hindered EU–Central Asia cooperation by delaying the adoption of the EU's latest long-term budget (the Multiannual Financial Framework for 2021–27). In the meantime, in May 2020 the European Commission announced an €8m (approximately US$9.4m) support programme for 'Sustainable Energy Connectivity in Central Asia' (SECCA).

Cloudy horizons

But the challenge that the pandemic presents to Central Asia's connectivity plans goes further. The economic crisis triggered by COVID-19 will drain these countries of the financial resources they need to build and upgrade infrastructure. Even before COVID-19 struck, poor governance had hindered efforts to foster domestic, regional and transcontinental connectivity in all five Central Asian states. As a result, several projects caused public scandal or were mired in protests and suspended or cancelled (such as the Naryn Free Economic Zone logistics centre and several mining projects in Kyrgyzstan). Uzbekistan was a fortunate exception, but it has not escaped the COVID-19 crisis.

The region is entering troubled times, with ambitious development strategies proving a mirage. Kazakhstan has invested more in connectivity projects than any other Central Asian country, but will find it hard to continue pumping money into infrastructure. In the best-case scenario, it will focus on improving policies, regulations and procedures – areas where there is plenty of room for improvement. In the worst-case scenario, governance will deteriorate while interest groups will push for borrowing from abroad to keep their businesses afloat. Kyrgyzstan and Tajikistan, already heavily indebted, will have to rely on the generosity of international donors. Uzbekistan has more autonomy and capacity to weather the storm, and is reaping benefits from international enthusiasm for President Mirziyoyev's economic reforms. As of June 2020, Turkmenistan denied having had a single case of COVID-19. As a result, it mishandled the pandemic more severely than any other country in the region. It faces a bleak future, with extremely poor governance aggravated by a drop in revenues due to low gas prices and China's reduced demand for Turkmen gas.

However, this dire situation could have a silver lining if Central Asian states now focus more on developing connectivity and supply chains within Central Asia and the wider region. Kazakh and Uzbek officials and experts argue that this will be necessary as once-global supply chains become increasingly local and regional. This is consistent with the advice of international donors, such as the World Bank, the US and the EU, that have long sought to encourage regional connectivity in Central Asia.

The external setting is also changing. While the interest of Chinese and other businesses in the Europe–Asia railway corridor spiked as a result of the pandemic, this could prove to be a temporary trend that will end once sea and air cargo shipments recover from disruptions. Given the status of the BRI in China's foreign policy, and its linkages with Xinjiang-related domestic policies, Beijing is likely to continue supporting and financing projects in Central Asia. But the level of subsidies and benefits that the Chinese government will provide to companies involved in BRI projects might decrease.

The EU is likely to continue promoting sustainable connectivity in the region, but the scale of this support is unclear. Two factors pull in opposite directions. On the one hand, the pandemic and other crises (such as developments in Belarus) that claim EU attention and resources could push Central Asia further down the EU's agenda. On the other hand, EU determination to be a more assertive geopolitical player vis-à-vis China could strengthen the incentive for greater engagement with the region.

In sum, the impact of the COVID-19 pandemic on Central Asia's connectivity plans is significant. Governments under stress will have fewer resources for the necessary infrastructure projects, and less attention to devote to improving the governance and regulations needed to make connectivity work and to deliver benefits.

External powers continue to engage. For China, this means building a 'Eurasian economic corridor', and for the EU, creating 'partnerships for sustainable connectivity'. This courting of Central Asian governments offers the latter opportunities and room for manoeuvre. They would be well advised to respond by focusing on developing needed, but neglected, connectivity within the region.

CHAPTER 9

Middle East and North Africa

MIDDLE EAST AND NORTH AFRICA

Drivers of Strategic Change

REGIONAL SHARE OF GLOBAL POPULATION, GDP AND DEFENCE BUDGET

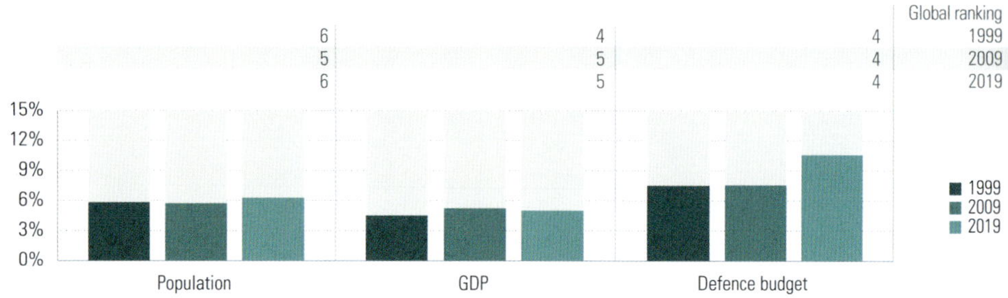

POPULATION

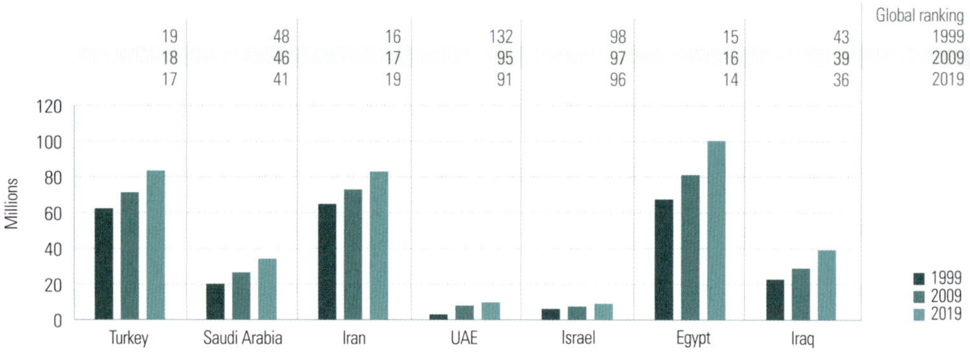

AGE STRUCTURE

(Percentage of national population)

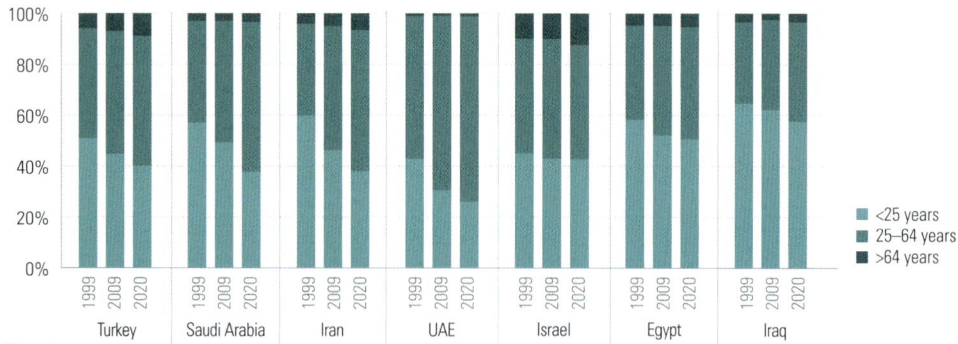

Median age	Turkey	Saudi Arabia	Iran	UAE	Israel	Egypt	Iraq
2000	24.9	21.3	21.2	28.1	28	21.1	18.2
2010	28.3	26	27	31.9	30.1	23.7	18.9
2020	31.5	31.8	32	34	30.5	24.6	20

GDP
(Constant 2010 US dollars)

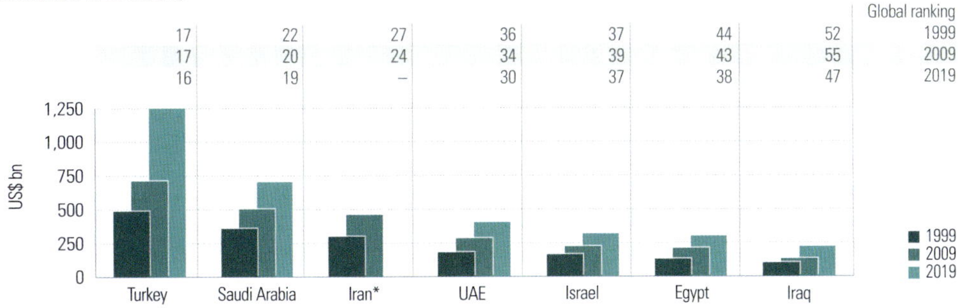

* No 2019 data available for Iran

GDP PER CAPITA
(Constant 2010 US dollars)

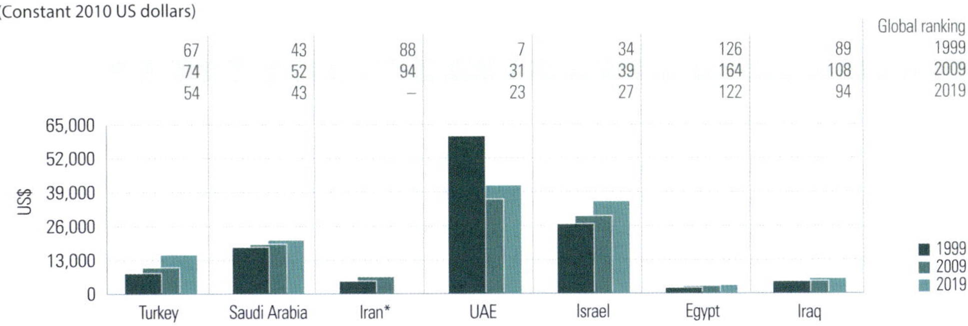

* No 2019 data available for Iran

DEFENCE BUDGET
(Constant 2015 US dollars)

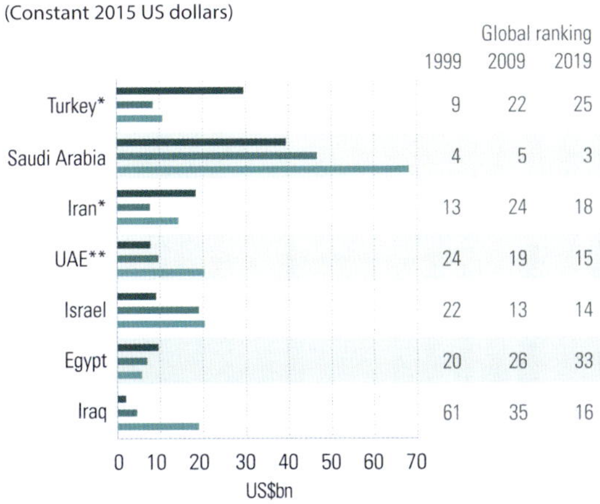

ACTIVE MILITARY PERSONNEL

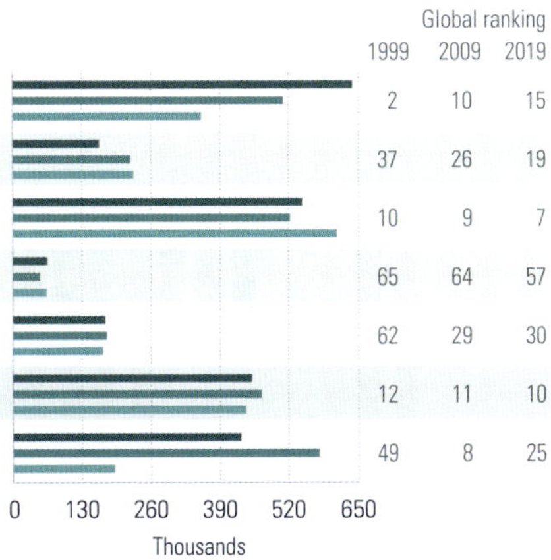

* 1999 defence budget values for Iran and Turkey are estimates, and may be distorted by high inflation rates in these countries

** No 2009 data available for UAE. 2010 data is therefore used in place of 2009 data

■ 1999 ■ 2009 ■ 2019

For explanation of drivers and sources, see page 7

HUMAN DEVELOPMENT INDEX (HDI)

(Score between 0 and 1, where 0 denotes a low level of development and 1 a high level of development)

	Turkey	Saudi Arabia	Iran	UAE	Israel	Egypt	Iraq
Global ranking 1999	80	47	74	31	14	91	92
2009	91	48	71	34	21	109	120
2019	58	36	65	35	21	116	120

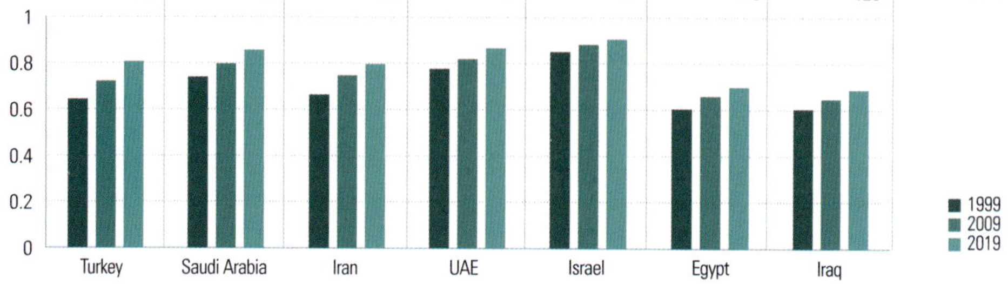

POLITICAL SYSTEM

(Score between 0 and 100, where 0 denotes no political freedom and 100 fully free)

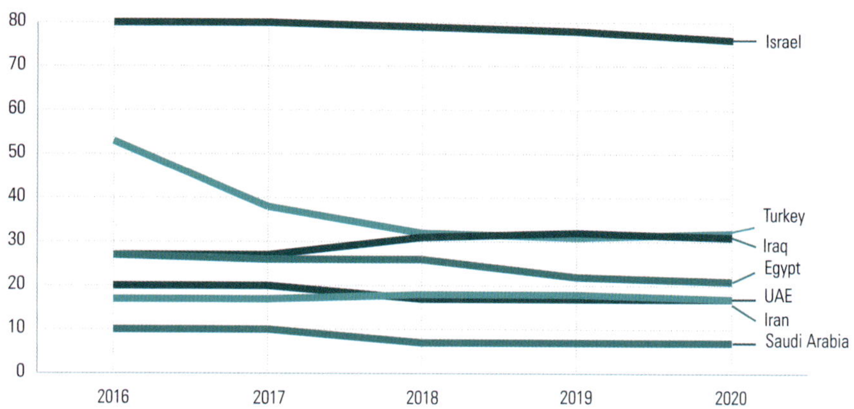

BREAKEVEN OIL PRICES (2015–20)

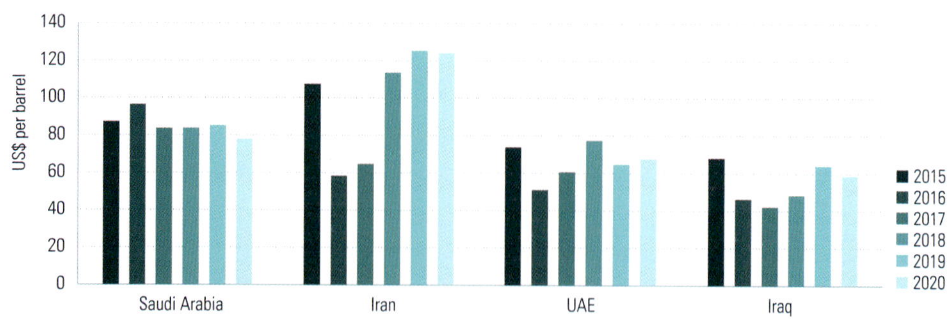

Average annual oil price 2015–19 (US$/barrel of Dubai Crude)

2015	2016	2017	2018	2019
$97.07	$41.19	$53.13	$69.51	$63.43

For explanation of drivers and sources, see page 7

2019–20 Review

Developments in the year to June 2020 amplified security threats and aggravated the systemic risks to a Middle Eastern order already reeling from a decade of destructive conflicts and unbridled regional competition.

A sense of disorder prevailed at both regional and local levels. The prospect of a US–Iran war, whether intentional or accidental, dominated an already complex regional agenda. Simultaneously, conflicts in Libya, Syria and Yemen remained intractable and sometimes grew more intense. Governmental dysfunction in Iraq and economic collapse in Lebanon revealed deficient governance in weak states and ignited popular protests. Israel faced political turmoil, with three elections in less than a year that produced weak governments. Elections in Iran resulted in a solid conservative-hardline majority, while municipal elections in Turkey demonstrated deep discontent with the rule of President Recep Tayyip Erdogan.

A tentative transition was taking place in Sudan, and Tunisia conducted a successful presidential election. Elsewhere, authoritarian rule remained unshaken and Algeria underwent a transition managed by the military. Turkey was involved simultaneously in two Arab states, Syria and Libya, with significant military consequences. Lasting discord among the Gulf states prevented the formulation of a regional agenda and frustrated Western security partners.

External powers often fed disorder instead of seeking stabilisation. Breaking with decades of US policy, the Trump administration acquiesced to Israel's expansionist plans and parted with the traditional US emphasis on a political settlement leading to a two-state solution. Russia remained both arbiter of and party to competition in Syria. It deepened its involvement in Libya, establishing itself as a power in the Eastern Mediterranean. In contrast, the European Union, focused on intra-European politics and divided over Libya and other conflicts, was largely absent from the strategic debate.

By March 2020, two new layers of uncertainty added to this troubled landscape. Rapidly falling oil prices and the repercussions of the

COVID-19 pandemic presaged another period of domestic unrest, economic dislocation and security upheaval, should the second-order effects of the crisis be mismanaged.

Rising tensions in the Gulf

The possibility of a conflict between Iran and the US remained the overriding geopolitical risk because of the sheer magnitude of such a war and the potential repercussions for global and energy security. The Trump administration's withdrawal in 2018 from the 2015 Iran nuclear deal, and its adoption of a maximum-pressure strategy aimed at compelling Iranian surrender, forced Iran to assume a posture of maximum resistance. These competing stances guaranteed an escalation of sorts and constant mutual provocations; the inherent preferences and limitations of both sides prevented, if only momentarily, a slide towards an all-out intentional war.

Reeling from US sanctions on oil exports and other severe measures, from early 2019 Iran conducted a series of escalating provocations in its immediate neighbourhood intended to impose costs on Arab countries allied with the US. Tehran deployed calibrated, deniable tactics including attacks on and harassment and seizure of oil tankers to emphasise its ability to disrupt its neighbours' oil exports if its own exports were sanctioned. The initial stages of this counter-pressure campaign inflicted no casualties and international reactions were subdued, partly because of general distrust of the Trump administration and fear about an accidental escalation into all-out war. Reflecting a growing consensus in Washington, the US administration itself was reluctant to risk a new intervention in the Middle East, which US President Donald Trump saw as needlessly consuming. The campaign culminated in September 2019 with spectacular drone and missile attacks against the Saudi oil-reprocessing facility of Abqaiq and the Khurais oilfield. The attacks reduced Saudi oil production by almost 50% for several weeks, but oil prices remained stable, reflecting ample supply and the fact that Middle Eastern instability had already been priced in.

These attacks demonstrated Iranian military skill. They also illustrated the difficulty of protecting critical infrastructure, including energy-sector installations, thus emphasising the exposure of Gulf countries to Iranian mischief. Strategically, the crisis revealed the gap between Washington's maximalist ambitions and its reluctance to risk war even if its closest partners were exposed to the repercussions of US policy, as long as US facilities or personnel were not targeted. Trump explicitly stated that the killing of a US citizen would result in US retaliation. Consequently, US reluctance to respond militarily fuelled confusion and concern among its Arab allies about the reliability of the US and the Trump administration's overall strategy. There were signs of tactical adjustments, such as Emirati dialogue with Iran, and more strategic hedging by the Gulf states, such as outreach to Asian powers, as a result.

Another round of escalation occurred in late December. Working through its militia partners in Iraq, Tehran began harassing US facilities located inside Iraqi military bases by firing rockets. Iraq was particularly vulnerable to US–Iranian tensions, which threatened its security and economic stability as it dealt with its own political infighting. The killing of a US contractor in a rocket attack on a base in Kirkuk province in December crossed the line Trump had articulated, and he ordered a US retaliation that killed several Iraqi militiamen. In turn, pro-Iran groups organised a protest that escalated into an attack on the US embassy in Baghdad. On 3 January 2020, a US drone flying in the vicinity of Baghdad airport fired missiles at a convoy carrying General Qasem Soleimani, the commander of Iran's Quds Force and architect of Iran's regional influence, and Abu Mahdi al-Muhandis, the de facto head of the Popular Mobilisation Units, the Iraqi coalition of pro-Iranian Shia militias. The attack killed both men.

The assassination of Soleimani stunned Iran and unnerved the international community over concerns of massive escalation in the Middle East. However, its aftermath illustrated the strategic limitations and preferences of each side. By way of retaliation, Iran conducted a one-off, direct and overt missile attack on a US facility in Iraq – the first time in 30 years that Iran had conducted a conventional attack on the US.

The attack did not kill any US personnel but injured many. Iranian diplomats announced that retaliation had been conducted in accordance with international law, while Supreme Leader Ali Khamenei vowed open revenge.

The contest subsequently played out primarily in the Iraqi arena: Iran's militia partners pledged to expel US forces while the US reduced and consolidated its presence to support Iraqi forces in their efforts against the Islamic State, also known as ISIS or ISIL. The end of the US presence in Iraq became a possibility: it was a strategic goal of Iran and a preference of Trump, despite the opposition of the rest of the US security establishment and many Iraqi officials. The absence of any direct channel or active mediation was felt keenly: after a French initiative to organise a meeting between Trump and Iranian President Hassan Rouhani on the margins of the UN General Assembly fell through, there was no attempt at negotiations. The death of Sultan Qaboos of Oman in January 2020 removed a traditional facilitator between the US and Iran.

Inside Iran, the escalation and economic pain further weakened Rouhani and benefitted more hardline groups. These groups dominated the new parliament elected in February and denounced diplomacy with the US as futile and dangerous. Even as war remained a possibility, the prospect of a Democrat candidate winning the US presidential elections in November 2020 preserved a tenuous hope that the US could re-enter the nuclear deal and jump-start talks in 2021. Much depended on whether Iran would remain in the nuclear deal or escalate the calibrated encroachment of it.

Regionalised conflicts

A combination of local and regional drivers meant there was no substantial progress towards a political settlement in any of the three major conflicts afflicting the Arab world. Indeed, Syria, Libya and Yemen each experienced phases of intense warfare and inconclusive peacemaking. Importantly, the COVID-19 pandemic did not stimulate any regional or local efforts at de-escalation. French attempts at the UN Security Council to encourage a regional ceasefire failed to gather momentum.

The fracture line between Turkey on the one hand and Saudi Arabia and the United Arab Emirates (UAE) on the other became as pronounced, potent and consequential as the one between the latter two and Iran. This competition played out through proxies across battlefields in Syria and Libya, but also in the media space, in Islamic organisations such as the Organisation of Islamic Cooperation and in other multilateral arenas.

In Syria, the Assad regime struggled to translate its defeat of the rebellion into a political victory and a full territorial reconquest of the country. Its ambitions were frustrated by weak military capabilities when not supported by Russian and Iranian forces, a collapsing economy, steadfast Kurdish control of northeast Syria and Turkish support for rebel factions in northwest Syria. Nonetheless, President Bashar al-Assad continued to benefit from competition among his enemies. In October 2019, Turkey launched a campaign to seize large portions of northeast Syria from Kurdish control and create a safe zone. In previous months, diplomacy between Ankara and Washington over security arrangements to prevent a Turkish–Kurdish showdown had stalled. Erdogan saw the growth and de facto autonomy of the People's Protection Units (YPG) as a threat to Turkish territory due to its affiliation with the Kurdistan Workers' Party (PKK). He capitalised on Trump's dislike of the US presence in Syria to push his agenda. The Turkish intervention threatened the ability and willingness of the YPG to fight ISIS remnants and stabilise the region, leading to considerable criticism from NATO members, notably France. Militarily, Ankara achieved its goals, but its operation cost the US credibility among its partners and complicated the fight against ISIS. Russian facilitation brought the Syrian regime and the YPG, both enemies of Ankara, closer together.

In northwest Syria, the Assad regime sought to capture the province of Idlib – the last one controlled by a mix of rebel and jihadi forces – in late 2019. Multiple ceasefires had collapsed in previous months as regime forces, backed by Russian airpower and Iranian help, strangled the area, which held more than three million people, including numerous internally displaced persons. Rebel forces, weakened by months of fighting, vulnerable to airpower and ostracised because of the large proportion of jihadi fighters in their ranks, lost ground rapidly, provoking a massive flight of civilians towards the border with Turkey. This was of great concern for Erdogan, who feared another wave of refugees and its domestic repercussions.

In February, Turkish forces engaged, for the first time, in battles against Syrian government forces to help the rebels hold their positions in Idlib. They inflicted heavy losses through air, drone and artillery bombings. This compelled Russia to seek another uneasy ceasefire and establish military coordination. However, the rebels lost around 40% of the territory they had controlled during summer 2019. They became ever more dependent on Turkish support and less able to counter jihadi groups in their ranks.

The Assad regime faced the prospect of lasting isolation despite its military successes. The US enacted the Caesar Syria Civilian Protection Act in December 2019, a comprehensive bill that imposed heavy constraints and sanctions on economic dealings with the regime. Combined with US diplomatic pressure and EU reluctance to restart relations with the Assad regime, this complicated any prospect of re-engagement by third parties. Neither China nor other actors seemed willing to invest in Syria. The humanitarian situation remained dire, with no prospect of return of refugees or improvement of conditions in their host countries. The US and the EU were unwilling to fund reconstruction without a political settlement. Together, the EU and the US provided 90% of humanitarian assistance to Syria in 2019, while Russia contributed a mere 0.3%.

Libya was another arena in which regional intervention shaped dynamics and outcomes. Since early 2019, a powerful Libyan faction led by General Khalifa Haftar and supported by Russia, Egypt, the UAE and France had captured oil-rich areas and the Fezzan region, and since April 2019 it had been laying siege to the capital Tripoli, where the UN-backed Government of National Accord and various militias, some of which were Islamist in character, operated. An international conference in

Berlin, organised by Germany, in January 2020 reached a notional agreement on de-escalation and a political process, but this was immediately breached, primarily by Haftar. In contravention of a UN embargo, Haftar benefitted from UAE-provided weaponry, notably drones, and Russian manpower in the form of mercenaries. Fearing a victory by its regional rivals, in January Turkey began deploying military advisers and advanced weapons systems to support Libyan factions opposed to Haftar. This intervention tipped the military balance: it broke Haftar's siege and momentum, forcing his troops to free Tripoli and retreat east. Further internationalisation of the conflict seemed inevitable, with Egypt and the UAE exploring military responses to contain the advance of the anti-Haftar forces.

In turn, the conflict in Libya fuelled competition in the Eastern Mediterranean, pitting Turkey and its Northern Cypriot allies against a large coalition comprising Greece, Cyprus, Israel and Egypt as well as Russia, France and the UAE. Enduring maritime-border disputes and the decades-long unresolved conflict over Cyprus were aggravated by the quest for oil and gas as well as pipeline projects. The possibility of armed conflict increased in the spring, emphasising the lack of US diplomatic engagement.

The war in Yemen also fluctuated wildly. The UAE drawdown from the country throughout 2019 following a decision to freeze a campaign to retake the strategic city-port of Hudaydah, combined with Saudi fatigue and inability to secure lasting military successes, reduced levels of violence. However, local drivers kept the conflict ablaze as the UN process failed to restrain the various players. In early 2020, seizing on the fatigue and disarray in the anti-Houthi coalition, the Houthi movement (Ansarullah) made significant advances towards the city of Marib, which had been spared until then. Moreover, these dynamics emboldened the UAE-backed secessionist Southern Transitional Council (STC): the STC challenged the weak rule of Abdu Rabbu Mansour Hadi, the Yemeni president in exile, despite an agreement signed in Riyadh in November between the STC and Hadi that sought to establish a power-sharing deal.

Impact of COVID-19

The COVID-19 pandemic hit the Middle East as it struggled with conflict, stabilisation and dire economic prospects. The first phase in spring 2020 affected Middle Eastern countries unevenly, magnifying the social, economic and capability gaps both between and within countries.

The first cases of COVID-19 infections in the region were reported in February in Iran, which quickly emerged as an epicentre of the crisis. There, the government appeared to understate the gravity of its spread, in part to preserve relations with China, a strategic partner that Iran needed to fend off US pressure, but also to prevent any disruption of its national elections. These were held in February, resulting in a sweeping victory for the conservative and hardline camps. Soon after, the death toll of the pandemic became clearer, especially as senior officials and clerics figured among the casualties.

The virus reached the Gulf states, Iraq, Jordan, Lebanon and other states in early March, primarily but not only by way of travellers and pilgrims from Iran. This forced a near-total suspension of travel, border closures, lockdowns, curfews and the imposition of other stringent measures to contain its spread.

The pandemic revealed significant vulnerabilities and gaps in the capabilities and preparedness of public health systems in the region. The Gulf states were particularly badly affected, with Asian expatriate workers living in precarious conditions especially vulnerable. But these states successfully deployed their financial power and international networks to manage and control the caseload by procuring tests, introducing technology and building health response centres. For them, the crucial consequences were economic: the pandemic, combined with low oil prices, threatened their economic plans and public finances. To protect their economies, they enacted financial measures, helped state-owned entities, and reviewed or froze some of their key projects. Hard-hit Dubai postponed Expo 2020, while Saudi Arabia suspended religious pilgrimages, tripled its VAT and considered suspending important elements of its Vision 2030 plan to diversify its economy away from oil. Some countries, mostly notably the UAE, conducted aid diplomacy to increase their soft power, sending shipments of aid and essential health products to allied or hard-hit countries. China's rapid containment of the pandemic allowed it to display its strengths and make inroads in the region: Beijing dispatched large quantities of tests, protective equipment and medicines to Gulf states and other Middle Eastern states, as the Gulf's traditional partners in Europe and the US struggled to adequately respond to the crisis.

2020 was expected to be a year of strategic rebound for Saudi Arabia, the host of the 2020 G20 Leaders' Summit. However, weakening global governance and multilateralism, and suspension of travel, marginalised the forum, denying it a role in jump-starting global cooperation against the pandemic or coordinating economic-recovery efforts. Simultaneously, the breakdown in March of an OPEC+ agreement led by Russia and Saudi Arabia to cut oil production led Saudi Arabia to announce a significant increase in its oil output to punish Russia for defecting from the agreement and to preserve its market share in Asia. This oversupply, and expectations that the pandemic would trigger a global recession – and, in particular, hit Asian oil consumers – sent oil prices crashing to historic lows. This created anger in the US, where higher-cost producers in the shale-oil industry were especially vulnerable to lower prices. Oil prices recovered slowly after Saudi Arabia and Russia agreed a new deal under US pressure in April, but they remained considerably inferior to the break-even price most oil exporters need to balance their national budgets, let alone fund vanity projects or arms purchases. Countries such as Bahrain and Oman faced adverse fiscal circumstances and adopted austerity measures.

In poorer Arab countries, there were concerns that the pandemic could overwhelm their precarious health systems but also that containment measures would amplify their economic woes. Jordan and Lebanon introduced severe measures to avoid being hit hard. These countries seemed at first to have managed the early phase of the crisis well, but fatigue among the populations and the growing pain associated with slow economic activity eroded governments' capacity to manage the pandemic.

Conflict-ridden countries such as Syria, Libya and especially Yemen appeared particularly exposed to the pandemic. All three relied on international assistance, and their collapsed health systems could not cope with or even monitor the pandemic's spread. Security and logistical obstacles made any humanitarian response even more complicated. By early summer 2020, Yemen and Syria were believed to be struggling with large numbers of COVID-19 infections, but local authorities were unable to properly assess its spread, let alone contain it.

The New Geopolitics of the Red Sea
Why is interest in the region resurging?

The Red Sea has re-emerged as an arena of strategic importance. Long regarded as a mere junction between the Mediterranean and the Indian Ocean, falling between two continents, the region is increasingly recognised by policymakers and researchers as a distinct geostrategic space with a logic of its own.

The interests of eastern African states, projection of Gulf ambitions and growing presence of major powers are creating a complex and dynamic landscape. But 2020 opens an era of uncertainty for the Red Sea region. After a recent rush for influence by regional and extra-regional players, the COVID-19 pandemic and subsequent economic crisis could act as a reality test for new commitments.

The re-emergence of the Red Sea as an area of strategic importance

The Red Sea is a major maritime choke point for global trade and connectivity, located at the centre of a region marked by protracted conflicts and instability. Some 12% of the world's trade volume, and 8% of seaborne-traded petroleum and liquefied natural gas (LNG), pass through the Suez Canal and the Suez–Mediterranean (SUMED) oil pipeline. The Red Sea is also a point of convergence of major fibre-optic cables linking Asia to Europe.

During the Cold War, the Red Sea was an arena of great-power rivalry. The collapse of the Soviet Union and curbing of piracy in the early 2010s led to political disengagement from the area. But in recent years the Red Sea has made a comeback in regional and global strategic agendas.

In the past five years, Gulf Cooperation Council (GCC) countries have dramatically increased their economic engagement in the Horn of Africa. Viewing East African countries as booming consumer markets and untapped reserves of natural resources, they seek to project influence and secure good relations with friendly regimes in this volatile but strategic region. A flurry of investments by Gulf states in ports and logistic infrastructure on Red Sea shorelines has also reinforced their positions on major global trade routes at a time when China is deepening its economic influence through its Belt and Road Initiative.

There is also a military and strategic rationale for engagement with the Horn. Gulf states have established military bases to support their involvement in the conflict in Yemen and help protect their regional security interests. Sudanese troops play a critical role in supporting United Arab Emirates (UAE) forces in Libya and Yemen. As the rivalry between Turkey and Qatar on the one hand, and Saudi Arabia and the UAE on the other, has ramped up in recent years, this competition for influence has extended into the Red Sea arena.

On the other side of the waterway, African states have been far from passive observers of these shifting dynamics. African players have competed to attract financial and political support from external powers to advance their domestic agendas. Ethiopia and Sudan are going through important political transitions. The appointment of Ethiopian Prime Minister Abiy Ahmed in Ethiopia in 2018, the peace deals between Ethiopia, Eritrea, Djibouti and Somalia the same year, and the collapse in 2019 of Omar al-Bashir's long-standing regime in Sudan have generated optimism about prospects for a greater political opening-up of the region.

These complex regional dynamics have been compounded by the resurgence of great-power involvement. China and other Asian powers have established a security presence in and around the region, which they conceive as part of their broader Indo-Pacific security strategies. The opening of a Chinese base in Djibouti in 2017 prompted considerable concern in Washington DC. Russia has reportedly made discreet political enquiries about securing a base in Somalia or Sudan. These developments have pushed the US and Europe to shift their focus on maritime security from piracy and trafficking towards potential military competition.

Efforts towards greater regional cooperation

Competing objectives hinder effective regional coordination. Yet the Red Sea region is not short of common challenges. All actors recognise the importance of maintaining stability around this strategic choke point. In mid-2020, there were five major conflicts in the region – in the Sinai

Peninsula, Somalia, South Sudan, Sudan and Yemen – as well as multiple state rivalries within the GCC or between Egypt and Ethiopia over the Grand Ethiopian Renaissance Dam (GERD). While piracy in the Gulf of Aden (an acute security concern in the 2000s) has been effectively curbed, illegal migration, arms trafficking and the spillover effects of conflicts in Yemen and Sinai continued to threaten freedom of navigation in the area. Environmental protection also warrants greater cooperation between Red Sea coastal states. For example, a rusting oil tanker loaded with more than one million barrels of crude oil lies off Yemen's Red Sea coast, threatening to trigger a major environmental crisis for neighbouring states if not addressed rapidly.

To address those shared challenges, Red Sea countries have recently attempted to foster better coordination through the creation of task forces and regional platforms. In February 2019, the Intergovernmental Authority on Development (IGAD), a grouping of eight East African countries, established a task force for the Red Sea and the Gulf of Aden. The African Union High-Level Implementation Panel (AUHIP) also organised a series of consultations with regional stakeholders. In January 2020, Saudi Arabia launched the Council of the Red Sea and Gulf of Aden Countries. This initiative aims to bring together coastal states (Djibouti, Egypt, Eritrea, Jordan, Saudi Arabia, Somalia, Sudan and Yemen) to discuss common challenges and shared interests on topics ranging from economic cooperation to maritime security and environmental protection.

Countries in the region have also sought to increase their cooperation in the maritime-security field. In January 2020, Saudi Arabia hosted the first such drill between Red Sea littoral countries, a few days before conducting a bilateral maritime training exercise with Egypt in the Red Sea. The same month, UAE leader Mohammed bin Zayed and Saudi Arabia's Deputy Minister of Defence Khalid Bin Salman were present at the opening of Egypt's Berenice naval base, reportedly the largest on the Red Sea.

However, regional cooperation continues to face numerous challenges. Debates within the Council of the Red Sea and Gulf of Aden

Countries reflect power dynamics and sometimes diverging interests. The Council's focus on coastal states allows Saudi Arabia to further delegitimise the presence of external powers (especially Iran and Turkey), while for Egypt it serves to exclude one of its main regional rivals, Ethiopia, which lacks direct access to the sea. In turn, Ethiopia used IGAD's consultations on the Red Sea in early 2020 to counter its exclusion from the Saudi initiative and reaffirm its relevance to Red Sea matters. While IGAD and the African Union are wary of meddling by Middle Eastern countries, their internal divisions hamper their ability to enforce their agendas in the region and rebalance their asymmetric relationship with Gulf powers.

A mosaic of national interests and objectives

Not all actors are invested in the region to the same degree. For Egypt, Ethiopia, Saudi Arabia and the UAE, and to a lesser extent Qatar and Turkey, investing and positioning themselves in the area is a way to support their regional ambitions, project influence, and also to meet or outmatch the involvement of their respective rivals. By contrast, countries with domestic difficulties such as Eritrea, Somalia and Sudan have been much less proactive in the Red Sea, primarily focusing on their internal challenges and on attracting regional help.

Egypt's policy towards the region is informed primarily by its strategic interest in the Suez Canal and pivotal position between the Middle East and Africa. Cairo considers the Red Sea to be part of its sphere of influence and is concerned by the growing ambitions of Ethiopia and Saudi Arabia. Its multi-billion-dollar project to enlarge the Suez Canal in 2015, as well as the opening of the Berenice base, served as Cairo's bid to remain the main guarantor of international trade flows passing through this strategic choke point.

Growing Saudi interest in the region has been motivated by a desire to marginalise extra-regional powers such as Iran, Qatar and Turkey, and also by a gradual rebalancing of Saudi Arabia's economic focus away from the Persian Gulf towards its Red Sea coast. Recent or planned infrastructure megaprojects – such as the futuristic Red Sea city of NEOM, the ports and industrial zones of Yanbu and Jizan, and the East–West pipeline – indicate that Saudi Arabia is betting on the strategic relevance of the Red Sea corridor for the future of its economy. For Riyadh, however, the Red Sea is as much a vulnerability as an opportunity, as it does not control its two main choke points of Suez and the Bab el-Mandeb Strait.

Access to the sea is a driver of many of Ethiopia's recent foreign-policy moves. In 1991, the war with Eritrea cut its maritime access, forcing Ethiopia to redirect its trade routes through the port of Djibouti. Such a dependency is unacceptable to Ethiopia, which has sought to develop alternative routes through Eritrea, Kenya, Somaliland and Sudan, sometimes in cooperation with the UAE, but with limited success. While Ethiopia seeks to position itself as an important middle power and a primary interlocutor for Gulf states, this source of weakness, combined with domestic economic and political difficulties, hampers its ambitions.

Small states like Djibouti, as well as the regional government of Somaliland, have attempted to capitalise on regional and international interests to compensate for their vulnerability and attract investment and protection. Djibouti's entire economy and political influence rely on its port – the main entry point for Ethiopia's maritime trade – and the international bases it hosts. Maintaining international interest in the Red Sea is therefore a vital interest.

External powers such as the UAE, Turkey and, to a lesser extent, Qatar have also been keen to secure a foothold in the Red Sea. The UAE has been particularly proactive on this front. Its base in Assab, Eritrea, was critical for its intervention in Yemen. In the longer term, it also pursues an agenda distinct from that of its main regional ally, Saudi Arabia. The UAE is conscious of the vulnerability of its location inside the Persian Gulf. Its attempts to control ports along the Red Sea and Gulf of Aden aim to position it at the centre of regional and global maritime trade. For Turkey, whose engagement in the Horn of Africa is even older and deeper than the UAE's, provision of aid to the region aims to reinforce its narrative of regional power.

Countries facing domestic difficulties, such as Eritrea, Somalia and Sudan, have been less proactive. While they perceive Gulf interest in the region as an opportunity to secure much-needed financial and political support, especially in the case of sanctioned regimes such as Eritrea and Sudan, this support is also seen as a potential vulnerability. In Sudan, for example, Gulf interference during the political transition provoked outrage in the streets of Khartoum. Nonetheless, Saudi and Emirati financial support has been essential to help the transitional government stabilise the economy.

The Red Sea remains a source of vulnerability for these countries. Although piracy has been curbed, the conflict in Yemen has exacerbated arms trafficking in the region. Weapons bought on the Yemeni arms market, some from Saudi Arabia, can be sold for three to ten times their price in Djibouti, Eritrea, Ethiopia and Somalia, which feeds instability in these fragile countries.

Increasingly interlinked rivalries across the Red Sea

As the Red Sea becomes more important on domestic and regional agendas, power dynamics are increasingly entangled across both sides of the waterway. However, after three years of a regional rush for influence, ambitions have met with harsh realities, forcing several countries to reassess and adjust their plans. Although political transitions in Ethiopia and Sudan present new opportunities for external engagement, some African regimes in the region remain unstable, making it harder for Gulf countries to consolidate their influence. Regional rivalries have also inflated the importance of these countries or led to hazardous and short-sighted investments. These new uncertainties will reshape Red Sea dynamics.

The projection of Middle Eastern rivalries into the Somalia–Somaliland dispute has attracted international attention. While Qatar and Turkey maintained close ties with the Somali federal government, the UAE drove a wedge between itself and Mogadishu by engaging with autonomist states such as Somaliland and Puntland, infuriating Somalia's President Mohamed Abdullahi Mohamed (popularly known as 'Farmaajo').

More recently, however, Saudi Arabia and the UAE have made gestures towards Mogadishu. Saudi Arabia included Somalia in its Red Sea Council but excluded Somaliland. During the COVID-19 crisis, the UAE sent medical aid to Mogadishu. Abu Dhabi's September 2019 decision to abandon its project to construct a naval base in Berbera, Somaliland, due to persistent logistical and political difficulties, could also help to ease tensions. Talks between Somalia and Somaliland in June 2020 – held in Djibouti with Ethiopia's support – are reviving hopes for dialogue between the two parties. If successful, this could also contribute to a relative improvement of Abu Dhabi's relations with Mogadishu, although the Somali government remains suspicious of the UAE for now.

The same Middle Eastern fault lines have been projected onto Sudan. In 2019, Riyadh and Abu Dhabi leveraged their connections with Sudanese military leaders, with whom they had previously collaborated in Yemen, to influence the transition process. In May 2019, they hosted the head of Sudan's military council General Abdel Fattah al-Burhan in Abu Dhabi and pledged US$3bn in financial support. They also coordinated with the US and the United Kingdom to broker a deal between military leaders and civilian protesters in August 2019 and promised to back Sudan's removal from the United States' list of state sponsors of terrorism. While Turkey and Qatar have courted Khartoum, Sudan's new administration has been unwilling to reciprocate for fear of compromising ties with the UAE and Saudi Arabia. As a result, Turkey appears to have suspended ambitions to build a military base on the island of Suakin in northeastern Sudan.

As part of their search for strong regional partners, Saudi Arabia and the UAE positioned themselves as supporters of Ethiopian Prime Minister Abiy after his nomination in 2018 and backed his move towards peace with Eritrea. Ethiopia is, however, careful not to take sides in Middle Eastern rivalries.

The GERD dispute is likely to put the triangular relationship between Egypt, the Gulf states and Ethiopia under increased pressure. The Arab League, whose secretary-general is Egyptian, has supported

Egypt's position in demanding a permanent water-management agreement. The Gulf states feel obliged to support their long-standing Arab partner. As tensions over the dam mount, it is uncertain whether Saudi Arabia and the UAE will manage to maintain a balance in their policies towards Egypt and Ethiopia, both important partners.

What future for Red Sea dynamics?

The COVID-19 pandemic created new uncertainties for the future of the Red Sea region. Fragile Horn of Africa states faced immense economic pressure, with Egypt, Ethiopia and Kenya receiving emergency financing from the IMF. As Gulf economies adjust to its painful economic ramifications, their commitment to the Horn of Africa will be tested. Should oil prices remain low, Gulf countries will likely lose some of the economic leverage necessary to influence the political landscape in the Horn of Africa.

Emirati and Saudi investments in the region have often been driven by political objectives rather than economic interests. As a result, many projects have turned out to be unprofitable and unsustainable, while others failed to materialise. The failure of Gulf investments in Sudan's agriculture – often referred to as the 'breadbasket gamble' – are a striking example of these limitations. Even in Ethiopia, often described as a land of economic opportunities, a foreign-exchange shortage currently prevents investors from repatriating their profits, and the underdevelopment of the private sector continues to weigh on the economy.

On the security front, several planned military bases have been slow to materialise. Difficult negotiations over basing agreements, and the cost of a military presence in such a volatile region, have hampered many, such as the UAE's planned bases in Djibouti and Somaliland. Saudi Arabia's base project in Djibouti has also been slow to take off. As Abu Dhabi reduced its involvement in Yemen, it also started withdrawing troops from its base in Assab, Eritrea. After the transition in Sudan, Turkey looked certain to abandon its base project in Suakin. Even the importance of the Turkish base in Mogadishu has been overestimated. With only 200 soldiers, and an objective limited to training the Somali Armed Forces, it is unlikely to herald significant Turkish power projection in the near term.

Gulf states privately express frustration about navigating Horn politics and uncertainty about their influence over their partners in the region. In Sudan, for example, the massacre of protesters by the Rapid Support Forces of General Mohamed Hamdan Dagolo ('Hemeti') – Riyadh and Abu Dhabi's closest ally in the country – on 3 June 2019, a few days after he had visited both states, came at a heavy political and reputational cost for the Gulf leaders. Similarly, Saudi Arabia and the UAE were increasingly struggling to balance their relationship with Ethiopia and Egypt as tensions around the GERD project mounted. The Ethiopia–Eritrea rapprochement they had sponsored in 2018 is stalling; the Eritrean government has even publicly attacked Saudi Arabia for claiming ownership of the peace agreement.

After throwing their support behind Abiy, Gulf states could be forced to review their calculations in case the Ethiopian prime minister loses the upcoming elections (initially planned for August 2020 but postponed due to the COVID-19 pandemic).

The future of Gulf–Horn relations will therefore depend on the GCC states' appetite and ability to continue investing in their African neighbours, despite the difficulties and the global economic crisis, and on how Horn states manage to address their own domestic challenges and intra-regional rivalries.

The strategic relevance of the Red Sea for regional actors will also partly depend on how its relevance for global powers evolves. The reduction in piracy and the gradual cooling of the war in Yemen have temporarily reduced international interest in the region. But with the ramping up of the US–China rivalry, the Red Sea could again become an arena of great-power competition. China has been gradually expanding its military base in Djibouti since its establishment in 2017. This could push the US and India to reinforce their security presence and shore up the strategic relevance of the area.

Selected military bases in the Red Sea and Gulf of Aden

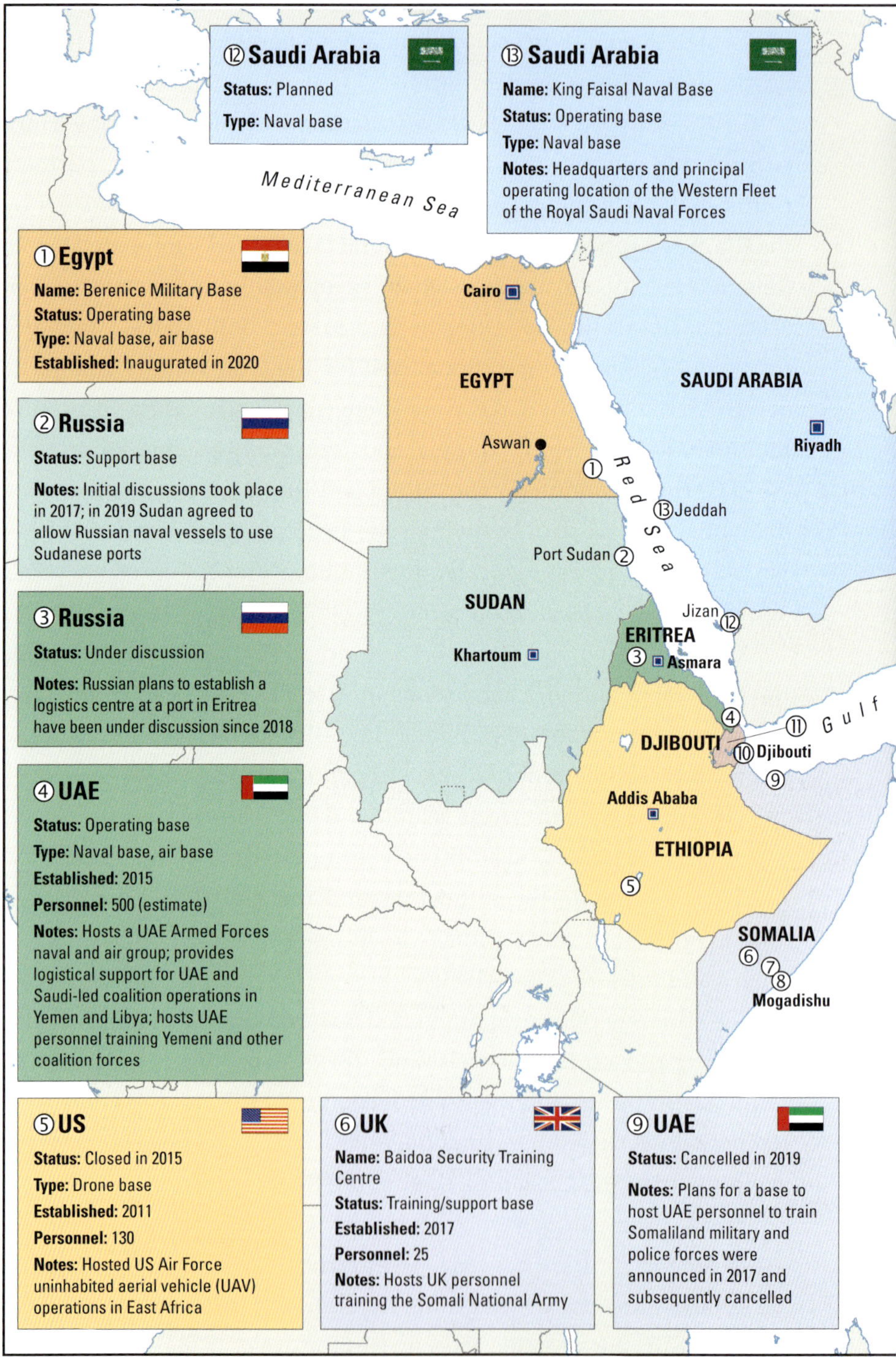

Sources: IISS; SIPRI

⑩ US

Names: Camp Lemonnier, Chabelley Airfield

Status: Operating base

Type: Air base

Established: 2001

Personnel: 4,000

Notes: Principal operating location of Combined Joint Task Force – Horn of Africa; hosts US Navy, Air Force and Special Operations Command air units on rotation for operations in the Horn of Africa, Gulf of Aden and Yemen; provides logistical support for other US Africa Command deployments in East Africa

⑩ Italy

Name: Italian Military Support Base (BMIS)

Status: Support base

Established: 2012

Personnel: 90

Notes: Provides logistical support for Italian national operations in the Horn of Africa and Gulf of Aden

⑩ France

Name: French Forces stationed in Djibouti (FFDj)

Status: Operating base

Type: Naval base, air base

Established: 1977

Personnel: 1,450

Also hosted: Germany, Spain, EUNAVFOR logistical support staff

Notes: Hosts French forces forward deployed in Djibouti, including an army combined-arms regiment and an air-force fighter detachment; hosts the German and Spanish air component of EUNAVFOR/*Operation Atalanta* counter-piracy mission; provides logistical support for French and European operations in the region

⑪ China

Name: PLA Support Base

Status: Operating base

Type: Naval base*

Established: 2017

Personnel: 500 (estimate)

Notes: Hosts a forward-deployed PLA Marine mechanised/special-forces company; provides berthing and logistical support for rotational PLA Navy counter-piracy task forces

⑩ Japan

Name: Japan Self-Defense Force Base in Djibouti

Status: Operating base

Type: Air base

Established: 2011

Personnel: 180

Other countries hosted: India (under discussion)

Notes: Hosts Japanese maritime-patrol aircraft conducting counter-piracy operations in the Gulf of Aden

UAE

Status: Cancelled in 2015

Notes: Intended to provide logistical support for UAE/Saudi-led coalition military operations in Yemen

Saudi Arabia

Status: Planned

Notes: In 2017 Djibouti approved a Saudi request to establish a military base in the country

⑦ US

Name: Baledogle Military Airfield

Status: Operating base

Type: Air base

Established: 2016

Personnel: 500**

Notes: Hosts US air operations in Somalia and US personnel training special-forces units of the Somali National Army

⑧ UAE

Status: Closed in 2018

Established: 2015

Personnel: 30–40

Notes: Hosted UAE personnel training the Somali National Army; closed following clashes with Somali security forces

⑧ Turkey

Name: Camp Turksom

Status: Training/support base

Established: 2017

Personnel: 200

Notes: Hosts Turkish personnel training the Somali National Army

* with heliport
** Figure represents total US deployment to Somalia, which also includes detachments to other parts of the country

Note: Locations of bases are not exact. Where location data is not provided, it is because this information is not available.

The Grand Ethiopian Renaissance Dam and Nile Geopolitics

Will Africa's huge new dam create a crisis?

On 30 March 2011, Ethiopia announced its plan to build the Grand Ethiopian Renaissance Dam (GERD), a massive hydroelectric dam on the Blue Nile tributary in the Ethiopian highlands close to the Sudanese border. Its purpose is to boost the country's economy by significantly increasing electricity production. GERD is the largest installation along the Nile since the Aswan Dam in Egypt was completed in 1970. With an expected capacity of 6.45 gigawatts, GERD will become the largest hydroelectric power plant in Africa and the seventh largest in the world.

The Millennium Reservoir that will serve the dam is designed to hold 74 billion cubic metres of water. The project will introduce a new era of Nile water usage that could affect allocations to the countries that share the resource, and it will upend policy enshrined in the 1959 Nile Waters Agreement (NWA) between downstream countries Egypt and Sudan, which apportioned water flows from the Nile between them and granted Egypt the lion's share. It has also created new tensions between Egypt, Ethiopia and Sudan (known as the 'Tripartite' in negotiations), the central issues being the effect of development of the river upstream and the effect of expected water shortages and reduced access to Nile water on downstream activities.

Egypt – the country furthest downstream and one of the most arid and water-stressed countries in the world – has long been defined as the Nile hegemon in the region. Since the end of the British–Egyptian mandate in Sudan in 1956, Egypt has both defined and designed Nile politics. Prior to the GERD announcement, Ethiopia rallied other Nile riparian states to join the Nile Basin Initiative (NBI), an intergovernmental partnership established in 1999 that looked to apportion access to the Nile and its waters more fairly among all states that share the river. In 2010, five of the upstream NBI member countries (Ethiopia, Kenya, Rwanda, Tanzania and Uganda) signed a water-cooperation agreement. Egypt and Sudan both refused to join.

The dam is the most overt challenge yet to Egypt's long-term hegemony over the river and has forced a shift in its foreign policy that reprioritises the Horn of Africa and neighbours after years of downgraded relations during the presidency of Hosni Mubarak. It has also become a symbol of Ethiopia's wider economic and political goals to enlarge its role within the Horn of Africa and redefine relationships with its neighbours. Egypt seeks to challenge these aspirations. It is not only resisting revision of the arrangements that govern Nile water allocation, it is also challenging power held throughout the Horn of Africa.

The result has been an increase in tensions among Nile riparian states over the past decade. This has coincided with domestic turmoil in all the Tripartite countries, which has diverted attention from the issue of water security. In Ethiopia, unrest has defined the decade, with ethnic and tribal tensions growing following the death of prime minister Meles Zenawi in 2012, through the turbulent years of Hailemariam Desalegn's leadership (2012–18) and into the tenure of Prime Minister Abiy Ahmed since 2018. Soon after Ethiopia announced it would build the dam, Egypt entered the throes of the revolution that ended Mubarak's decades-long rule. Egypt's 2013 counter-revolution ushered in military rule under President Abdel Fattah Al-Sisi, who was elected president in 2014.

In Sudan, the overthrow of Omar al-Bashir in April 2019 began a transition under a new civilian–military leadership that is balancing internal forces but coming under increased pressure. Furthermore, increased engagement with the Horn by global powers such as China has threatened the dominance of traditional Western allies, especially France, the United Kingdom and the United States. As the Gulf crisis between Saudi Arabia and the United Arab Emirates (UAE) on one hand, and Qatar on the other, begins to spill over into the Horn of Africa, the politics of the wider Middle East are also redefining relationships and priorities, directly inserting themselves into the Nile dispute.

Impact of GERD on the Tripartite countries

All three parties have remained relatively entrenched in their respective political positions regarding the Nile as the dispute has developed.

Each portrays the dispute domestically as an existential crisis, exacerbating the stalemate and rallying citizens to a protectionist and nationalist agenda.

Ethiopia

Ethiopia has been one of the fastest-growing economies in the world, with an average annual growth rate of 10% over the past decade. As Ethiopia seeks to become a middle-income country by 2025, GERD will not only help to meet its rising energy demands but will also provide a valuable commodity – power – to sell across the continent. GERD is a largely self-funded endeavour, owned by the Ethiopian Electric Power Corporation and supported mainly through government bonds and private donations. It has become a national symbol of Ethiopia's economic ambition and political prowess.

Egypt

GERD threatens Egypt's water security, with Egyptian officials arguing that it has the potential to reduce its access to water by up to 30% if the reservoir is filled within the shortest suggested time frame. Egypt has also alleged poor structural design and lack of consultation among riparian states and demanded managed water flows to adjust to and respond to prolonged drought periods. Since the 2011 announcement, Egypt has sought to engage water experts, technical assistance and international diplomacy in order to achieve a negotiated water-management agreement.

Sudan

GERD may have benefits and drawbacks for Sudan. In the long term it would give Sudan access to reduced and managed water flows, preventing the flooding of arable land that can then be cultivated, but water-flow releases could also risk significant flooding and raise questions over safety and related environmental issues. Sudan has had a security relationship with Egypt for decades (not always as an ally), but the Bashir presidency was just as heavily defined by fluctuating relations with Ethiopia. The transition that followed Bashir's ousting in 2019 remains fragile, as the country struggles with civilian–military relations, and ethnic and religious differences. Sudan's priority is to avoid conflict of any form and ensure that it is not caught in the middle of the dispute between Egypt and Ethiopia, both of which have direct interests in the country, with respect to Nile water access and the political system that emerges from transition.

Tripartite relations and negotiations, 2011–19

Tripartite relations have fluctuated since construction of the dam began in 2011. There was little multilateral engagement among the Tripartite in the early years, largely because of Egypt's domestic upheaval until Sisi was elected president in 2014. Although it has now relaxed its position, Ethiopia originally proposed to fill the reservoir in three years, which would have seriously harmed Egypt's domestic water access and usage. This set the tone for continued animosity in diplomatic negotiations. Until 2015, Egypt and Sudan (treaty-bound by the 1959 NWA to negotiate as a bloc with Egypt) were a united front in arguing that Ethiopia's unilateral decision to build the dam was not only illegal, but also a national-security threat to downstream states. Egypt found little sympathy among the international community in the early years of the dispute, although that has shifted in more recent times. Ethiopia has recently been portrayed by some international partners, as well as by Egypt, as a difficult negotiating party that refuses to acknowledge the implications of GERD for other states.

The 2015 Khartoum Agreement (or 'Declaration of Principles') between the Tripartite countries was hailed as a breakthrough in the dispute over the dam and the sharing of Nile waters. It committed the three parties to conduct technical studies and incorporate international experts to agree guidelines on filling (article 5), provided a loosely worded mediation mechanism (article 10) and required them to ensure 'no significant harm' (article 2). But although the declaration appeared to represent an agreed approach to Nile water management, a political stalemate ensued after technical disagreements and differing interpretations stalled the negotiations.

After the Khartoum Agreement was signed, Sudan's public position on Nile relations shifted from generally supporting Egypt's defence of its water allocation to allying with Ethiopia. It became more aware of the benefits of access to GERD-generated power and more balanced water flows to its flood-prone arable land. In 2016 Sudan's leadership even offered its apportionment of Nile waters (as defined in the 1959 NWA) to Ethiopia to contribute to filling the Millennium Reservoir, although this offer is no longer considered valid. Sudan's decision to abandon Egypt's GERD policy at the time, in favour of supporting Ethiopia, caused the biggest fracture between the two states in recent years.

The parties have built new alliances in the wider Horn region to support their positions. Under Sisi, Egypt has embarked on a flurry of diplomatic excursions to renew its role in Africa, building new political and security alliances with Eritrea, South Sudan and Uganda and exerting soft-power influence by funding several medical and educational facilities of smaller powers such as Burundi. Ethiopia has responded in kind by deepening its political and economic partnership with the UAE. Furthermore, Abiy has sought to serve as a peace broker between other Horn countries such as Kenya and Somalia, as well as becoming the designated mediator in the talks that resulted in the creation of the Sudanese Transitional National Council following the ouster of Bashir. Sudan has flip-flopped in its alliances with the other Tripartite powers, upsetting Egypt by signing security agreements with Turkey in 2017, as well as engaging in disputes over border issues between the two countries, internal policy towards the Muslim Brotherhood and security in Libya.

In 2017, EU Special Representative to the Horn of Africa Ambassador Alexander Rondos urged the Tripartite countries to broaden their negotiations to include security representatives. Considering the national-security defence raised by Egypt and Ethiopia and Egypt's new form of military rule, it was deemed reasonable to allow security officials to take part in negotiations. In practice, however, their involvement further stalled progress, as domestic turbulence in each country diverted attention from the wider water-security issue, delaying construction of the dam. The parties did not use this extra time to reach an agreement, however. Egypt in particular did not press the issue as urgently as it might have. Ethiopia's announcement in October 2019 that it had almost completed the construction of the dam and was ready to begin filling the Millennium Reservoir as early as summer 2020 focused minds on the issue and revived diplomacy.

The Washington and African Union processes, 2019–present

Ethiopia's October 2019 announcement came as a surprise to Egypt. In response, Egypt called for urgent talks after months of quiet, appealing to both the US administration and the World Bank to support mediation. Despite Ethiopia's long-standing defiance of such mediation (it believes the US and the World Bank to be more supportive of Egypt), Tripartite negotiations took place in Washington DC between November 2019 and February 2020. The US and World Bank were both observers, with the US hosting and facilitating the talks. Egypt and Ethiopia continued to trade antagonistic remarks. Ethiopia claimed that Egypt's allies had bullied it into participating, while Egypt accused Ethiopia of behaving obstructively and trying to sabotage progress.

The talks collapsed in February at the final stage over disagreement on a longer-term water-management agreement. Egypt demanded a permanent agreement with a fully transparent and binding arbitration mechanism, while Ethiopia would agree only on the fill schedule for the reservoir (around eight years) and refused to consider long-term commitments on drought considerations for water management. The new leadership in Sudan sought to mediate, defending both Egypt's right to use the Nile to meet its water-security needs and Ethiopia's right to construct the GERD and produce electricity to support its growing economy.

Egypt has appealed to the UN Security Council (UNSC) for urgent mediation, which is unlikely to occur as the UNSC has partners that have their own water-management issues (China and water allocation in the South China Sea, for example), but also because it tends not to interfere in non-military disputes. However, the appeal afforded some space for renewed urgent mediation on the issues, this time from the African Union (AU). The three heads of state – mediated by South African President Cyril Ramaphosa – met for the first time in over a year in the week of 23 June. Despite differing messages of success conveyed to their own citizens, Ethiopia did agree a stay of at least two weeks on its threats to fill the dam regardless of an agreement being in place. Egypt and Sudan both interpreted this publicly as a commitment not to fill the reservoir without an agreement. Although materially untrue, Egypt and Sudan both hope that continuing to show Ethiopia acting – in their terms – in bad faith may corner the country into making concessions to reach an agreement before the reservoir fill begins.

At an urgent meeting on GERD convened on 29 June, the UNSC confirmed that there was no appetite among the Tripartite to mediate or engage in diplomatic negotiations, dealing a blow to Egypt's efforts to elevate the issue and engage more international partners. The UNSC did commit to support the AU process and the Tripartite talks. Pushing the parties to reach an agreement also acted to diminish Ethiopia's political stance of attempting to fill the reservoir without a mutual agreement in place.

The key issues of water flow and control over the Nile have remained unchanged since GERD was announced. A water-management deal has always depended on the willingness of the parties to make concessions. For Egypt, this would require acknowledgement of its weak bargaining power as the country furthest downstream, as well as acceptance of the need to allocate Nile water more equitably and share in the benefits of developing the river. For Ethiopia, an agreement would require understanding that unilateral activity by upstream countries threatens the security of dependent downstream ones. Sudan is caught in the middle, trying to balance long-term economic development that could follow from GERD with the short-term support it can earn from Egypt, Gulf allies and international partners. The Ethiopia–Sudan relationship has fluctuated, with years of strong relations between Meles and Bashir upended by the secession of South Sudan and subsequent competing security interests between the two countries.

A looming 'water war'?

As the dispute has dragged on, Egypt has periodically hinted at its readiness to consider military action should the filling of the reservoir begin before a water-management agreement is signed. In 2013, then-president Muhammad Morsi was caught on videotape in a closed-door cabinet meeting threatening war with Ethiopia over this issue. Egyptian officials privately acknowledge that this sowed mistrust between the parties and set back negotiations. Nonetheless, Egypt has always maintained that 'all cards remain on the table, including military action'.

Under Sisi, Egypt has softened its position somewhat and retreated from any direct threats of military action. However, it continues to send covert messages of such threats, mainly through media messaging. Sisi has also expanded the Egyptian military and developed new maritime, army and air bases along its borders, including the recently opened Ras Banas base near the southern border with Sudan, opened in late 2019. In response to the collapse of the talks in Washington, Ethiopia began amassing large numbers of troops and military equipment around the site of the dam in March 2020 and near the Sudanese border, in apparent preparation for retaliation against an Egyptian military strike on the dam.

While military options remain a possibility, it is increasingly unlikely that Egypt will deploy them, largely because it would find it difficult to significantly change the realities on the ground with a military strike. The ability of limited airstrikes to cause significant damage to the dam structure is heavily in question: such a strike would be unlikely to damage the large and dense structure severely enough to halt the project. More generally, Egypt's military apparatus remains averse to direct confrontation as a legacy of the Arab–Israeli wars.

Furthermore, Egypt has no international or regional support from allies to undertake any military activity. Such an attack would invite severe international criticism. It might legally require US approval, based on foreign military funding guidelines on end-line monitoring of equipment. Notably, Sudan has continued to downplay the threat of any military action, not only because of the state of its fragile transition, but also because of genuine fears of domestic upheaval should an 'Arab–African' war become a material threat.

Egypt is more likely to seek to delay completion of the dam to create more time for negotiations than to attack it. To this end, Egypt could attempt to exacerbate the current domestic crisis within Ethiopia through support for opposition challengers to Abiy, thereby aiming to avert Ethiopian domestic attention away from the dam.

The future of the Horn: domestic and international challenges

Ethiopia has faced a multitude of problems since Abiy came to power. The cancellation of the 2020 elections set off a series of events that has left the country on the brink of civil conflict. Internal divisions in the reshaped ruling party, the Ethiopian People's Revolutionary Democratic Front (EPRDF – reformed as the Prosperity Party), and the broader ongoing power struggle between the ruling party and the Tigray regional leaders, the Tigray People's Liberation Front (TPLF), has been brewing for years. Abiy also faces a possible rebellion in the Oromia Region. Ethiopia's domestic struggles will shape the way it projects power in the region. Indeed, it is the general public's staunch support for GERD, and the need for the prime minister to curry political favour domestically, that sparked the government's drive to complete the dam. GERD is the brainchild of the TPLF and – even with Abiy's lukewarm approach to the national project – he has been unable to wrest power from the TPLF by raising nationalist rhetoric towards the dam.

Sudan is arguably more threatened than other countries by continued insecurity in the Horn region. The benefits for Sudan of more efficient management of the Nile by GERD could be significant if it develops the political will to effectively build its agricultural sector, including through partnership with more experienced partners like Egypt, to help it realise the value of GERD. On the other hand, concerns have been raised over the years about the structure of the dam. A leak or breach would threaten Sudan with severe flooding that could reach Khartoum.

Egypt has been forced to make significant concessions in the rewriting of Nile politics and to accept that the reality on the ground vis-à-vis water allocation has moved faster than diplomatic negotiations were able to shape or control. The completion of GERD is no longer in question. Barring attempts to destabilise Ethiopia domestically, the dam will be ready to begin filling in 2020. The reality of GERD completion will likely force Egypt's eventual retreat from the battle to be the outright Horn hegemon, though it will likely impel it to work harder to infiltrate domestic Horn developments, causing further instability for countries that have threatened its interests. Military conflict along the Nile is unlikely. Some experts predicted better-than-expected rainfall during the 2020 rainy season, meaning that the GERD could begin to fill without any material effect on Egypt's water access for the time being. Nevertheless, urgency around the need to agree a deal will no doubt remain – both for GERD and for other envisaged projects.

The politics of the water dispute have escalated far beyond the three main riparian states. China's fast-expanding footprint amid a confused and largely absent US policy across the Horn of Africa, coupled with Russia's incursion in the Libya conflict, have created a new terrain for competition among superpowers. The Gulf crisis spillover into the Horn has added new security threats that are likely to affect Eritrea, Ethiopia and Sudan. Furthermore, Egypt faces security crises of historical proportions along all its borders, the most serious of them in Libya. Turkey's movement into Libya threatens Egypt's security interests along the Mediterranean coastline and into the Red Sea, as well as the security of its western land borders with Libya and its southern border with Sudan.

Amid these domestic issues surrounding the Nile and the wider international issues affecting the region, geopolitics will continue to evolve rapidly – regardless of whether water agreements are signed, or if relations among the Tripartite are repaired. The multitude of relationships, foreign engagement and shifting regional ties leaves the region in disarray, with few signs of tensions abating. Domestic unrest, and wider regional and great-power involvement, will only complicate politics across the Horn region in the coming years.

Grand Ethiopian Renaissance Dam and Tripartite states in Nile waters negotiations

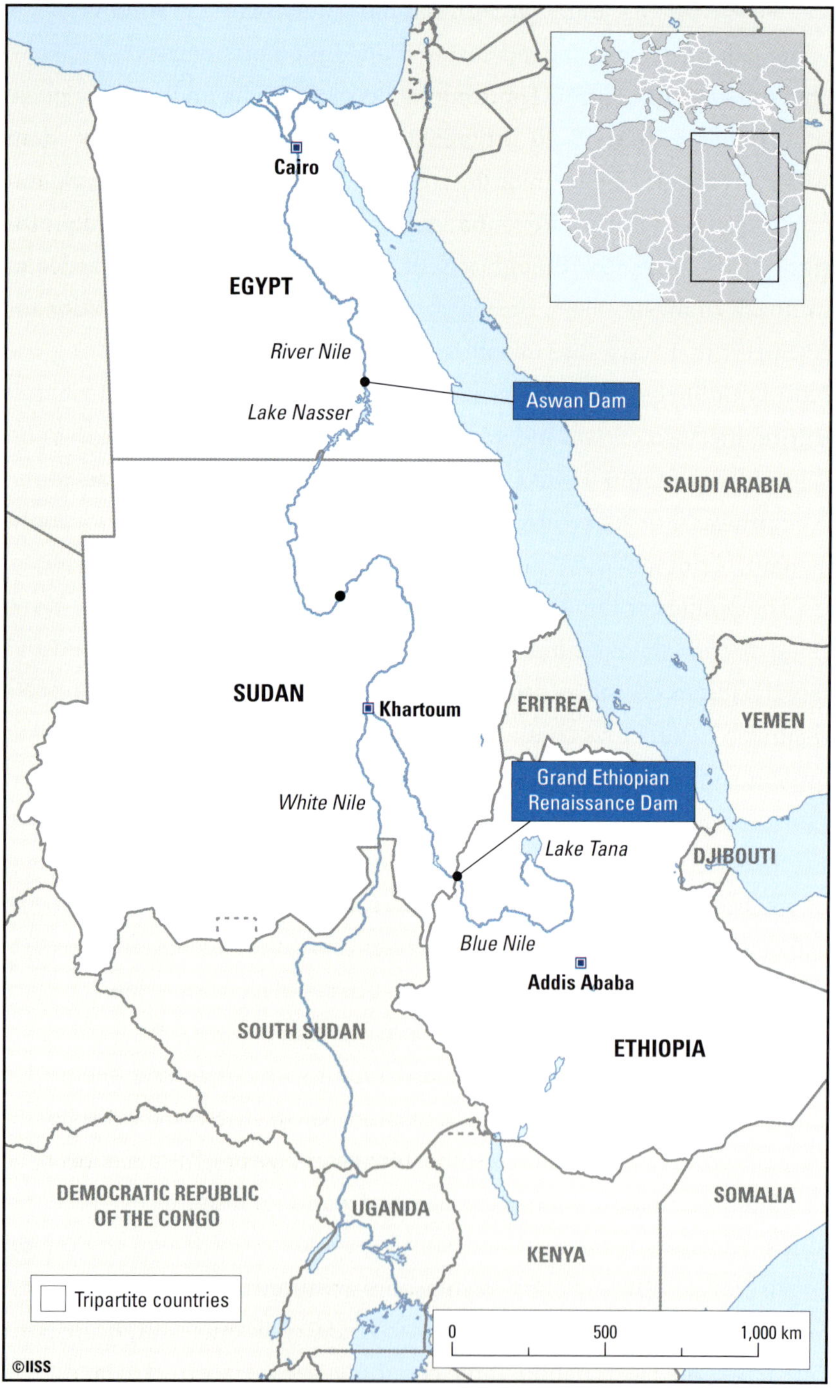

Turkey's Force Generation Abroad
How and why is Erdogan projecting power?

In a speech in early 2016, Turkish President Recep Tayyip Erdogan set out a proactive approach to security policy that broke with the country's past military restraint. To counter the separatist threat of Kurdistan Workers' Party (PKK) operations from bases in northern Iraq, and to contain the growth of the People's Protection Units (YPG) – the PKK's Syrian affiliate and a key partner of the United States-led coalition fighting the Islamic State, also known as ISIS or ISIL – Erdogan announced that:

> After today we won't wait for border problems to break out. We won't wait until the last minute, until we fully sink into a morass. From now on, we will confront problems. Is there a problem with terror? We won't wait for terror outfits to strike us. Wherever they are, we will go find them and hit them hard.

In the next three years, Turkey mounted three major cross-border ground interventions in Syria:

- *Operation Euphrates Shield* in northeast Syria to help its Syrian allies capture territory from ISIS and pre-empt a YPG attempt to do so (August 2016–March 2017)
- *Operation Olive Branch* in northwest Syria to seize the YPG-held pocket of Afrin (January–March 2018)
- *Operation Peace Spring* to control a vast strip of land along the Turkey–Syria border from Tel Abyad to Ras Al-Ayn and expel YPG fighters and their allies in the US-backed Syrian Democratic Forces (October–November 2019). This has been the most ambitious operation under Erdogan so far.

In May 2019, Turkey also began conducting *Operation Claw*, a series of aerial and ground operations against PKK targets and bases in northern Iraq to counter its mobility and operations across Iraq, Turkey and Iran.

But Turkey's military operations are no longer confined to the traditional goal of countering the PKK. They have widened to include actions against state actors in support of broader strategic ambitions. By July 2020, Turkey was conducting:

- *Operation Spring Shield* in the Idlib region in northwest Syria to support its Syrian rebel allies against an Assad-regime campaign backed by Russia
- an operation in Libya in support of the Tripoli-based Government of National Accord (GNA) against a coalition of Libyan forces backed by Egypt, Russia and the United Arab Emirates (UAE)
- aerial and naval patrols in North Africa and the Eastern Mediterranean for limited anti-access/area-denial (A2/AD) purposes and to protect its own hydrocarbon seismic research and deep-sea drilling efforts.

The new 'Erdogan Doctrine' holds that Turkey must adopt a proactive and preventive military approach to external security challenges. Why has Turkey developed this more muscular approach? What capacities does it involve? What are the benefits and risks to Turkey?

Until the 2010s, Ankara used only limited military force to manage a complex and multi-threat environment. Its main priority was the four-decade conflict with the PKK. Diplomacy and deterrence were used to freeze rivalry in the Eastern Mediterranean and on the Cyprus issue. This reluctance to generate and project force abroad began to change in the summer of 2015, when a breakdown in the two-year-long negotiation process with the PKK led to the re-initiation of full military clashes, first inside Turkey and then in the immediate neighbourhood.

Turkey's embrace of muscular methods is rooted in profound transformations in its external environment and domestic dynamics. Externally, Ankara's threat perceptions shifted east and south during the previous decade owing to growing security risks in the Eastern Mediterranean, Iraq, North Africa and Syria, and to strategic competition with Egypt,

Russia, Saudi Arabia, the UAE and other regional powerhouses. Ankara was particularly unnerved by NATO's passivity on its southern flank during the Syria crisis, which contributed to a security vacuum there. In relying on the YPG, the PKK's Syrian franchise, to counter ISIS, Western countries ignored or dismissed Turkey's well-known concerns. Finally, there is a pervasive and enduring sense among the Turkish ruling elite that NATO failed to adequately support Ankara during and after the failed coup of July 2016.

Domestic factors have also driven the growing militarisation of policy. Firstly, since the executive-presidency system, which granted the president significantly greater powers, took effect in June 2018, foreign policy has become a crucial plank of Ankara's political agenda. Military actions abroad enjoy strong popular support and help sustain Erdogan's popularity. His embrace of a more nationalist political discourse at home has helped consolidate his ad hoc coalition with the ultra-nationalist National Movement Party (MHP). Secondly, sending the Turkish military abroad is popular with the armed forces themselves. It boosts morale and motivation through extra wages and opportunities for promotion, and it provides valuable experience in joint-force operations.

A third domestic driver of Turkey's more militaristic approach is its role in harmonising relations between military and civilian elites. These elites agree on the importance of equipping Turkey with new military capabilities and strengthening its national-defence industry. The Turkish Armed Forces (TAF), more concerned with the technical dimensions of this consensus, is pursuing a transformation and restructuring process with its 'TAF 2033 Vision'. Civilian politicians want to use this new capacity and energy in domestic and foreign policy. They also hope that keeping the military busy abroad will make civilian control easier because it will divert the generals' attention to external rather than domestic affairs.

Finally, the boom in the Turkish defence industry enables Ankara to pursue a more independent strategy and allows it to display its defence systems for the purpose of international marketing. As a result of these domestic and external factors, Turkey has sought greater autonomy in its relations with regional actors and has challenged constraints it had long observed in its relations with the US and NATO. This has strained Turkey's transatlantic ties, as demonstrated by the row over its acquisition of the S-400 air-defence system from Russia, its intervention in northeast Syria in 2019 and its involvement in the Libyan conflict.

Turkey's military activism

Turkey's military activism manifests itself in two ways: an expanding military footprint and wider military cooperation. Turkey's military presence in the Middle East is now extensive. In Libya, since December 2019 Turkey has engaged in train-and-equip programmes, proxy warfare and direct operations. In early February 2020, Turkey established an operations centre, headed by a Turkish army general and comprising officials from the army, navy and air force, with the aim of coordinating Tripoli's air and ground defences. As of July 2020, there were around 350 Turkish military personnel in Tripoli as well as intelligence elements under the command of a major-general from the Turkish General Staff. Turkish liaison teams from the National Intelligence Organization coordinated around 3,000 Syrian National Army (SNA) fighters who were transferred from Turkish-controlled areas in northern Syria to Libya to reinforce the defence of Tripoli. The Turkish Navy's *Goksu* and *Gokova* frigates have also provided air and naval protection to the forces of Turkey's local partner, the Tripoli-based GNA, off the coastline between Tripoli and Sabratha since early February 2020, for limited A2/AD purposes.

Libya's geographical location and the capabilities of Turkey's regional adversaries – especially Egypt's and the UAE's air forces and Egypt's naval elements – make for a complex and risky environment for Turkey and the GNA in their fight against the General Khalifa Haftar-led forces controlling the eastern and southern parts of Libya. But after the involvement of Turkish military and intelligence elements and transfer of SNA fighters in early 2020, the GNA and its local allies secured Tripoli and even developed offensive capabilities that led to operational achievements. In late March 2020, GNA forces launched a counter-offensive, *Operation Peace Storm*, to repel Haftar's forces, drawing on Turkish military support.

In September 2017, Turkey opened Camp TURKSOM in Somalia, south of Mogadishu. It has used this for partner-capacity building, the assistance of and involvement in domestic-security operations, and humanitarian assistance. It also runs a military academy with 300 Turkish military personnel. As of June 2020, 10,000 Somali military personnel had been trained there. The camp has played a significant role in renewing the Somali military. In Qatar, Turkey has assisted with security, building alliances and power projection and beefing up its presence since the 2017 crisis that pitted Saudi Arabia and the UAE against the small emirate. Turkey's Khalid Ibn Walid military base is on course to expand by the end of 2020, with the addition of air and naval elements. As a result, the Turkish contingent will grow into a brigade task force of about 2,000 soldiers, including army, air-force, navy and special-forces elements. The base will become the second-biggest foreign military base in Qatar after the United States' Al Udeid base.

In Iraq, Turkey has engaged in counter-terrorism operations; intelligence, surveillance, target acquisition and reconnaissance (ISTAR); and train-and-equip missions. In northern Iraq, the Turkish military has around ten liaison offices run by special-forces elements, operating in liaison with the Iraqi Kurds and engaging in intelligence gathering. As of July 2020, Turkey had deployed around 20,000 soldiers in four zones in northern Syria: the *Euphrates Shield* pocket (the Jarablus–al-Bab–al-Rai triangle), the Afrin region, the Tel Abyad–Ras al-Ayn–M4 highway region and Idlib province. Turkey's main enemy in the first three zones was the YPG, while in Idlib it faced the Assad regime, its Iranian and Russian allies, and associated militias.

In the rebel stronghold of Idlib, Turkey faced a major dilemma after its March 2020 deal with Russia led to a fragile lull. Turkey's priority was to avoid a new influx of refugees across its southern border. It appeared stuck between the demands of Moscow and those of Syrian factions and residents in Idlib, where radical jihadi groups held sway. The Assad regime was intent on capturing this territory while displacing many of its residents and had used the ceasefire to consolidate its positions and replenish its ranks. For its part, Turkey reinforced its troops and positions in Idlib and assisted its rebel allies in their attempt to isolate or moderate jihadi groups. Turkey was also conducting joint patrols with Russian forces along the front line. But a breakdown of the ceasefire and a return to fighting were likely. Ankara had to choose between difficult options, from outmatching any escalation to abandoning its presence there. Its decision will impact the security rapprochement in Syria and the Eastern Mediterranean that Turkey has developed over the past four years, culminating in its procurement of the Russian S-400 mid-range air-defence systems.

In the Eastern Mediterranean, Turkish naval activity has focused on securing gas fields in contested waters, inflaming regional rivalry and poisoning intra-NATO politics. During exploratory expeditions, Turkey's deep-sea drill ships are always escorted by naval elements. In February 2018, for instance, Turkish frigates stopped a drill ship operated by the Italian company Eni from accessing Cyprus's claimed Block 3, which overlaps with Northern Cyprus's claimed Area F, where the Turkish Cypriots have granted a licence to Turkey's main exploration company, TPAO. It did so on the pretext that live-fire exercises were under way in the area.

In January 2020, Ankara announced it would send naval escorts for seismic-research ships to the Turkey–Libya maritime zone to the southeast of Crete, where Greece claims an exclusive economic zone (EEZ). It is therefore likely that the escalation of a naval military presence will spread to the waters around Crete in late 2020. To assert its claims in the EEZ, Turkey has deployed a significant naval force of two submarines and 14 warships in the Eastern Mediterranean since the summer of 2016. As of July 2020, almost half the Turkish Navy inventory was deployed in the south Aegean and Eastern Mediterranean. But the need to share ports with civilian vessels limits the resupply, maintenance, repairs and support these ports can provide. The nearest main naval base is at Aksaz, on the southeast coast of the Aegean Sea. Ankara is considering establishing such a base in Northern Cyprus, which would provide significant facilities for the Turkish Navy and reduce the time required for Turkish warships to travel to operational zones. Turkey is also considering reactivating a disused civilian airport in Gecitkale, Northern Cyprus, for the use of F-16 fighters. This already serves as the main air base for Turkish military activities in the Mediterranean with the transfer of TB2 uninhabited aerial vehicles (UAVs) for ISTAR tasks.

Turkey has also invested in vigorous defence diplomacy. In Azerbaijan, it has military bases for alliance-building and train-and-equip missions. In late 2017 it signed a military-cooperation agreement with Sudan. Reports that Turkey planned to build a military base and naval facility in Suakin island in the Red Sea alarmed the Saudi and Egyptian governments. This would have been the third Turkish foreign military base in the Red Sea/Gulf region after those in Qatar and Somalia. But following the ouster of president Omar al-Bashir in 2019, the new administration in Khartoum, which maintains close ties to Saudi Arabia and the UAE, has apparently rejected such plans.

Turkey has also internationalised its military activities through cooperation. The TAF has signed cooperation agreements with 68 countries and is negotiating agreements with 41 more. Turkey trains a growing number of foreign-military personnel and has significantly increased its number of military attachés overseas. In addition to these deployments, Turkey continues to take part in multilateral missions. It maintains 600 soldiers as part of NATO's *Operation Resolute Support* mission in Afghanistan, training, advising and assisting the Afghan National Defense and Security Forces. In Lebanon, a naval element serves the United Nations Interim Force.

Turkey's military role and its defence industry

The Erdogan government believes that Turkey cannot become a major regional power without military deterrence and capabilities. But it does not feel that it can trust weapons suppliers from traditional Western allies because their systems come to Turkey with planning, implementation and geographical restrictions. The desire to overcome these constraints has prompted the development of a vibrant Turkish arms industry, both for national use and for export. The Erdogan government has offered major incentives to the defence industry in an effort to develop indigenous production, national identity and strategic autonomy.

Countries in Central, East and South Asia, and in the Middle East, have shown interest in Turkey's defence industry. During his visit to Ankara in October 2017, Pakistani Prime Minister Shahid Khaqan Abbasi personally participated in a test flight of a T129 attack helicopter manufactured by Turkish Aerospace Industries. The following year Pakistan and Turkey negotiated a deal worth US$1.6 billion for the former to replace its worn-out Bell AH-1F and AH-1S *Cobra* attack helicopters with 30 T129Bs, though the proposed sale had not been completed as of June 2020 as Turkey had failed to secure the necessary export licence from the US. In November 2019, Turkish defence companies participated in the Tri-Service Asian Defense and Security Exhibition in Bangkok, touting T129s, frigates developed under the MILGEM warship-development programme, armed and unarmed UAVs, *Hurkus* training aircraft and armoured fighting vehicles. Drawing on its own experience, Turkey has developed weapons and systems for counter-terrorism and counter-insurgency operations. These are of interest to other countries facing similar challenges. There is particularly strong interest in Turkey's devices to counter roadside bombs, such as jammers and compressors, as well as its armed and unarmed UAVs, border-surveillance systems, and armoured utility vehicles and personnel carriers. Ankara appears to be pleased with the marketing success stories and sales of Turkish defence products abroad. So far, it judges that the growth of its defence industry is a strategic success.

The pitfalls of Turkey's ambitious power projection abroad

Ankara's force-generation and force-projection ambitions, and its defence procurement, carry risks and limitations as well as opportunities. Firstly, Turkey's force-projection capacity is incomplete. It is essential to provide close air support to ground forces with combat aircraft, attack helicopters and armed UAVs. During the al-Bab siege during *Operation Euphrates Shield* in late 2016 and early 2017, and in Tripoli in 2020, the lack of close air support was an operational weakness. Turkey also lacks the capacity to airlift battalion-sized units and heavy military equipment. The absence of a strategic airlifter such as a C-17 *Globemaster* or Ilyushin Il-76 *Candid* in the Turkish Air Force inventory is becoming a serious liability.

Secondly, Turkey lacks a medium- and long-range air-defence umbrella. On 27 February 2020, an airstrike on a Turkish commando battalion in the south of Idlib killed 36 Turkish soldiers. This showed the risks of deploying Turkish army units on the ground without effective medium- and long-range air defence. This liability is visible in Libya as well. Turkey's inability to impose air supremacy through fighters or missile systems is another operational weakness. Ankara seeks to fill this gap with TB2 *Bayraktar* and *Anka* armed UAVs, but this is not sufficient to create a deterrent effect.

Thirdly, Turkey's ambitions are a drain on a troubled economy. For instance, new orders in the Turkish defence industry in 2019 totalled US$10.7bn, a decrease of more than 12% from the previous year. The drop, which came amid domestic economic turmoil, ongoing quality and trust problems in the industry, and various crises stemming from Ankara's foreign policy, shows that Turkish defence-industry firms are struggling to find foreign markets, and that the sector has a structural export problem. The sector now supplies 90% of domestic requirements. Without increasing foreign sales, Turkey will struggle to sustain its expansion.

Power projection abroad may help promote the defence industry, but it is costly. A knowledgeable security official interviewed in Ankara in May 2020 suggested that Turkey's military activities in northern Syria and northern Iraq cost around US$200 million and US$120m per month respectively, and that the cost of the two-month-long *Operation Peace Spring* (October–November 2019) was around US$1bn. He assessed the total annual cost of Turkish military activism at around US$8bn, on top of the US$22bn 2020 defence budget. Total defence outlay of US$30bn constitutes around 3.6% of GDP.

Conclusion

Several domestic and external factors are driving Turkey's growing force generation and projection. These include the country's gradual drift away from the Western security bloc; the role of force projection in bringing together military and civilian elites; the domestic popularity of military actions abroad, especially ahead of the 2023 presidential elections; and the need to advertise Turkish defence-industry products in international markets. Erdogan also justifies his executive presidency as enabling Turkey to conduct a proactive and assertive foreign policy.

More immediate factors amplify this. These include conflict with the PKK and affiliated armed groups in Iraq and Syria; the need to protect the maritime deal signed with Libya's GNA; the need to maintain gunboat diplomacy in the Eastern Mediterranean and North Africa to deter the anti-Turkish alliances in the region between the Republic of Cyprus, Greece, Egypt and the UAE; and the drive to open new defence/security and energy markets in Africa and South Asia. Turkey's more assertive military posture, a striking development over the past four years, is therefore likely to develop further.

CHAPTER 10

Sub-Saharan Africa

SUB-SAHARAN AFRICA

Drivers of Strategic Change

REGIONAL SHARE OF GLOBAL POPULATION, GDP AND DEFENCE BUDGET

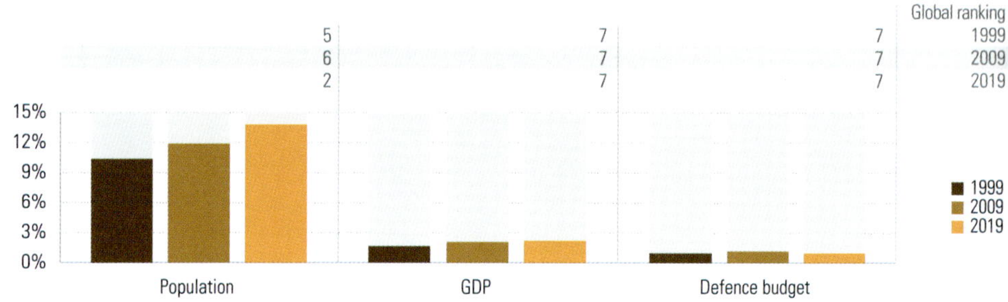

POPULATION

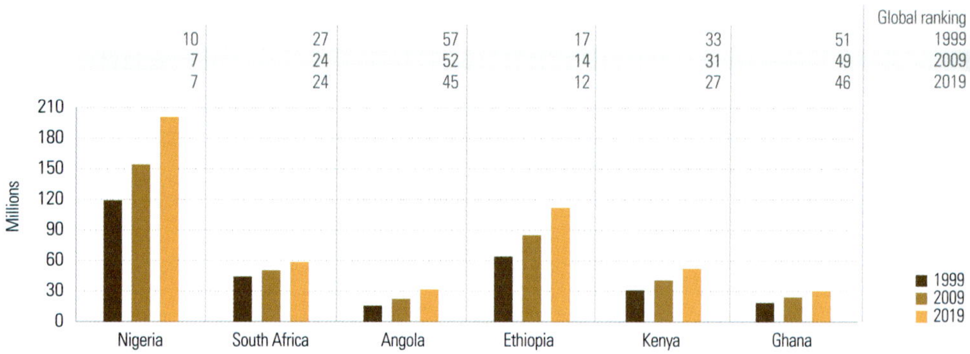

AGE STRUCTURE

(Percentage of national population)

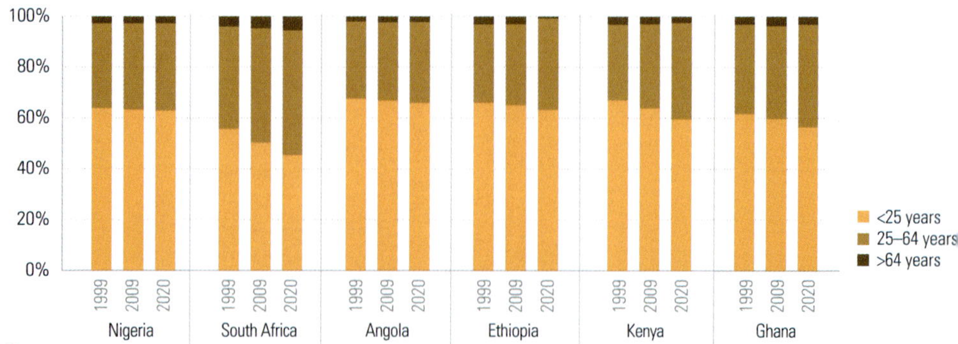

Median age	Nigeria	South Africa	Angola	Ethiopia	Kenya	Ghana
2000	17.9	22.6	16.3	16.6	17	18.8
2010	17.9	25	16.4	17.5	18.1	19.9
2020	18.1	27.6	16.7	19.8	20.1	21.1

GDP

(Constant 2010 US dollars)

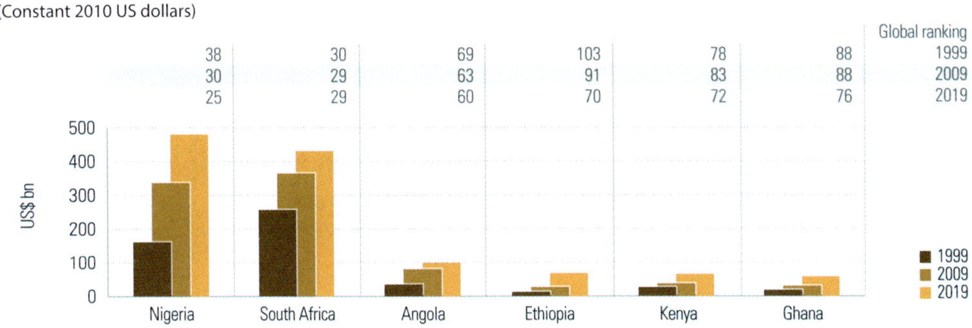

GDP PER CAPITA

(Constant 2010 US dollars)

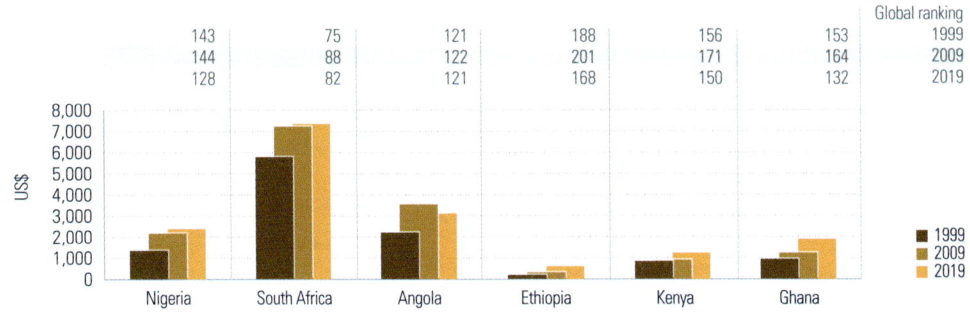

DEFENCE BUDGET

(Constant 2015 US dollars)

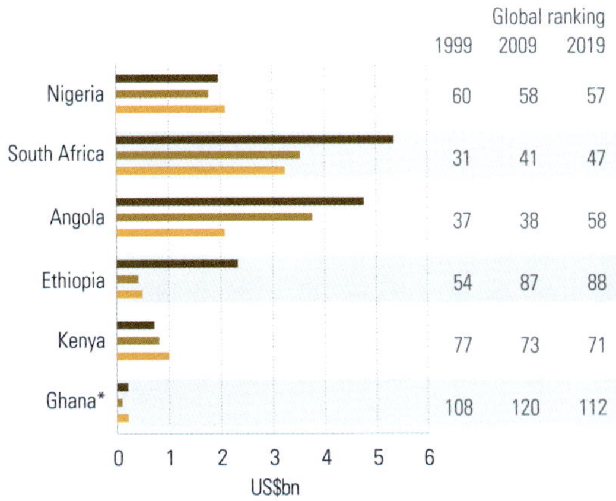

ACTIVE MILITARY PERSONNEL

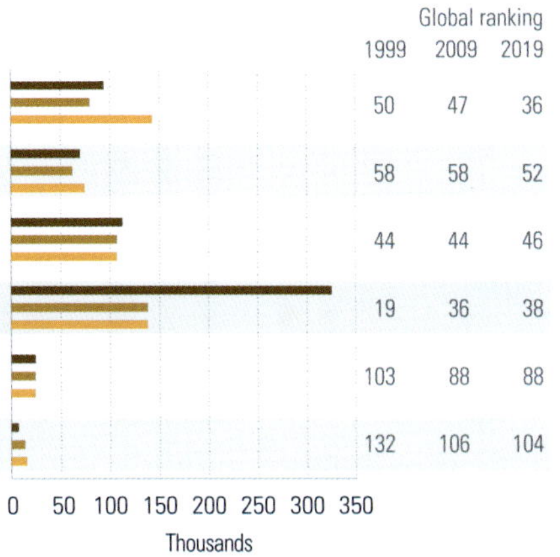

* 1999 defence budget value for Ghana is an estimate, and may be distorted by high inflation rates

■ 1999 ■ 2009 ■ 2019

For explanation of drivers and sources, see page 7

HUMAN DEVELOPMENT INDEX (HDI)

(Score between 0 and 1, where 0 denotes a low level of development and 1 a high level of development)

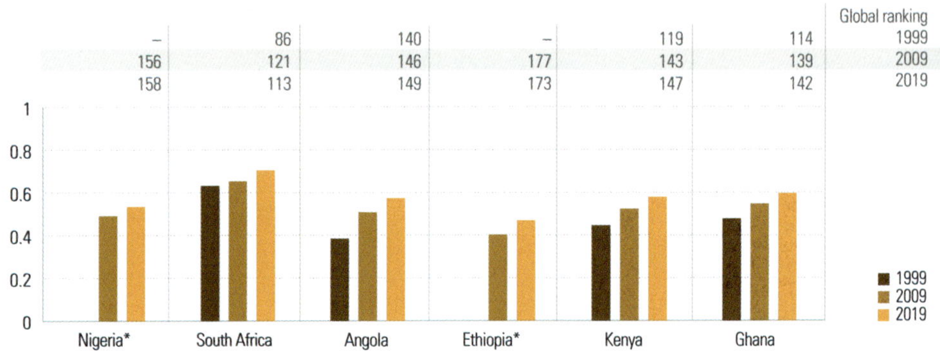

* No 1999 data available for Ethiopia or Nigeria

POLITICAL SYSTEM

(Score between 0 and 100, where 0 denotes no political freedom and 100 fully free)

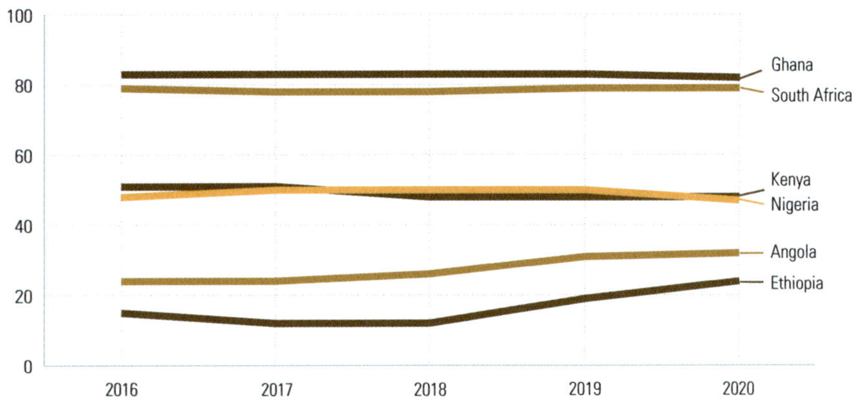

PERCENTAGE OF CHILDREN IN EDUCATION (GROSS)*

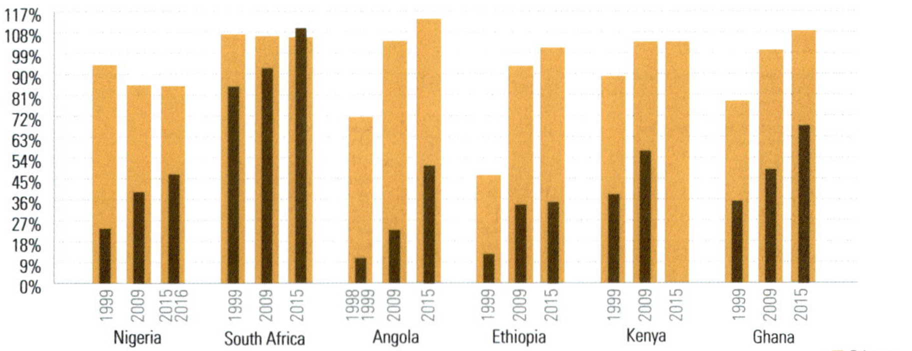

Primary data for Angola is 1998 and for Nigeria 2016. Secondary data for Kenya 2015 not available

* The data shows gross enrolment, which can exceed the population of the age group that officially corresponds to the level of education because of repeaters or late enrollers. Therefore some values are greater than 100%

For explanation of drivers and sources, see page 7

2019–20 Review

Around a dozen sub-Saharan African states held national elections between July 2019 and June 2020. Incumbents performed well in almost all cases, though this did not necessarily reflect their overwhelming popularity. In Cameroon and Guinea, for example, there were full or partial opposition boycotts of polls. Several other long-standing presidents and ruling parties made the most of the advantages of incumbency, further hindering underfunded and poorly organised opposition groups.

Two elections stood out. In Guinea-Bissau, José Mário Vaz, president since June 2014 but standing as an independent candidate, secured just 13% of the vote in the first round of presidential polls in November 2019 and was forced to drop out. The eventual victor was Umaro Sissoco Embaló, who won 54% in the second round, despite securing only 28% in the first – well behind Domingos Simões Pereira of the ruling Partido Africano para a Independência da Guiné e Cabo Verde (PAIGC), who had polled 40%.

Guinea-Bissau has been suffering a political crisis since 2015, in part due to a power struggle between Vaz and members of the PAIGC, which the elections have not resolved. Embaló's February appointment of one of his political allies, Nuno Gomes Nabiam, as prime minister provoked strong opposition from the PAIGC, which in turn triggered military intervention in support of President Embaló. Army units occupied government buildings and the national radio and television stations. Embaló shored up his position, gaining recognition of his election victory from the Economic Community of West African States, the African Union, the European Union and the United Nations. However, he lacked a strong legislative majority and planned fresh elections in the second half of 2020 – though these were at risk of postponement owing to the COVID-19 pandemic. Embaló's decision to bring the military back into politics is potentially troubling in a state that suffered a coup during elections as recently as 2012, and which only returned to civilian rule in 2014.

Ethiopia may also see political stability weaken as the result of a vote – and of the postponement of another. The Sidama Zone in the south overwhelmingly voted in favour of autonomy in November 2019 and was due to become the tenth state of federal Ethiopia. More than a dozen other ethnic groups are thought to be considering demanding similar votes. This could make the federal structure unwieldy and exacerbate growing ethno-nationalist tensions.

In February, Ethiopia announced that the 2020 parliamentary election would be delayed from May to late August because of COVID-19. While opposition parties such as the Oromo Liberation Front and the National Movement of Amhara supported this delay, other groups, such as the Oromo Federalist Congress, did not. Resultant protests could add to security concerns. Moreover, there is a risk that public expectations of the election are unreasonably high. Ethiopian elections traditionally saw the then-ruling Ethiopian People's Revolutionary Democratic Front (EPRDF) win overwhelming majorities. However, the holding of freer and fairer elections is unlikely to result in a new political dispensation, since years of democratic constraints mean that opposition groups lack the funding and organisation capable of attracting national support. The newly formed Ethiopian Prosperity Party, comprising three of the four parties in the EPRDF, together with five regional parties, is thus likely to win a substantial majority.

COVID-19 presents a threat – and an opportunity – for China

China persisted with its long-standing dual approach to sub-Saharan Africa: to protect its economic and security interests while expanding its influence in the region. But it faced setbacks arising from the COVID-19 pandemic. With the next triennial Forum on China–Africa Cooperation (FOCAC) scheduled for 2021, China focused on building on the pledge made by President Xi Jinping at the 2018 FOCAC to channel investment towards a China–Africa Peace and Security Fund. It also addressed security issues, including military assistance, law and order, peacekeeping, anti-piracy and counter-terrorism.

In July 2019, Beijing hosted the first China–Africa Peace and Security Forum, with delegations from the African Union and 50 of 54 African states. Attended by 15 defence ministers and chiefs of defence forces, the forum focused on maritime security and peacekeeping. As of July 2020, about 2,000 Chinese personnel served in UN peacekeeping missions in the Democratic Republic of the Congo, Mali, South Sudan, Sudan and Western Sahara, while Chinese operations conducted from its base in Djibouti have included humanitarian assistance, peacekeeping and counter-piracy operations. However, the division between such operations and more traditional military activities is not always clear-cut. For example, Chinese naval vessels have visited a number of African ports as part of anti-piracy 'escort' missions, while live-fire exercises have been conducted at the Djibouti base.

In mid-2019, General Thomas D. Waldhauser, then head of United States Africa Command (AFRICOM), told Congress that while Chinese efforts in Africa were unlikely to impede US military access and operations in the short term, 'China could gain that capability within the next decade'. An April 2020 report by the US Naval War College's China Maritime Studies Institute suggested that the Djibouti base was capable of 'supporting a wide range of operations' beyond just logistical support and had the capacity to station as many as 10,000 personnel – more than double the 4,000 based at the United States' Camp Lemonnier, just a few miles away. The report also underscored the People's Liberation Army Navy's ability to increase the length of deployments, with the possible use of Djibouti as a shore-leave location potentially allowing deployment times to the Gulf of Aden to be extended by as much as one-fifth.

The COVID-19 pandemic presented both an opportunity and a threat for China in Africa. China moved quickly to provide aid in the region, leveraging the Jack Ma Foundation (set up by the co-founder of the Alibaba Group) to deliver 1.1 million testing kits, 6m masks and 60,000 protective suits and face shields in March. Set against this, videos circulated on social media showing discrimination against Africans in the southern Chinese city of Guangzhou almost certainly dented Chinese soft power. Heavy-handed measures – including forcible testing and eviction from homes – provoked strong criticism in Africa, and some unusually public political reproaches. For example, the speaker of Nigeria's House of Representatives, Femi Gbajabiamila, posted a video of himself reproaching the Chinese ambassador over alleged mistreatment of Nigerian citizens in Guangzhou.

The longer-term impact on Sino-African relations will probably be determined by Chinese attitudes to debt relief. According to the China Africa Research Initiative at Johns Hopkins University, the Chinese government, as well as Chinese banks and contractors, extended more than US$150 billion to African governments and state-owned enterprises between 2010 and 2018. Many African states now have high financial exposure. China gave its backing to a G20 plan suspending all bilateral-government-loan repayments due from eligible low-income countries from 1 May to the end of 2020. In all, 77 low-income countries will see their debt repayments to China deferred, but not forgiven.

China has traditionally preferred debt renegotiation to debt relief, although it has offered forgiveness on some (usually small) loans. If it fails to offer more generous debt relief, China will likely face accusations of profiting from African states' poverty and using the G20 to support its debt issuance. In contrast, taking the lead on debt could secure it substantial political capital. However, China's position could ultimately be determined by its domestic economic performance. Already more risk-averse in its engagement with sub-Saharan Africa, if its growth slows due to COVID-19 it may be less willing to provide large-scale debt relief.

Russia's courting of Africa

Russia has developed an increasingly bold approach to the region. General Stephen Townsend, who became head of AFRICOM in July 2019, has described Russia as the biggest threat after terrorism to US interests in Africa, while adding that he expected it to be eclipsed by China. Russia's growing role in Africa was exemplified by the October 2019 Russia–Africa Summit in Sochi. Around 10,000 delegates from 54 African countries took part, including 43 heads of state. Some US$12.5bn-worth of agreements were signed – though many were not legally binding, and it remains to be seen how many will be realised.

But Russia's efforts to find new markets and partners remained at a nascent stage, particularly when compared with China. Even if fully implemented, new deals and investment are less than one-quarter of the US$60bn of investment agreed around the Forum on China–Africa Cooperation in 2018. That year, Russia's exports to sub-Saharan Africa totalled US$3bn and imports just US$1.7bn. Russia has relatively little to offer sub-Saharan economies.

Armaments remained the major exception. Russia has signed more than 20 military-cooperation agreements since 2015, and this was one of the key priorities of the Sochi summit. Between 2013 and 2017, Russia supplied 39% of Africa's imported arms. The relatively low cost (much of the equipment is second-hand) and lack of political conditionality are selling points, particularly for states facing Western criticism for their human-rights records. Before the Sochi summit, Russian Minister of Foreign Affairs Sergei Lavrov began negotiations with the Rwandan government for the sale of a missile-defence system and with the Eritrean government to establish a logistics centre in the Horn of Africa. The latter would give Russia access to a Red Sea port – a key strategic aim of the Putin government.

While Russia's formal presence in Africa on the ground was small compared to the US or France, it offered support via private military contractors such as the Wagner Group (reportedly owned by Yevgeny Prigozhin, a businessman with close links to Russian President Vladimir Putin), as well as Russian intelligence agencies. According to AFRICOM, Russian private military companies were active in 15 African states, notably the Central African Republic and Mozambique. In the latter, Russian private contractors assisted local security services in tackling an insurgency in Cabo Delgado province that claimed to be affiliated to the Islamic State, also known as ISIS or ISIL. Seven Russian personnel were reportedly killed in October 2019. Nonetheless, the use of private contractors is likely to continue, since this will enable the Russian government to offer relatively low-cost (and in some cases diplomatically deniable) military support.

By addressing security and defence deficiencies in African countries that are suffering, or recovering, from conflict, Russia positioned itself for future defence contracts and broader trade gains at a time when the

West appeared to be focusing more on migration and related issues. This approach also enabled the Putin administration to cultivate allies – both within the UN at a time of growing tensions among the five permanent members of the UN Security Council, and in its challenge to the current Euro-Atlantic-dominated security order.

Mixed messages from US undermine its position

AFRICOM maintains a light and relatively low-cost footprint on the continent. Many of its operations are focused on providing air support for European and African troops carrying out counter-terrorism missions, and most of the around 5,200 US troops and 1,000 Defense Department civilians or contractors carry out training or exercises with local forces. But, in contrast to Russia and China, there are signs that the US is reappraising its security presence. In December 2019, it was reported that US Defense Secretary Mark Esper – carrying out a global troop review to free resources for deployment in Asia – was considering a major reduction, or even complete withdrawal, of US forces from West Africa. Other suggestions for a US drawdown included the abandonment of a US$110m drone base in Niger – which only began operating in November 2019 – and halting assistance to French forces tackling militants in Burkina Faso, Mali and Niger. This was in line with the 2018 National Defense Strategy, which mooted a reorientation away from counter-terrorism operations against groups with no demonstrable ability to attack the US and towards competition with China and Russia.

However, these mooted plans prompted concern from allies – notably France, which relies on US intelligence and logistics to support its 4,500 troops in the Sahel – and domestic opposition. A bipartisan congressional group sought to introduce legislation preventing the administration from cutting AFRICOM forces without a detailed report to Congress on the potential impact. And, while not a direct response to Esper's suggestions, General Townsend appeared to endorse legislators' contentions that the removal of troops from Africa would be counterproductive given peer competitors' deepening military footprints in the region. In January 2020, he told Congress that Africa remained a key theatre for competition, with China and Russia 'aggressively using economic and military means to expand their access and influence'. He noted that a joint Sino-Russian naval exercise off South Africa in 2019 was the 'first visible sign of cooperation we've seen' on the continent.

Esper's force-reduction proposals also appeared to contradict a new State Department initiative to tackle ISIS in West Africa. In November 2019, Secretary of State Mike Pompeo stated that 'ISIS is outpacing the ability of regional governments and international partners to address that threat', and announced the creation of a special task force to focus on the deteriorating security situation in sub-Saharan Africa. The threat posed by terrorist groups in West Africa, particularly given growing signs of collaboration between affiliates of al-Qaeda and ISIS, was one reason the US sponsored the February 2020 *Operation Flintlock* military exercises in Senegal and Mauritania, involving more than 1,500 personnel from 34 African and Western states.

These mixed signals contributed to the sense that President Donald Trump's administration lacked a coherent strategy towards the continent. Pompeo's visit in February, his first as Secretary of State, took place less than a month after Trump announced that the US would no longer issue immigrant visas to Nigerians and Eritreans, and that citizens of Tanzania and Sudan would be excluded from the United States' green-card visa lottery scheme. However, proposed military cutbacks, combined with slow progress on the Trump administration's Prosper Africa scheme (announced in December 2018) and a lack of coordinated assistance during the COVID-19 pandemic, suggested that the US risks being outmanoeuvred in Africa in both the military and soft-power spheres.

The regional impact of COVID-19

As of the end of June 2020, sub-Saharan Africa was the region least affected by COVID-19. It was not yet clear whether this reflected a lack of testing – Senegal and South Africa had more extensive testing programmes and higher numbers of cases – or whether underlying factors gave sub-Saharan states some protection. Only 3% of the population in sub-Saharan Africa is aged 65 or above, while 43% is younger than

15 years old. The hot climate and limited connectivity may also have limited infection rates. Set against this, less well-resourced healthcare systems meant lower survival rates for hospitalised patients.

Nevertheless, COVID-19 was already hitting Africa's economy through its effects on major trading partners, commodity prices and tourism. The World Bank predicted the continent would suffer its first recession in 25 years, with economic growth contracting to between −2.1% and −5.1% against population growth of 2.7%. Urban areas, dependent on the cash-driven informal sector, were particularly hard-hit by lockdown measures. There was a risk that efforts to enforce these by security forces could prompt serious unrest. At particular risk were Zimbabwe (already suffering serious macroeconomic imbalances); Sudan and Ethiopia (in the midst of political transition); and South Africa, where in March the government deployed 70,000 troops after crowds launched a series of attacks on trucks delivering food and other provisions.

Debt Crisis in Sub-Saharan Africa
How severe – and will Washington and Beijing help?

Sub-Saharan Africa faces a debt crisis. Public debt in Africa has been rising slowly but steadily – although not evenly across the region – since the mid-2000s. With some countries already facing debt distress at the start of 2020, the COVID-19 pandemic and ensuing collapse in oil and other commodity prices are likely to cause many more states in the region to struggle to meet their debt obligations. How exactly this debt crisis plays out will depend on the speed of the region's economic recovery from the pandemic and the extent to which it receives timely liquidity support from international financial institutions and debt relief from creditors.

The debt crisis has strategic implications for the continent. Debt distress brings economic hardship, raising the prospect of political turmoil and a surge in emigration. African governments' relations with China will change, as borrowing for infrastructure projects will be curtailed. And, stung by capital flight during the crisis, governments may resort to controls on cross-border portfolio investment. The crisis will therefore be an important moment in sub-Saharan Africa's political and economic development.

Rising debt risks and economic shock

External public debt in sub-Saharan Africa has been high before. In 1994, it reached 58% of gross national income. However, between 1996 and 2012, the Heavily Indebted Poor Countries (HIPC) Initiative and the Millennium Debt Relief Initiative (MDRI) wrote off around two-thirds of the public debt of qualifying countries.

Since then, sub-Saharan African governments have renewed their borrowing, in particular to finance their huge infrastructure needs. With the Paris Club of major creditor countries scaling back lending, African governments turned to China and private-sector lenders. In the early 2000s, China began directing some of its vast current-account surplus towards developing countries by offering to finance infrastructure projects carried out by Chinese construction companies. Sub-Saharan Africa quickly became the largest recipient of Chinese loans and now has the second-largest stock of debt to Beijing of any region except Latin America and the Caribbean (see Figures 1 and 2). Then, in the wake of the 2007–08 global financial crisis and resulting recession, with fixed-income investments in the West offering poor returns, private-sector investors looked to developing countries for better rates of return, leading to unprecedented portfolio inflows into sub-Saharan African government debt. Several sub-Saharan African governments also took the opportunity to issue eurobonds.

This coming-together of willing lenders and capital-hungry governments has resulted in a surge in public debt with a very different composition to that of past debt. Most African countries have become more indebted, especially commodity-exporting countries including Angola, Mozambique, the Republic of the Congo and Zambia (see Figure 4).

Figure 1: Estimated outstanding debts to China by region, 2000–17

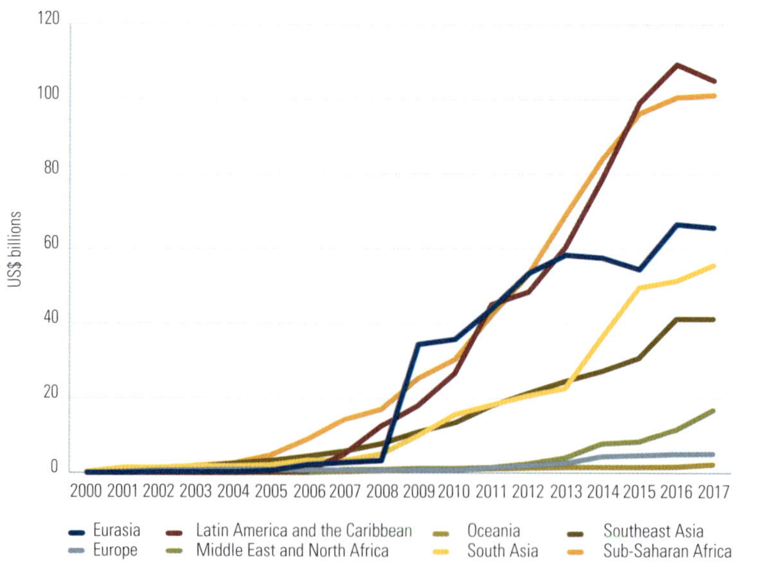

Source: Horn, Sebastian, Carmen M. Reinhart, and Christoph Trebesch. 2019. 'China's Overseas Lending'. NBER Working Paper No. 26050

Figure 2: Estimated outstanding sub-Saharan African (SSA) debts to Paris Club creditors and China, 2008–17

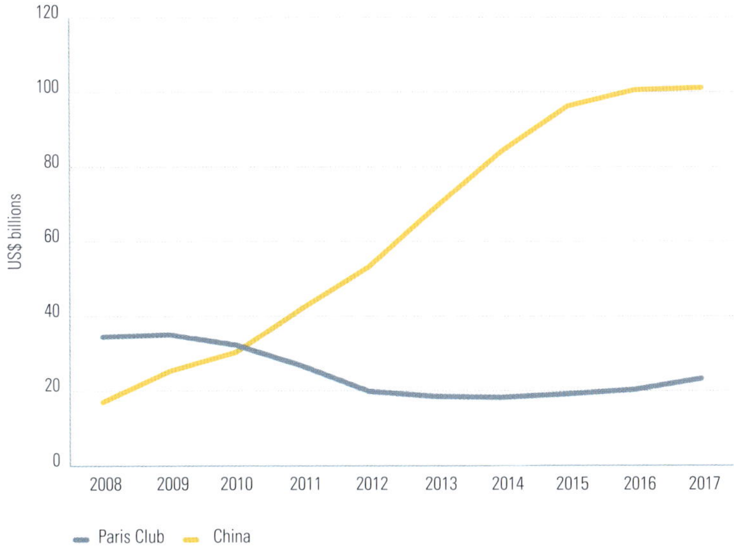

Sources: Horn, Sebastian, Carmen M. Reinhart, and Christoph Trebesch. 2019. 'China's Overseas Lending'. NBER Working Paper No. 26050; Paris Club, 'Paris Club Claims', Press Releases, 2008–17

©IISS

Figure 3: SSA external public and publicly guaranteed debt and bonds, 1978–2018*

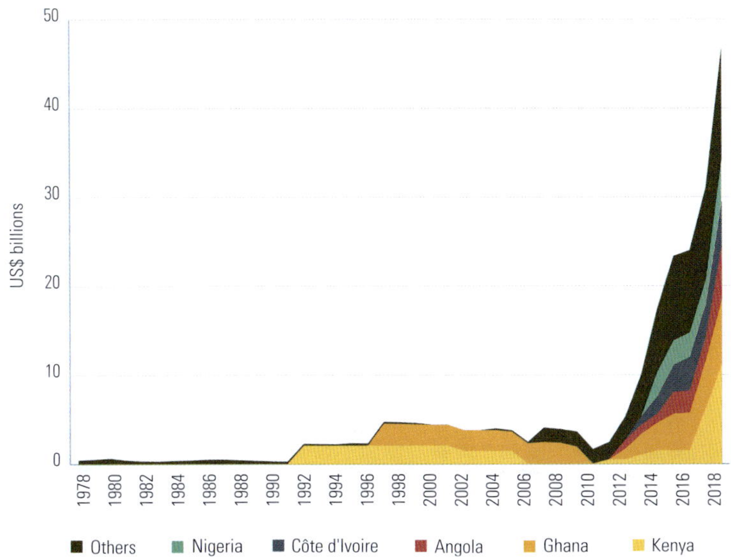

Source: World Bank

** Excludes South Africa*

©IISS

Figure 4: Gross government debt of ten largest SSA economies, shown as percentage of GDP

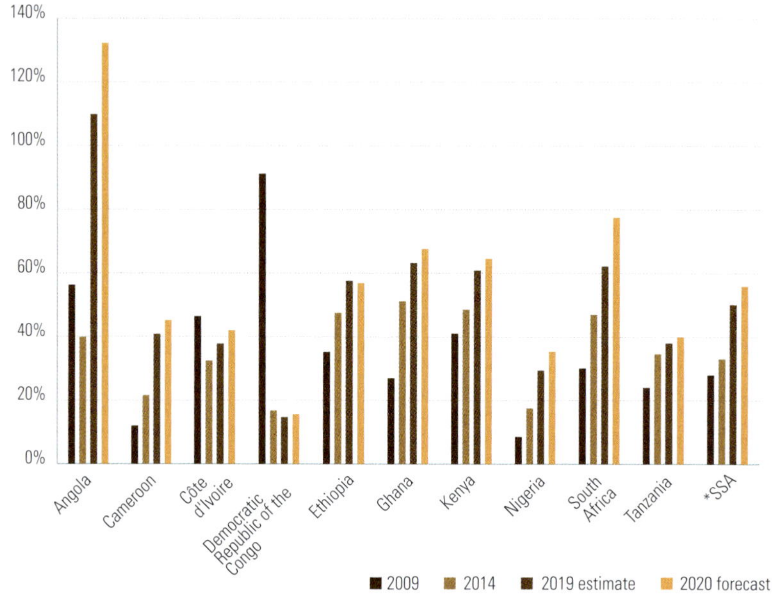

Source: IMF

** Excludes Djibouti, Mauritania, Somalia and Sudan* ©IISS

Figure 5: Domestic–external composition of general government debt, shown as percentage of GDP, for the ten largest SSA economies, 2008 and 2018

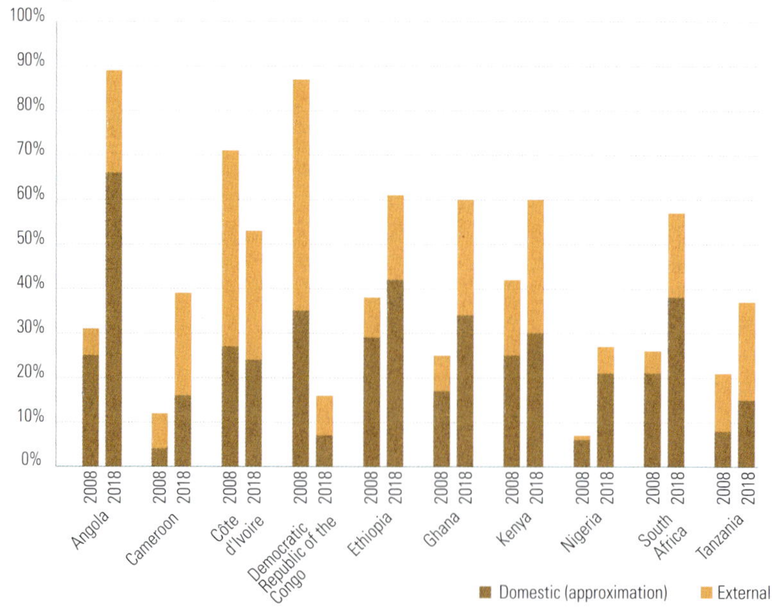

Sources: IMF; World Bank

©IISS

The extent and composition of this debt has created vulnerabilities to economic shocks. By relying on external creditors (see Figure 5), sub-Saharan African governments are less able than governments elsewhere to use monetary policy to reduce their debt. Since much of their debt is issued in foreign currencies (see Figure 6), debt-servicing costs are vulnerable to exchange-rate fluctuations. On top of these long-standing vulnerabilities, by relying more on China and foreign private-sector creditors, governments in the region face higher interest rates on debts and greater difficulty negotiating debt relief.

Furthermore, few sub-Saharan African countries have established controls on cross-border capital flows. For three decades, the IMF has been firmly opposed to their use, arguing that they limit developing countries' access to much-needed finance. Most sub-Saharan African governments have therefore either relaxed controls on portfolio inflows and outflows (such as South Africa's liberalisation of outflow controls) or not imposed them. Having accumulated large stocks of cross-border portfolio capital on which there are few capital controls, many countries are therefore vulnerable to destabilising capital flight.

Figure 6: Denomination of general government debt, shown as percentage of GDP, for select SSA economies, 2010 and 2019

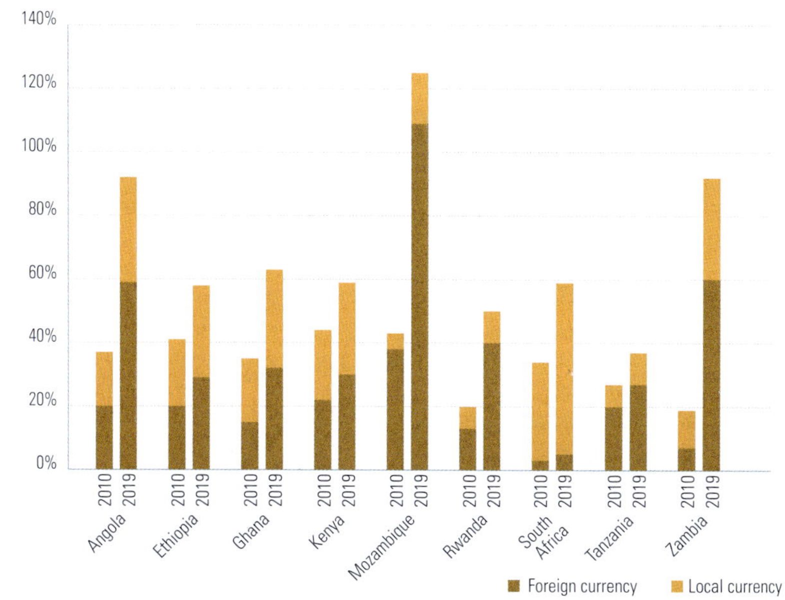

Source: IMF, 'Global Financial Stability Report: Markets in the Time of COVID-19', April 2020, p. 49

©IISS

China's appetite for imported commodities and investing overseas in the extractives industry has made countries like Angola and Zambia increasingly reliant on commodity exports to China. While exports to China fuelled economic growth, these governments paid little attention to increasing intra-African trade despite the long-term benefits anticipated from doing so. With 43% of trade within Africa in 2015 comprising manufactured goods, compared to 19% of its exports to countries outside Africa, focusing on intra-regional trade would have helped sub-Saharan African economies move up the value chain and diversify their exports and export partners. Instead, growing trade with China increased dependence on commodity exports to a single export partner.

As a result, economic growth, government revenue and the value of African currencies have become more vulnerable to external shocks, and especially to slowdown in China. Few governments in sub-Saharan Africa have amassed sufficient foreign reserves to buffer against such shocks. (Nigeria and Tanzania are exceptions.)

As a result, by late 2019 eight countries in the region were already unable to service their current debt obligations, according to the IMF. It classified a further 11 (including Ethiopia and Ghana) as at high risk of debt distress. Of those, Zambia had already defaulted on debt payments to Italian bank Intesa Sanpaolo and run up arrears with its Chinese creditors.

The COVID-19 pandemic and collapse in oil prices have made debt crises very likely for some and near inevitable for others. Compounding the disruption of lockdowns within Africa, those imposed elsewhere have hit demand for commodity exports, manufacturing inputs, tourism and transport. The World Bank also expects remittances to Africa to fall by over 23% in 2020.

The global dip in investor confidence also triggered record portfolio capital outflows from emerging markets and developing countries, including from sub-Saharan Africa. The cumulative portfolio outflow from South Africa in February and March 2020 amounted to 1% of GDP. Furthermore, inflows of foreign direct investment (FDI) are expected to fall by 25–40% across Africa as a whole in 2020.

Figure 7: Percentage change in forecasted 2020 GDP growth for ten largest SSA economies

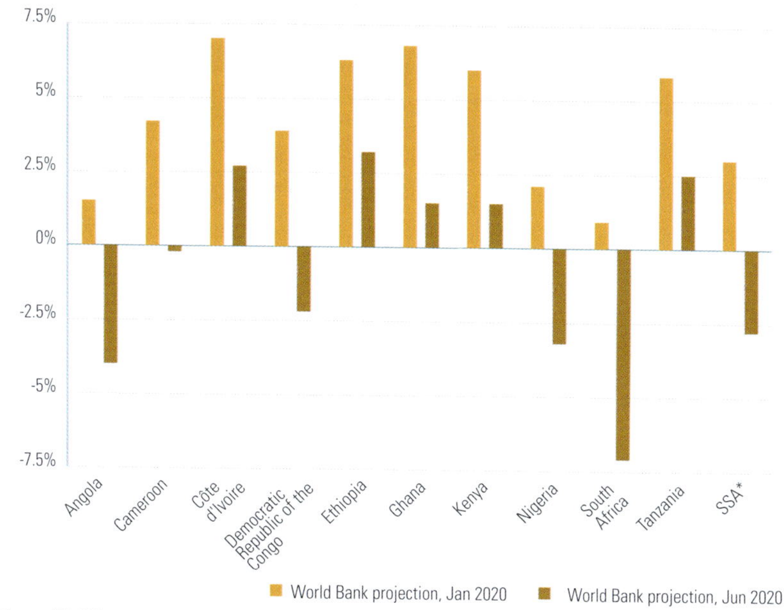

Source: World Bank

** Excludes Mauritius* ©**IISS**

Figure 8: Change in forecasted 2020 fiscal balance for the ten largest SSA economies, shown as percentage of GDP

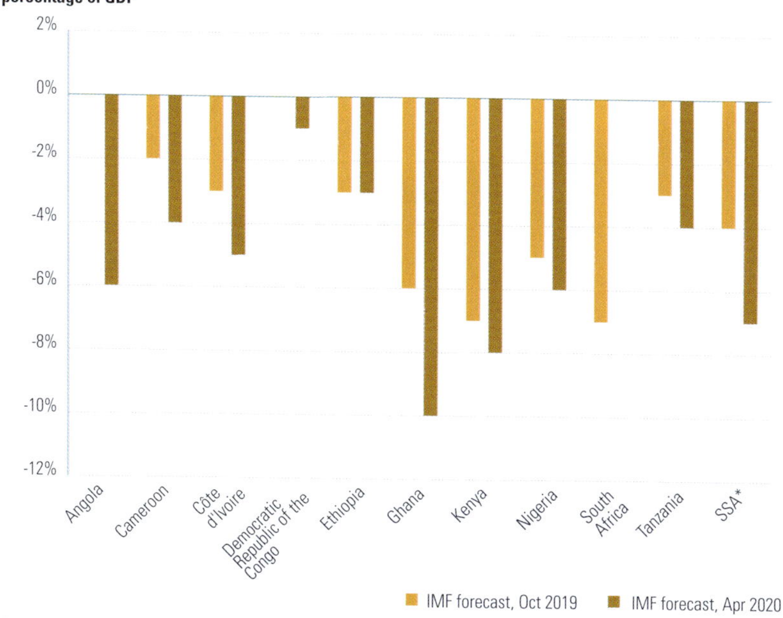

Source: World Bank

** Excludes Somalia* ©**IISS**

Against the background of a global recession, government deficits will grow (see Figure 8) even as borrowing and debt servicing costs rise (see Figure 9). Many African governments will find it increasingly difficult to meet their debt obligations – especially for foreign currency-denominated debt.

Vulnerability to debt distress varies across the region. Côte d'Ivoire, Kenya and Senegal will fare comparatively well, maintaining GDP growth and manageable budget deficits, and benefiting from (comparatively) low debt-to-GDP ratios. But in other countries, the combination of higher borrowing and debt-servicing costs, shrinking tax revenues and additional COVID-19-related spending make it likely that they will be unable to service their debts. This is true of Ethiopia, Ghana and South Africa in particular, other commodity exporters, and smaller tourism-dependent economies including Cabo Verde, Comoros and Seychelles. By June 2020, the crisis had already tipped Angola and Zambia definitively into debt distress to the point where they lacked the resources to service their debts on their original terms.

Figure 9: Basis points increase in borrowing costs for select SSA economies, 20 Jan 2020–20 May 2020

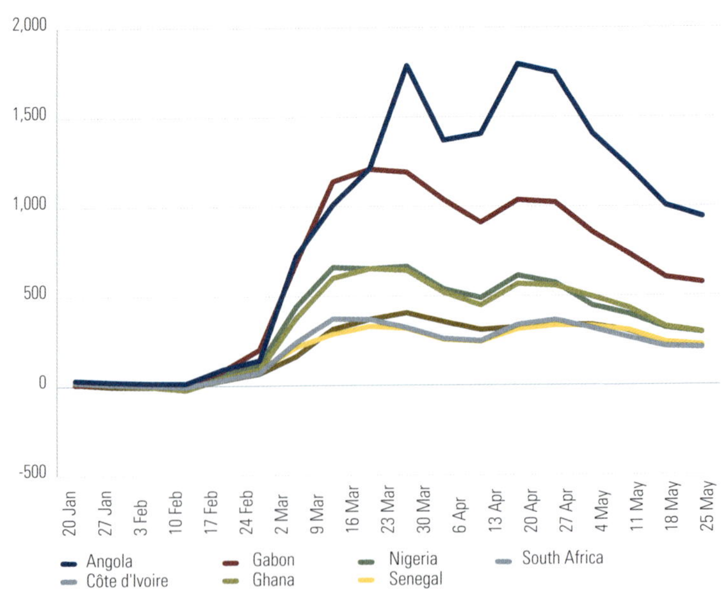

Source: World Bank

Any further economic shock could make the situation worse. Large swarms of desert locusts in East Africa are a serious threat to agricultural production in Ethiopia, Kenya and to a lesser extent Uganda. The need to provide food assistance to their populations will strain governments' budgets still further, especially in Ethiopia.

The debt crisis is set to unfold over the next two to three years as budget deficits accumulate and large debt repayments (especially bond principals) fall due. And unlike the occasional, isolated sovereign defaults witnessed on the continent since the HIPC and MDRI, this crisis could affect tens of countries in the region.

Curbing the crisis

The extent and severity of this debt crisis will depend on two factors: the speed of economic recovery in Africa and globally, and the response to the crisis from African governments, creditors and international financial institutions.

Governments' ability to respond to the crisis is limited. Expansionary monetary policies would devalue foreign-owned, local-currency debt and encourage further capital outflows. Debt denominated in foreign currencies would also become more expensive to service. Most countries in the region have little fiscal space to use government spending to boost economic growth, and for those that do, borrowing costs have risen dramatically. For those struggling to cover the cost of debt service from existing government revenues, few options are available but to cut spending and increase revenues. Nigeria, for example, has announced a 15% cut in government expenditure. The IMF advised vulnerable countries to begin early conversations with creditors about rescheduling debts and securing refinancing for debts that are due to mature.

The IMF and World Bank reacted quickly to the crisis, but the tools currently at their disposal are expected to prove inadequate. At the beginning of April 2020, both institutions announced funds to cover the immediate costs of dealing with the pandemic and its economic impact. African governments began immediately to make use of this facility. By May, IMF lending to the region spiked to unprecedented levels (see Figure 10).

However, African governments' need for hard currency is likely to exceed the current lending capacity of the IMF and World Bank.

One strategy to address temporary shortages of dollars is for the United States Federal Reserve to introduce currency-swap arrangements with African central banks. A second strategy, which would help in the case of a longer-term shortage of hard currency, is for the IMF to issue new special drawing rights (SDRs). This involves creating new assets in a bundle of reserve currencies and assigning these to IMF member countries via the SDR allocation mechanism, allowing African countries to draw down hard currency to meet their obligations. Rich countries could draw down on their own allocations and make the funds available to developing countries' governments.

However, the US government seems disinclined to pursue either strategy. It has so far opposed the issuance of new SDRs by the IMF, without giving clear reasons. Without US support the IMF cannot act. Despite providing currency-swap lines to Brazil and Mexico, the Federal Reserve has made no moves to provide such a facility for African central banks.

Figure 10: IMF credit outstanding for SSA, May 1984–Jun 2020

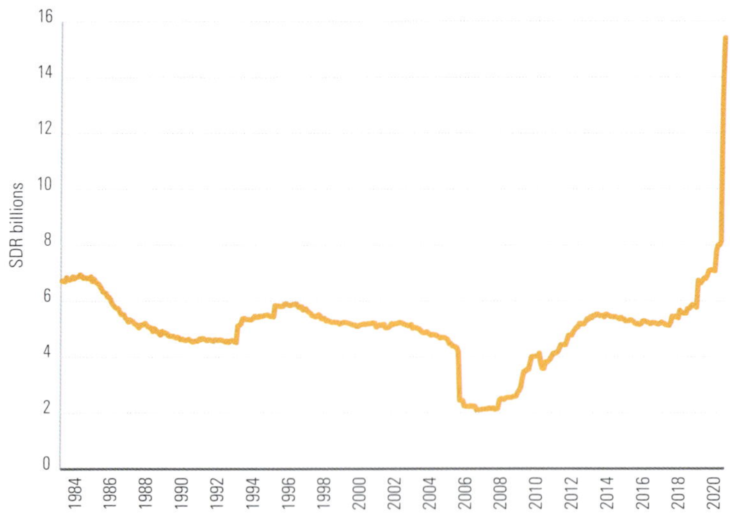

Source: IMF

China could also offer currency-swap lines. The People's Bank of China (PBoC) has previously offered currency swaps to South Africa's and Nigeria's central banks, and could expand these arrangements to other African countries. But for Beijing, this involves a trade-off. On the one hand, if Beijing pushes Chinese policy banks and Chinese exporters to redenominate their US dollar-based contracts with African partners in renminbi (RMB), then providing swap lines to African central banks will ensure those partners have access to the hard currency needed to pay for imports from, and debt repayments to, China. On the other hand, the effect would be nullified if African borrowers instead converted the renminbi into US dollars, taking advantage of the relatively stable USD–RMB exchange rate, with little benefit to China. And although currency-swap lines usually involve little credit risk, lending to central banks in African countries that face a realistic prospect of debt distress would involve non-negligible risks for the PBoC.

African countries will also need to negotiate debt relief with multilateral lenders, bilateral lenders and private creditors. Both the US and China will affect how this plays out. Multilateral lenders face their own financing risks, with debt write-offs threatening to affect their credit ratings and borrowing costs. They will be looking to rich nations to recapitalise them in equal measure. It is far from clear that the US will be willing to do so.

Several African countries will also seek relief on their Paris Club debt. Granting this requires a consensus of all Paris Club members, but US support for debt relief is uncertain. To write down bilateral debts, the US Treasury needs to appropriate funds through Congress. In the latter years of the Obama administration, Congress blocked appropriations to offer debt relief through the MDRI; it is unlikely to approve appropriations for debt relief if the Republicans remain in control of the Senate or White House – particularly if the US has no guarantee that China is offering equivalent debt relief.

China, meanwhile, is reluctant to offer debt relief, and its approach to renegotiating debts complicates the process. Loans from Beijing are opaque, often secured against commodity-export revenues and only partially reported to the IMF and Bank for International Settlements. Beijing conducts debt renegotiations bilaterally, and on a loan-by-loan basis. This makes it difficult for debtors to renegotiate with other official (bilateral and multilateral) creditors. Those other creditors demand a thorough assessment of a government's total debt obligations and ability to pay before agreeing on a write-down on the whole parcel of bilateral debt – so the burden of debt relief is evenly shared. Unless Beijing provides transparency on debts it is owed by African governments – and writes down debts to the same degree as other creditors – African countries may be unable to negotiate debt relief in a timely fashion. Furthermore, their treasuries do not always have a clear picture of their own countries' total sovereign debt.

The US and China therefore have considerable influence over how the debt crisis will evolve – in particular, whether sufficient hard-currency liquidity is provided to avoid debt distress, and whether debts can be renegotiated before governments reach the point of default.

Furthermore, although most debt maturing in 2020–21 is owed to China, from 2022 onwards African governments face large principal repayments on eurobond debts to private creditors, which they will need to refinance or renegotiate. Most governments have adopted aggregated collective action clauses (CACs) in bonds issued since 2015, allowing a supermajority of bondholders to agree renegotiated payments on behalf of all bondholders across all bond issuances. However, these aggregated CACs are less common in earlier bonds. Governments such as Zambia's, which issued eurobonds without aggregated CACs, therefore face the risk of holdout creditors who prolong the process of debt renegotiation and block access to international finance.

Outlook

The effects of the debt crisis will be felt for several years. Government spending, especially capital spending, will be depressed both in countries currently in debt distress and in those at risk of it, including Ethiopia, Ghana and South Africa. The continent's infrastructure-investment gap, estimated by the African Development Bank even before the crisis to be between US$68 billion and US$108bn per year, will increase further, meaning slower advances in access to electricity, transport and digital connectivity.

The political and social impact of economic and financial problems on a continent with a youthful, rapidly growing population could be especially significant. Falling incomes and growing joblessness could fuel protests and even turmoil over cuts to government spending, rising taxation and economic hardship. Debt distress will also draw government attention and capacity away from other issues, such as dealing with conflicts (in Mozambique and Nigeria, for example) or pushing ahead with regional economic-integration plans. The effects will be felt outside the region, in Europe in particular, as emigration from debt-stricken countries grows, even as humanitarian and development assistance comes under pressure from rising debt in donor countries.

There are few simple fixes to the factors that left African governments vulnerable to debt distress. It will be a long time before Africa's commodity exporters manage to diversify their economies. And it will take many years for domestic credit markets in Africa to deepen to the extent that governments no longer rely on external borrowing.

However, the debt crisis, and the broader global economic crisis, may trigger changes to rules on international capital flows. If large outflows of portfolio capital are perceived to be a major contributor to the debt crisis in Africa, the case for imposing controls on cross-border portfolio capital flows will strengthen. Aside from the current crisis, two factors open the door to wider use of capital controls. Firstly, the IMF's longstanding opposition to them is beginning to soften. In April 2020, the IMF acknowledged that they might be of use in responding to the current economic crisis, albeit temporarily. Secondly, the use of capital controls by China and other leading emerging-market economies has engendered wider acceptance of such measures. Despite limiting access to external finance, capital controls offer greater financial stability. Widespread debt crises in sub-Saharan Africa (and in other emerging-market and developing economies) could therefore nudge African governments towards their use.

So far, African countries have been reluctant to introduce capital controls for fear of losing access to external finance. A green light from the IMF and US would provide them with some reassurance. However, under a Republican administration, the US will undoubtedly remain opposed, primarily reflecting US investors' interests. Wider uptake of capital controls would mark a move away from the Washington Consensus and could be seen as a move towards a new 'Beijing Consensus'. The prospects for capital controls therefore depend in part on the policies of the US administration elected in November 2020, and in part on African governments' determination to pursue them regardless.

High African indebtedness will also encourage evolution of the region's relationship with China. Many countries no longer have the fiscal space to take on additional debt for new turnkey infrastructure projects. As a result, Chinese companies may look instead to public–private partnership or concession models for infrastructure projects in Africa. The latter is already in use in Nigeria: in 2019, the China Harbour Engineering Company signed a deal for a build–own–operate–transfer contract for Lekki Deep Water Port with funding from the China Development Bank. However, although projects of this kind circumvent the immediate constraints on fiscal space as the implementer bears the upfront cost, fiscal risks to African governments often remain. They will still need to offer revenue guarantees for the infrastructure built and operated by Chinese companies. This could prove costly further down the line, especially if African economies grow more slowly than forecast.

The scale, severity and consequences of the African debt crisis will depend in part on how the pandemic develops on the continent and beyond. Much will also depend on the US and China, whose policies will determine whether sufficient and timely liquidity support and debt relief are available to the continent's most indebted countries. This will itself depend on whether there is a change in US policy following the US presidential election in November 2020. At its worst, economic hardship associated with debt distress could cause political upheaval in numerous African countries.

Sub-Saharan Africa will continue to struggle to balance the need to bridge its infrastructure gap with the risk of running up debts to finance capital projects. This may involve rebalancing the current mix of concessional loans from multilateral development banks, non-concessional loans from China and bonds sold to external creditors. Deeper domestic debt markets would allow governments to borrow more domestically and avoid some of the risks associated with external credit.

Boosting intra-regional trade would also help diversify Africa's exports and export partners and reduce exposure to external shocks. Although the launch of the Africa Regional Free Trade Initiative, scheduled for July 2020, was delayed due to COVID-19, the pandemic is prompting a wider reassessment of sub-Saharan Africa's economic dependence on other regions, including at the African Development Bank and United Nations Economic Commission for Africa. A lesson of the debt crisis is that African governments should press ahead with regional economic cooperation.

South Africa's Difficult Choices
Does 'Ramaphoria' have a future?

May 2020 marked the first anniversary of Cyril Ramaphosa's election as post-apartheid South Africa's fifth president. He had initially taken office following Jacob Zuma's resignation in February 2018, before being elected in his own right. A wave of 'Ramaphoria' swept the country as he promised a 'New Dawn' that would tackle the corruption, crony capitalism and decade-long economic decline of Zuma's presidency. The way seemed open to substantive political and economic reform.

Zuma, his family and associates, notably the Gupta family of Indo-South African businessmen, had sought to capture and undermine key state institutions to facilitate the flow of economic rents from corrupt state-procurement deals, and to prevent legal consequences of this. Zuma replaced the leadership of the South African Revenue Services (SARS), once a model agency in the developing world; of the National Prosecuting Authority; and of the police and security agencies.

Zuma also replaced the leadership of the key state-owned enterprises (SOEs), whose major infrastructural projects offered tempting targets. Instead of providing an enabling environment for investment and private-sector activity, these state-owned network industries provided unreliable and ever-more-expensive services, inhibiting rather than promoting growth while requiring ballooning state subsidies and guarantees. These included, most notably, the electricity-supply commission Eskom, the rail-and-port authority Transnet and South African Airways.

The only key state agencies that managed to hold out against a sustained assault were the Treasury, the independent Reserve Bank (the country's central bank) and the public protector, whose role is to investigate improper conduct in government. The last of these, Thuli Madonsela, courageously published a seminal exposé of state corruption, 'Report on the State of Capture', at the end of her term in October 2016. Finally, Zuma tried but failed to subvert the judiciary, the free press and non-governmental organisations, all of which proved to be significant thorns in his flesh.

Previous African National Congress (ANC) governments had weakened the capacity of the public sector with a policy of cadre deployment, which favoured political loyalty over professional competence, and rapid race-based affirmative action. President Thabo Mbeki (1999–2008) had abolished the independent and effective anti-corruption unit, the Scorpions. What was new under Zuma was the open pursuit of rents.

All this constituted a formidable agenda for institutional and economic reform. Yet even in Ramaphosa's camp, significant factions and vested interests inimical to reform remained. In the ruling Tripartite Alliance, the Congress of South African Trade Unions – whose strength lay principally in the public-sector unions – and the South African Communist Party (SACP) – though opposed to corruption – remained hostile to market-based reforms that they labelled 'neoliberal'. Within the ANC, the statist bias of the liberation movement persisted in the doctrine of the 'developmental state', first outlined by Mbeki, who had otherwise pursued fairly orthodox economic policies, in 2004. This was originally, in effect, a version of the Asian developmental state blended with Scandinavian social democracy, but without either the effective Asian state administration or the resources and cohesion of the Scandinavians.

Powerful vested interests retained their influence in the private sector. The policy of black economic empowerment under its various iterations, while politically understandable and defensible, has been captured by elite interests and in many instances morphed into a form of crony capitalism. For example, the reallocation of coal contracts for Eskom power stations from established multinational miners to nascent black interests drove up energy costs, added to Eskom's financial woes and created a coal lobby hostile to the more competitive and clean renewable-energy sector.

Established businesses, including global consultancies and multinationals, also participated in the corruption of the Zuma era. In several high-profile cases companies were caught enabling state capture, directing rents from inflated contracts to politicians and officials. There were major corporate scandals, such as the multibillion-dollar Steinhoff fraud. This behaviour weakened large businesses' legitimacy amidst an already difficult and racialised business discourse.

The official opposition party, the Democratic Alliance, faced leadership woes in late 2019 and has struggled to cast off the perception that it represents white interests, despite winning significant support from mixed-race, Indian and black voters and four major metropolitan centres in 2016. This has limited its effectiveness in promoting a reform agenda. More challenging for Ramaphosa is the Economic Freedom Fighters party under the charismatic and populist Julius Malema. Expelled from the ANC in 2012 by a Ramaphosa-led committee, Malema has been able to outbid Ramaphosa by using inflammatory racial rhetoric that appeals to the large youth constituency impatient for change and critical of the 1994 Mandela-led constitutional compromise.

Ramaphosa's response: institutional reform

Some argue that Ramaphosa should have used his first 100 days, when his popularity was at its height and his opponents at their weakest, to drive through key political and economic reforms. But the nature of his victory, the complex factional politics of the ANC and associated vested interests, and his style of leadership were obstacles. The forces associated with Zuma remained strong and in key positions. Ramaphosa began his presidency by purging his cabinet of half a dozen supporters of Zuma deemed least effective or likely to support reform, but most of them remained in leadership positions in the party and its organs. The cabinet itself was a mix of Ramaphosa and Zuma loyalists, as were the top six officials at the ANC headquarters, Luthuli House.

Ramaphosa had signalled his intent to probe corruption by appointing in August 2018 a judicial commission of inquiry under former deputy chief justice Raymond Zondo. The revelations presented over two years have confirmed the allegations of the Madonsela report and numerous exposés by investigative journalists. But, in the manner of such enquiries, progress has been ponderous.

A focused judicial inquiry into SARS quickly produced the ammunition needed to eject Zuma's man, Tom Moyane. An investigation into the Public Investment Corporation – which invests the substantial Government Employees Pension Fund monies – also presented sufficient evidence to change out its leadership. A new director and team installed in early 2019 at the key National Prosecuting Authority took 18 months to bring its first major corruption case. However, no high-ranking official has yet been brought to book.

The economic crisis

The COVID-19 pandemic hit South Africa at a time when the economy was in considerable distress. The February 2020 budget optimistically projected a budget deficit of 6.8% and a GDP growth rate of 0.9% after three successive years of declining growth of, according to the World Bank, 1.4%, 0.8% and 0.2% respectively. In March 2020 Moody's became the last of the three major rating agencies to downgrade South Africa's credit rating to junk. This was owing to the low growth outlook, rising debt (62.5% in 2019 after the historic pre-global financial crisis low of 26.5% in 2008, an era of fiscal prudence under Mbeki) and questions about the ability of the government to constrain spending.

In an attempt to signal its intent to maintain some discipline, and as part of its efforts to cut spending and stabilise debt, the government proposed in the budget to renege on the final year of increases in a three-year wage deal with the public sector. This was an unprecedented step after years of loose policy and well-above-inflation increases won by the strong public-sector unions, key political allies of the ANC. The move elicited union fury and, as of the end of June 2020, was the subject of court challenges and ongoing negotiations.

Ramaphosa has been praised for moving rapidly against the COVID-19 pandemic. In late March 2020 he imposed a total lockdown and curfew, enforced by calling up the South African National Defence Force (SANDF). The SANDF has been accused of excesses, however. In April SANDF personnel allegedly beat a man named Collins Khosa to death in Alexandra township, Johannesburg. The health impact of the lockdown has been mixed, with outbreaks concentrated chiefly in the densely populated townships. By the end of June there had been close to 144,464 cases, according to the World Health Organization, and some 2,529 deaths. The economic impact has been devastating. With the country shut down for two months and a slow phased opening beset by complex, contradictory and heavy-handed regulations, GDP is forecast to decline by 7–10% in 2020. Unemployment, which reached 30% in the first quarter of 2020, is certain to increase dramatically.

The Ramaphosa administration announced in April a package of support interventions. But, as in other developing economies, these were constrained by limited resources – and in South Africa's case the situation was exacerbated by chronic economic mismanagement over the preceding decade. A support package comprising a R500 billion (US$28.7bn) fund (much of it re-prioritised funding), including a state-guaranteed private-bank lending fund of R200bn (US$11.6bn) and a variety of furlough support programmes, was small given the scale of the challenge. Weak delivery capacity, corruption risks and the contradictory policy objectives of siloed ministries will further vitiate its impact.

Domestic support programmes were to be supplemented by a proposed US$7bn of international borrowing and support from a variety of agencies, most importantly discussions with the IMF to draw down the US$4.2bn to which it is entitled under the IMF's rapid-financing instrument (RFI). Until now, borrowing money from the IMF has been ideological anathema to the Tripartite Alliance. It is a significant victory for Ramaphosa and Finance Minister Tito Mboweni to have persuaded their cabinet colleagues to accept such an arrangement – even if an RFI programme does not bring the stringent commitments of an IMF structural-adjustment programme.

On 24 June 2020 Mboweni delivered an emergency budget. Against the February budget it projected increases in the deficit from 6.8% to 15.7% and in debt to GDP from 66% to 82% in 2020–21, stabilising at around 87% at the end of the three-year cycle in 2023–24. All three ratings agencies have already cast doubt on these numbers and 100% seems more realistic. Around the budget Mboweni reiterated themes set out in the August 2019 Treasury plan for the reform of the economy: tough fiscal consolidation and an end to subsidies for SOEs, which should support themselves, be privatised or be allowed to fail. The much-delayed roll-out of broadband spectrum, the rapid expansion of the renewables programme and the implementation of Eskom reforms were immediate priorities. Mboweni warned that unless these measures were put in place South Africa would require a full IMF structural-reform programme within four years.

The chances of reform

What are the prospects for such fiscal consolidation and reform? The forces supporting economic reform centre around Mboweni and include the Treasury, the Reserve Bank, SARS and key individuals in the cabinet and the ANC. Ramaphosa is generally thought to support Mboweni, but until now has been unwilling to force change at the expense of party unity – or even to defend him publicly. He has preferred to work behind the scenes to build consensus between the members of the Alliance and within the ANC, and then to try to forge compacts with business (pro-reform) and labour (acute sceptics of reform). In the past such a strategy has often yielded high-flown rhetorical 'compacts' which failed to translate into much on the ground, since they ignored both profound differences in the detail of what each side wanted and the limits of state capacity. Crucially, this form of corporatism neglects the interests of small and medium enterprises and the unemployed.

While the new management of Eskom is restoring some stability, major reform has been delayed by another two years owing to contestation over the nature of such reform and to the extent of the institutional damage wrought during the Zuma years. Furthermore, Minister of Mineral Resources and Energy Gwede Mantashe, a senior SACP official, has given his support to the coal lobby, which employs unionised workers, and dragged his feet on the key renewables programme. He has also resuscitated the idea of a nuclear deal, which would face significant funding challenges but, as with a previous deal – with Russia – proposed under Zuma, is also attractive from a rent perspective. A nuclear deal would narrow the scope for wind and solar projects, which to date have been conducted transparently and competitively.

Leading party and government officials continue to support ANC resolutions calling for the formation of a new state pharmaceutical company, a state petroleum company and a state bank from the consolidation of existing assets. The government is also committed to a National Health Insurance scheme that is desirable in principle but would be costly. It will also face pressure to extend the temporary rise in welfare payments and expand them to cover the growing numbers of unemployed workers.

Such projects and initiatives require significant state expenditure when the country's fiscal circumstances are increasingly constrained. The developmental state, reiterated by Ramaphosa, and ongoing patronage requirements mean there is little appetite for private-sector-led solutions. Opponents of fiscal consolidation (that is, economic austerity) propose removing the Reserve Bank's independence and compelling it to engage in far-reaching quantitative easing (QE). However, attempting QE in a developing country such as South Africa would likely weaken the rand (down by over 50% against the dollar over the past decade) further and stoke inflation.

Although its economy faces serious and growing difficulties, South Africa has resources to support a vigorous reform programme. It has a strong and resilient finance sector with well-capitalised banks. Its insurance and savings industries, and financial-market infrastructure, can imbue global investors with confidence that their investments are safe from the contagion of national policies. It also has an extraordinarily resilient private sector that has weathered misrule in both the apartheid and post-apartheid eras.

Much therefore rests on the willingness and ability of Ramaphosa and the reformists to take a strong lead. The ANC and its allies will need to make painful choices rather than use domestic (privileged whites, unpatriotic minorities) and international (the West, the destructive Bretton Woods institutions) scapegoats as excuses for delay and deflection.

Foreign policy

In early 2019 Ramaphosa ordered a review of foreign policy. He promised a pivot to focus on Africa in both economic and political diplomacy. In February 2020 he began a year-long term as chair of the African Union

(AU), and he exercises influence by virtue of his personal reputation as well as the relative strength of South Africa in the continent.

Ramaphosa has used this position, and South Africa's 2019–21 tenure on the United Nations Security Council, to try to progress the AU's aspirations of 'silencing the guns' in Africa by the end of 2020. About 70,000 of the UN's 82,000 uniformed and civilian peacekeepers are deployed in Africa, and South Africa has 1,107 troops in the UN Stabilization Mission in the Democratic Republic of the Congo (DRC) and small numbers of non-military personnel in the UN missions in Darfur and South Sudan. The continent's hotspots – Burundi, the Central African Republic, the DRC, Libya, South Sudan and Sudan – were mired in a range of conflicting interests and alliances that made the goal of ending conflict in 2020 extremely ambitious. The case of Libya – where the AU is struggling to play a part amid interventions by Egypt, France, Russia, Saudi Arabia, Turkey and the United Arab Emirates – illustrates the challenges of piloting a meaningful AU role. Finally, South Africa continues to ignore the dramatic implosion of Zimbabwe, notwithstanding the very large numbers of economic refugees fleeing to South Africa.

South Africa has the potential to benefit substantially from the African Continental Free Trade Agreement, which has been signed by all AU member states except Eritrea. Its July 2020 implementation, though, has been delayed owing to the pandemic. The history of African regional free-trade agreements, in South Africa's case the Southern African Development Community, illustrates just how slow progress from formal agreement to implementation can be.

The country's continental foreign policy faces significant headwinds. Its limited military resources are depleting fast as defence budgets have been relentlessly slashed, and the profile of its troops is ageing. Some African countries such as Kenya and Rwanda are reforming their economies faster than South Africa, leaping ahead particularly in education and internet connectivity. Township riots against economic competition from entrepreneurial African migrants have damaged the country's reputation in the rest of Africa. The legacy of Nkosazana Dlamini-Zuma, who was an unpopular chair of the AU Commission (equivalent to the European Commission) from 2012 to 2017, has not helped.

South Africa's major businesses have also fared patchily in their expansion into African markets, experiencing resentment at their presence and associated tax and regulatory problems. Mobile-telecommunications giant MTN and Standard Bank are two cases in point. Economic challenges, such as the inability to remit profits (usually as a result of foreign-exchange shortages), and lack of consistent growth in target markets, have not helped. Having burned their fingers in a number of countries, many companies have retreated, and few are currently contemplating new forays beyond South Africa.

South Africa's wider foreign policy continues to exhibit a shift away from the West, reflecting liberation solidarity with former Eastern-bloc countries and China. Notwithstanding Ramaphosa's business experience and the continued importance of Western nations in trade and development, the BRICS nations (Brazil, Russia, India and China) – as well as Bolivia, Cuba and Venezuela – enjoy priority. South Africa's voting record at the UN and other multilateral agencies continues to reflect this positioning.

China remains an alluring model for ANC party ideologues, who have made a number of trips to the country to study party systems and thinking. Relations between South Africa and China remain close, though those with Russia have cooled somewhat since Zuma's failure to steer through the nuclear deal. Senior ANC figures continue to receive medical treatment in Russia. South Africa's support for the principle of sovereignty, even when states are openly committing human-rights abuses, often puts it in coalition with China and Russia. Under Zuma, South Africa took steps to withdraw from the International Criminal Court, but Ramaphosa has halted this process.

In sum, it seems likely that, if South Africa's economy continues to decline and its domestic preoccupations increase, its diplomatic influence both in Africa and beyond will wane.

CHAPTER 11

Latin America

LATIN AMERICA

Drivers of Strategic Change

REGIONAL SHARE OF GLOBAL POPULATION, GDP AND DEFENCE BUDGET

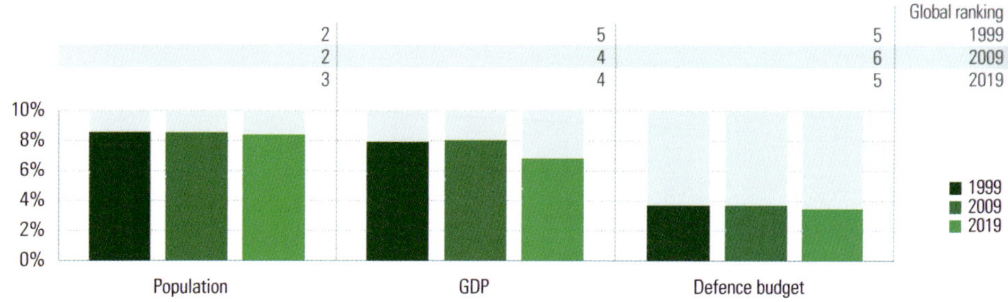

POPULATION

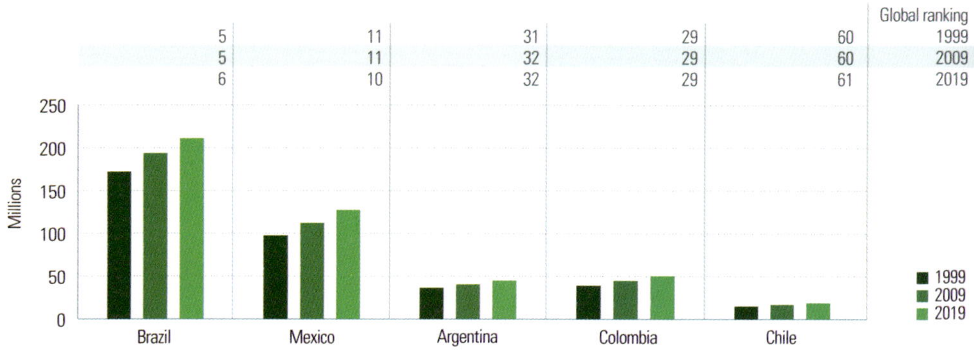

AGE STRUCTURE

(Percentage of national population)

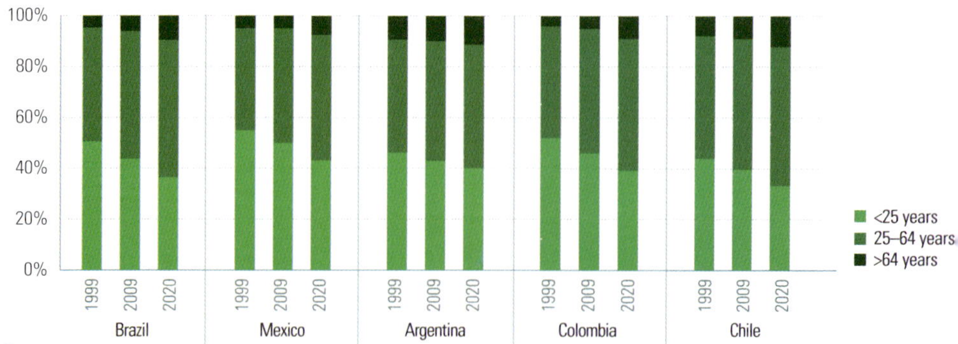

Median age	Brazil	Mexico	Argentina	Colombia	Chile
2000	25.3	22.9	27.6	24.1	28.7
2010	29.2	26.2	29.6	27.5	32.2
2020	33.5	29.2	31.5	31.3	35.3

GDP
(Constant 2010 US dollars)

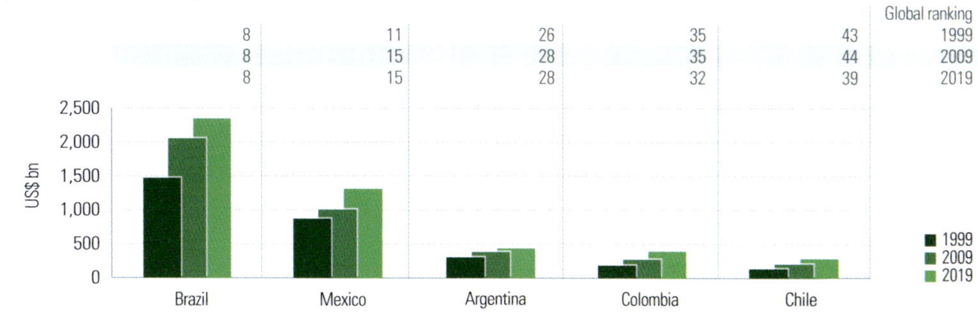

GDP PER CAPITA
(Constant 2010 US dollars)

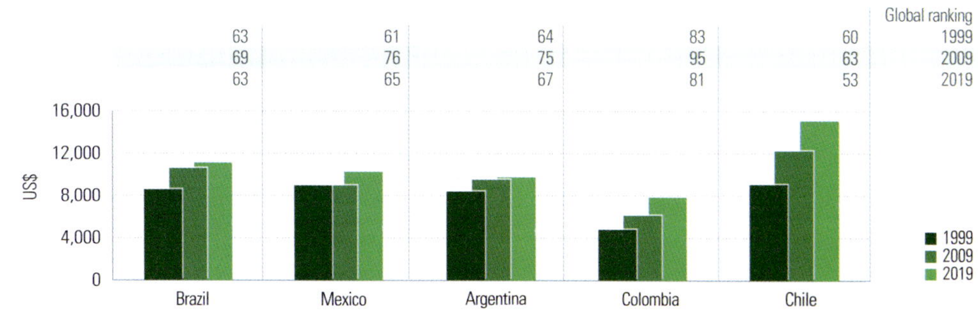

DEFENCE BUDGET
(Constant 2015 US dollars)

ACTIVE MILITARY PERSONNEL

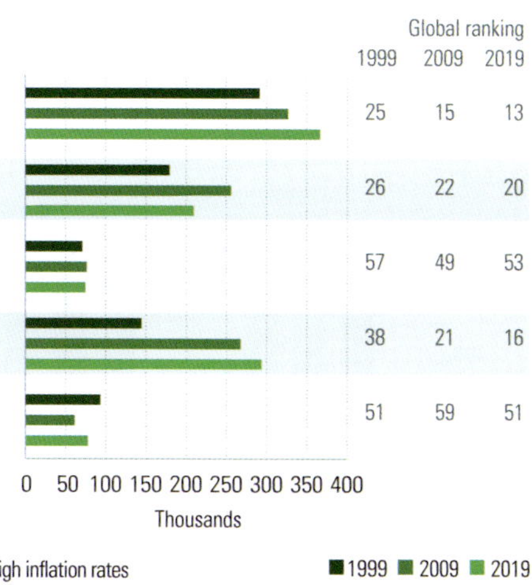

* 1999 defence budget value for Argentina is an estimate, and may be distorted by high inflation rates

For explanation of drivers and sources, see page 7

HUMAN DEVELOPMENT INDEX (HDI)

(Score between 0 and 1, where 0 denotes a low level of development and 1 a high level of development)

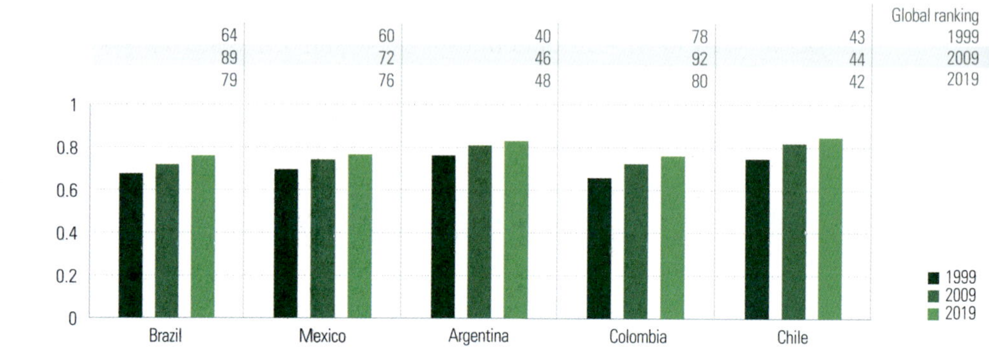

POLITICAL SYSTEM

(Score between 0 and 100, where 0 denotes no political freedom and 100 fully free)

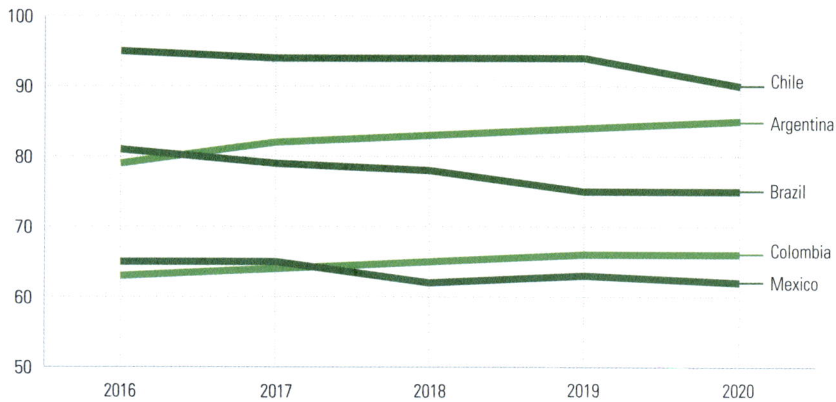

TRUST IN INSTITUTIONS

(Average level of trust)

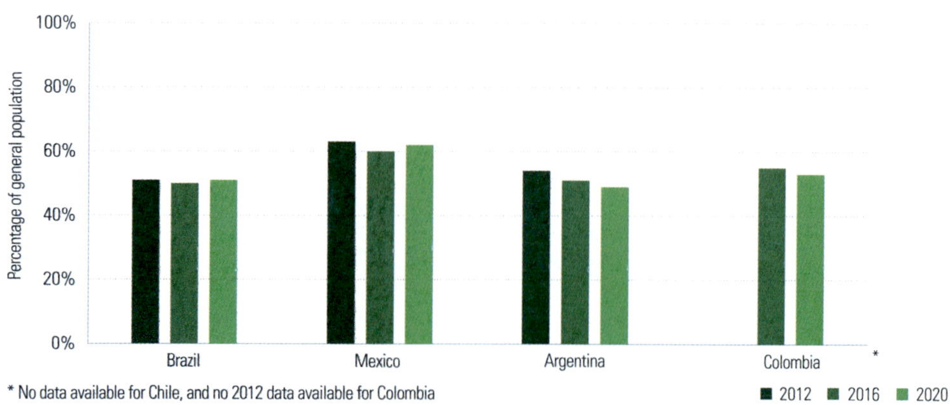

* No data available for Chile, and no 2012 data available for Colombia

For explanation of drivers and sources, see page 7

2019–20 Review

Even before the COVID-19 pandemic hit Latin America in February 2020, with the first case reported in Brazil, much of the region was struggling with a series of chronic problems: stagnant economic growth, over-stretched public finances, widespread income inequality, high rates of crime, pervasive corruption and rising social discontent. Latin America had experienced years of economic stagnation as it continued to adjust to the end of the commodities price bubble in 2014. Governments had failed to invest earlier windfalls in education, infrastructure and economic diversification. The resulting shrinkage of state coffers led to austerity measures.

In 2019, Latin American output per capita (at purchasing power parity) declined by 1.2%. Performance varied from 3.3% real GDP growth in Colombia, the best of any of the region's major economies, to a devasting contraction of 35% in Venezuela. This exacerbated political polarisation and popular frustration with political parties and institutions, and sparked demonstrations, some of them violent, in many countries across the region, with some of the worst unrest in Bolivia, Chile and Ecuador.

Corruption scandals, political dysfunction and, in some cases, a breakdown in democratic norms fuelled instability in several countries in the second half of 2019. In Bolivia, violent protests and a police mutiny forced the increasingly authoritarian president, socialist Evo Morales, to resign in November at the behest of the military. This occurred after allegations, contested by some international observers, that he had stolen the October election through fraud – an election in which he ran after his allies on the constitutional court overturned the results of an earlier referendum that would have banned him from seeking a fourth term. An interim government led by a senator, Jeanine Áñez, a Christian conservative, acted swiftly to shift policies to the right. For example, she distanced Bolivia from previous allies Cuba and Venezuela, stacked her cabinet with conservative ministers and injected heavily religious messaging into her government. Elections were postponed by several months from May owing to the pandemic. This prolonged uncertainty and raised the risk of further political polarisation and social turmoil in a country with a long history of racial divisions.

In Peru, citizens weary of corruption and political impunity took to the streets in September 2019 to support the decision of President Martín Vizcarra to dissolve the legislature, which triggered a constitutional crisis. Although snap legislative elections in January 2020 strengthened the president's hand, long-running political conflicts looked set to re-emerge, as social inequities deepened in tandem with a weakening of the economy in 2020.

In Ecuador, labour unions and indigenous groups led large-scale demonstrations in October against a reduction in fuel subsidies, forcing President Lenín Moreno to reinstate the subsidies. But, in a fragmented political system in which consensus is difficult to reach, grievances ran much deeper over political, racial and regional divisions. With the government already struggling to meet unpopular commitments made under a three-year IMF lending programme, Moreno's approval rating (around 20%) was already low, and his political capital weak, when COVID-19 slammed the country.

Chile, considered the region's most stable democracy and the one with the most orthodox economic policies, faced an unexpected surge of violent protests. In October, a transport-fare hike triggered demonstrations that attracted up to one million people in the capital city of Santiago alone. The deeper roots lay in dashed middle-class expectations and frustration with stark income disparities, stagnating living standards and low-quality public services, including the pension system. There were also demands for a new constitution. President Sebastián Piñera took steps to address some of these grievances and to reform the constitution, although COVID-19 delayed the reform referendum. But challenges to stability looked set to persist.

Colombia confronted challenges to its peace process with former left-wing guerrillas and the increased burdens of absorbing and supporting more than 2m refugees from Venezuela, as that country's economic collapse and political turmoil continued. Despite relatively strong economic performance in 2019, nationwide strikes against the government coalesced around a range of issues, including class disparities, high unemployment, and the government's lethargy in pushing through anti-corruption reforms and implementing the peace agreement.

The crisis in Venezuela showed no signs of bottoming out, with the economy contracting for the sixth consecutive year, oil production plummeting and living conditions continuing to deteriorate – and no democratic transition to a new government in sight. With a 35% collapse in GDP in 2019, Venezuela remained mired in the worst economic depression in Latin America's modern history. The opposition leader, Juan Guaidó, who had declared himself the interim president at the start of the year and garnered the support of dozens of countries around the world, including the United States, proved incapable of dislodging the incumbent, Nicolás Maduro. Anti-government protests lost momentum during the year as security forces clamped down and the president retained the backing of the military. International efforts to try to resolve the Venezuelan crisis waned.

The region's three largest economies – Brazil, Mexico and Argentina – faced complex political and economic situations. In Brazil, Jair Bolsonaro, the far-right anti-establishment president and former army captain who had taken office in January 2019, became engulfed in scandals and political crises. The Supreme Court launched an investigation into Bolsonaro's dismissal of the federal-police commander, who was investigating one of his sons for wrongdoing. Bolsonaro also packed his administration with military officers, some of whom launched threats against the judiciary, Congress and the media. Critics accuse him of endangering Brazil's democracy. Bolsonaro's confrontational approach to governing only heightened after COVID-19 arrived, with the president minimising its importance.

Mexico's left-leaning president, Andrés Manuel López Obrador, failed to deliver on his promises of robust economic growth, narrower income inequality and lower crime rates. Despite the creation of a National Guard to combat drug trafficking, violence increased in López Obrador's first year in office – with the murder rate hitting a new record of 23.24 per 100,000 inhabitants for the calendar year 2019. Nevertheless, the president's populist messaging still garnered him a firm approval rate of 59% in December 2019, according to pollster Mitofsky. But his erratic style of governing and nationalist policies, particularly towards the energy industry, alienated many foreign investors, contributing to a 4.9% contraction in real gross investment in 2019. With the president's party, MORENA, holding majority control of Congress, there seemed little threat to his authority. But the fragile economy and debilitated public finances – combined with his denial of the severity of the pandemic – left the government with little firepower to combat the health emergency.

In Argentina, the conservative administration of Mauricio Macri found itself mired in economic crisis and once again on the precipice of debt default – the second since 2000 and ninth in Argentina's modern history. Public debt stood at around US$324 billion, equivalent to 90% of the country's GDP. Despite an IMF loan programme intended to help correct long-standing fiscal imbalances, Macri was unable to stabilise the economy, the peso or the public finances. GDP contracted for the second consecutive year, poverty levels increased, and Macri lost the presidential election of October 2019 to a left-leaning opponent, Alberto Fernández. Upon taking office, Fernández reversed some of the policies of his predecessor and entered talks with creditors to restructure US$65bn in foreign debt. Yet Argentina formally defaulted in May, and in June negotiations were at a standstill.

Meanwhile, smaller economies in Latin America faced their own set of difficulties. Central American and Caribbean countries, struggling with high rates of crime and deep social inequities, saw an important outlet – emigration to the US – effectively cut off because of the Trump administration's immigration policies. The US government increased deportations and practically shut down the border, making it difficult even for asylum-seekers to have their cases heard. The clampdown also contributed to a slowdown in the growth of family remittances to Latin America – which became a sharp contraction as the US economy went into lockdown during the pandemic. This dealt a particularly harsh blow to countries such as El Salvador, Haiti and Honduras owing to their outsized reliance on such inflows for foreign exchange and household-income support.

Pandemic impacts immediately felt

COVID-19 arrived against the background of these political, social and economic challenges. In May 2020, the World Health Organization (WHO) declared Latin America the new global epicentre of the disease. Many governments – Brazil and Mexico being the notable exceptions – had quickly imposed strict lockdowns and social-distancing measures. However, these measures were haphazard and uneven. A high degree of urbanisation and labour informality, and in some cases leaky enforcement of quarantines, helped to fuel the virus's spread. By June, Brazil registered the second-highest number of infections and deaths in the world.

Pressures on healthcare systems, economic dislocations and the disruption of international trade once again exposed the region's structural, political and economic deficiencies. They stress-tested response mechanisms, fiscal resources, social safety nets and political institutions already under strain. The fallout from COVID-19 was both internal and external. Lockdowns smothered domestic demand, the economies of major trading partners contracted, commodity prices and inflows of remittances fell, global travel and tourism were frozen, and supply chains were disrupted.

Output was already down severely in March, the first month of the lockdowns. According to the Economist Intelligence Unit, the month-on-month contractions ranged from 10.4% in Peru, 9.8% in Argentina, 8.2% in Colombia, and 5.8% in Chile, to a lower 4.3% in Mexico and 3% in Brazil. The differences relate to the timing of containment measures and their degree. The first four countries imposed strict measures early, whereas Mexico and Brazil were lax. Indeed, both Bolsonaro of Brazil and López Obrador of Mexico downplayed the seriousness of the pandemic and did not impose strong federal measures, partly to prevent their devastating impact on the economy. Bolsonaro left the responsibility to state and local authorities, while López Obrador put some measures in place while failing to enforce mandatory protections. Regardless, economic data for the subsequent months showed even steeper drops in economic activity across the board.

Political pressures intensify

The combination of the pandemic and sharp economic contraction exacerbated political tensions. This was most clear in Brazil, where the Bolsonaro government's denial of the pandemic's severity and sabotage of quarantine measures adopted by state-level authorities led to a massive increase in cases and a severe death toll. Two consecutive health ministers departed and in June, the government stopped disclosing comprehensive COVID-19 statistics. Bolsonaro's approval rating plummeted, although he retained the support of his core base of around 30% of voters. The government found itself on the brink of institutional crisis. Conflict with the judiciary, Congress and state governors deepened amid growing calls for Bolsonaro's impeachment. Many middle-class voters who had applauded Bolsonaro's anti-establishment message began turning away. Although his removal from office did not look likely as of June 2020 – impeachment is a lengthy process that would damage efforts to deal with the health and economic crises – political tensions were poised to continue. With polarisation between the left and the right increasing, and Bolsonaro relying more on his support base in the army, concerns about the state of Brazilian democracy deepened.

In Mexico, López Obrador was widely criticised for failing to introduce adequate testing and was accused by healthcare experts of concealing the true COVID-19 statistics. He devoted few resources to economic support measures. Even as the disease seemed to be peaking in June, the government began to relax its quarantine restrictions. This appeared likely to spread COVID-19 further. In May the president's approval rating fell below 50% for the first time since he took office in December 2018. With his party controlling Congress, he was not at any immediate political risk. But the health crisis, dwindling fiscal resources and poor economic indicators looked set to erode his public standing further, sideline his populist plans to 'transform' Mexico and widen the socio-economic divide.

In Argentina, Fernández acted quickly to impose strict containment measures, which earned him short-term political gains and an unusually high level of social cohesion around government policy. The country posted fewer victims of the virus than elsewhere in the region. Nonetheless, the economy reeled from both the pandemic and a deep fiscal and sovereign-debt crisis, and looked set to suffer the longest recession since Argentina's 1998–2002 financial meltdown. Lasting support for the government was not assured, and political divisions between left- and right-leaning parties seemed likely to re-emerge.

Ecuador's problems deepened in early 2020 with the double blow of COVID-19 and a steep drop in global oil prices. Oil-export dependence left the government no fiscal space to cushion the impact of the pandemic. These difficulties emboldened political opposition to the Moreno government and heightened the risk of social discontent.

Other countries looked set to weather the health emergency somewhat better. Colombia, which strictly adhered to WHO guidelines to combat the pandemic, suffered fewer deaths than Brazil or Mexico. This boosted President Iván Duque, whose approval rose to 52% in April, the highest since he took office in 2018. Even so, the economic downturn and the government's limited fiscal capacity threatened to fuel social discontent in the second half of 2020. They also further detracted from the government's ability to meet its obligations under the 2016 peace accord with the FARC leftist guerrilla group, and to combat other insurgent groups still active in the field. While the peace accord did not appear to be at risk, implementation looked likely to be even slower and more complex.

Chile and Peru suffered high mortality from COVID-19 but enjoyed greater fiscal resources to mitigate the pandemic's economic repercussions. This put them on a potentially better path to recovery, though dependence on exports kept them vulnerable to the vagaries of external demand. Meanwhile, Uruguay, a small country with stronger institutions and a robust social safety net, benefited from firm popular confidence in government policy, which encouraged Uruguayans to voluntarily adopt social-distancing measures that kept its infection rate low.

Venezuela remained an outlier owing to the length and severity of its economic and political crisis. Despite a historic economic depression, its inability to confront COVID-19 owing to the collapse of its healthcare system, and the massive exodus of what the Inter-American Dialogue estimated would be 5.7m refugees, asylum-seekers and migrants by the end of 2020, the Maduro government seemed entrenched. Even as the economy was headed for another double-digit contraction in 2020, the government was preparing to call new legislative elections before the end of the year, under terms that would further undermine the position of opposition parties in the already impotent National Assembly. Yet international fatigue with the Venezuelan crisis, and the harsh repression of domestic opposition and protests, meant that no reversal of the country's democratic breakdown was on the horizon.

Table 1: **Depleted resources: public-sector fiscal deficits (percentage of GDP)**

Country	2019	2020*
Argentina	-3.8	-8.4
Bolivia	-7.2	-12.8
Brazil	-5.9	-16.3
Chile	-2.8	-14.0
Colombia	-2.4	-7.8
Ecuador	-2.8	-7.0
Mexico	-1.7	-4.6
Peru	-1.6	-11.5
Uruguay	-3.0	-7.0
Venezuela	-28.4	-21.5

* Forecast
Source: Economist Intelligence Unit

Flattening the curve: challenges ahead

As of June 2020, Latin America had not yet seen a flattening of the COVID-19 curve. Nonetheless, some countries began to loosen restrictions and reopen parts of the economy. The road to recovery was set to be bumpy. Whereas some countries, such as Chile and Peru, seemed poised to bounce back in 2021, others, including Bolivia, Brazil and Mexico, will likely see a very tepid rebound. Venezuela will struggle the most.

Political pressures were bound to increase against this backdrop. And, even amid the pandemic, the region's endemic corruption did not abate. Scandals erupted in several countries – including Brazil, Colombia, Ecuador and Mexico – involving public officials and local entrepreneurs committing fraud or price gouging in contracts for emergency medical supplies.

Looking beyond the short term, it seemed that much-needed structural reforms and social agendas would once again be put on the back-burner amid fragile economies, deeper fiscal imbalances and increased political polarisation. It was unclear when Latin American countries would address chronic problems such as substandard educational outcomes, infrastructure needs, severe income inequality, high levels of crime and corruption, and weakening democratic practices. If they do not, economic and social pressures will intensify and popular unrest will resurge.

The Expansion of Brazilian Organised Crime

How serious a threat to Latin American governance and beyond?

Brazil has long been an important transit route for cocaine trafficked from production hubs in the Andean countries of Bolivia, Colombia and Peru. It is now a major transit point for drugs originating in South America and destined for Africa and Europe. But over the past four years, Brazilian organised-crime groups have expanded their operations, spreading their recruitment and violent tactics to neighbouring countries and beyond. They have even begun probing direct involvement in drug trafficking in sub-Saharan Africa. This has shifted South America's organised-crime landscape, undermining governance and security in vast areas ranging from Colombia to Paraguay. These trends have continued despite the COVID-19 pandemic. Indeed, criminal groups have exploited the pandemic to enhance their influence and standing.

The expansion of Brazilian organised-criminal activity began after a two-decade truce between the country's two largest criminal groups, First Capital Command (Primeiro Comando da Capital, PCC) and the Red Command (Comando Vermelho, CV), unravelled in 2016. The PCC, headquartered in São Paulo, is at the forefront of this expansion. It now has the largest geographical reach of any criminal group in South America, with drug-trafficking operations stretching from Bolivia to Africa and Western Europe. In early 2020, its major rival, the CV, began to expand from its traditional base in Rio de Janeiro to the northwestern Amazon region, eyeing a lucrative drug route from the Andean countries.

The PCC's main targets in South America have been Bolivia and Paraguay, two of Latin America's poorest states. The PCC also incorporates local criminals, who previously operated in a more fragmented manner, into the larger criminal organisation. The closer relationship between South American criminal groups operating in the profitable Europe-bound cocaine traffic also amplifies the regional challenge of organised crime at a time when South America lacks comprehensive security-cooperation mechanisms. Modest attempts at regional inter-state cooperation came to an end in recent years when the Union of South American Nations (UNASUR) bloc ceased to meet regularly amid criticism over a lack of action against the Nicolás Maduro regime in Venezuela.

The PCC's continental expansion

The PCC's expansion has caused tectonic shifts in South America's organised-crime scene. It began in spectacular fashion with the assassination in 2016 of an infamous drug kingpin, Jorge Rafaat Toumani (by then known as the 'king of the border'), in the Paraguayan city of Pedro Juan Caballero. By 2017, the PCC had a presence in nearly all of Brazil's 26 states, and a burgeoning international position. The PCC maintains a robust presence in Bolivia and Paraguay and, to a lesser extent, small yet functional outposts in Argentina, Colombia and Peru. From these positions, and with its more than 30,000 members in Brazil, the PCC exercises considerable control over the flow of drugs and weapons across Brazil's often porous and lightly guarded borders. From there, it ferries illicit goods to the many ports where it has infiltrated and has corrupted officials, such as Fortaleza, Santos and Suape, from where drugs are smuggled on cargo ships to Africa and Europe.

Bolivia served as a natural expansion point for the PCC's international campaign. The group expended considerable resources to build a footprint in the country, sending members of its leadership team to spearhead the effort. The PCC has used Bolivian equipment and explosives to carry out hundreds of bank robberies and loot thousands of ATMs in Brazil. Most importantly, the PCC has forced its way into the local supply chain of coca production in Bolivia's Chapare region. It even operates cocaine-processing laboratories near the border with Brazil and has sent manpower to the region to protect the coca crop's yearly cultivation.

Colombia's coca cultivation is also an attractive target for criminal groups. In 2017, the PCC established a position near the Brazil–Colombia–Peru tri-border area, another major zone for cocaine production and transit, though this position was challenged and weakened by the CV in 2019–20. Following the historic 2016 peace deal between the Colombian state and the FARC guerrilla group, the PCC hired FARC dissidents who had rejected the demobilisation agreement, helping the PCC to build knowledge of trans-shipment routes for cocaine and weapons in the northern Amazon.

Brazil's Amazon region, with its network of poorly patrolled rivers, has also become a significant area of contestation, rising in importance for the CV as a route to bring in cocaine and cannabis from both Colombia and Peru. This region is crucial for the CV to continue its lucrative drug-selling business, especially in Rio de Janeiro, given that the PCC now directly controls most of the cocaine produced in Bolivia and the drug-trafficking route through Paraguay. By expanding here, the CV seeks to use an alternative route through the Solimões, a section of the Amazon river, which, alongside its tributaries, forms a riverine transportation network connecting Colombia and Peru.

The 'Solimões Route' for drug trafficking

In late May 2019, several gruesome prison massacres heralded novel developments in Brazil's criminal underworld. Fifty-five inmates were killed with smuggled and improvised weapons over two days in the prisons of Amazonas state – most of them in its capital, Manaus. This was the opening salvo in a war between criminal groups that has altered the criminal dynamics of northern Brazil, with implications for Colombia and Peru, both of which border Amazonas. Manaus is the headquarters of another powerful faction, the Northern Family (Família do Norte, FDN), now weakened by the territorial encroachment of rival groups, mainly the CV.

The CV's gradual victory in Amazonas and neighbouring Acre state brought a dangerous security dynamic to the northern parts of South America. This violent criminal group now possesses a strong territorial base in several key cities. Part of the CV's interest in the two states lies in their proximity to the two largest coca-growing countries in the world, Colombia and Peru. By taking over Acre, for instance, the CV sought to displace the PCC and its local ally, Bonde dos 13, from a drug route coming from neighbouring Peru, according to a local prosecutor cited by Brazilian news magazine *piauí*.

Colombian authorities are aware of dealings between Brazilian and Colombian criminal groups, including reports of involvement by demobilised guerrillas from FARC. In 2019, FDN members were reportedly introduced to a leader of the Los Pelusos criminal group in Colombia by a former FARC member. The FDN contingent that took part answered to Gelson Carnaúba, alias Mano G, who later switched sides from the FDN to the CV. The CV may therefore be able to take over FDN cross-border relationships with relative ease.

Brazilian criminal groups do not yet have any significant presence on Colombian territory other than ferrying drugs to Brazil. But they have a strong impact on the riverine traffic used by locals in the region to transport people and goods. Allan de Abreu, a reporter covering organised crime for *piauí* magazine, noted that the heightened criminal presence 'has rendered the Solimões route very insecure for those that conduct licit trade there', with many ships having to hire armed escorts. Abreu said criminal groups boarding commercial and passenger ships are often more interested in stealing fuel than money, presumably because they possess plentiful amounts of the latter already.

During its expansion operations, the CV brought its trademark brutality and confrontational tactics to the streets of northern Brazil. In 2019 and 2020 – even during the height of the COVID-19 pandemic – several cities in Amazonas and Acre witnessed decapitations or multiple homicides as the CV ruthlessly displaced local groups for control of urban territories.

The arrival of the CV came, literally, with fireworks. The sound of fireworks has become an ominous symbol used by the CV to celebrate victories over its rivals in specific neighbourhoods. Manaus's inhabitants have come to fear the sound owing to the CV's violent methods and intimidation tactics. Videos have circulated online of CV members painting the group's initials on walls and executing men suspected of being FDN members on sight. Fear of the CV is so great that its arrival induced a wave of conversions to local evangelical churches during the second half of 2019 in Acre, following orders by the criminal group. The CV issued a binary ultimatum to members of rival factions: convert to the CV or join local evangelical churches, which are some of the only institutions in Brazil that criminal organisations consistently respect owing to their presence on the ground in many *favelas* and a salvation message readily applicable to former criminals.

Institutional corruption

The expansion of Brazilian organised crime through South America poses new risks to national and local governments with little experience in dealing with large and well-resourced armed groups. It also heightens known risks, such as the corruption of law-enforcement agents and the gradual transformation of prisons into criminal headquarters (see text box). One of the main effects the PCC and similar organisations have had as they have expanded through Brazil has been the corruption of police officers, prison-system personnel and local politicians. South American countries have long faced these challenges, but the PCC's ability to completely dominate the criminal landscape in Brazil's richest state, São Paulo, indicates it plays in a different league from many local groups. Its organisation and resources may further undermine the region's rule of law and institutional stability.

Organised crime and the flawed prison system in Brazil

Because many of Brazil's criminal groups are prison-based in nature – they began, matured and continue to operate mostly from behind bars – prisons are an important node in the region's organised-crime landscape and a critical barometer of shifting dynamics in the criminal underworld. Most of the PCC's and CV's top leaders are incarcerated, and the prison experience is central to these groups' identities. Prison-based groups fill gaps in the Brazilian state's provision of goods and services as, throughout the country, prisoners rely on friends and family to provide essentials such as food and clothing. Criminal groups step into the void and not only offer physical protection and material goods but also recycle the profits from illicit activity to provide families with the opportunity to visit incarcerated members in far-flung prisons. They also offer legal protection of members when needed and negotiate better conditions with prison authorities.

In Paraguay, which has long served as a hub for drug trafficking towards Brazil, groups like the CV and PCC seem to have infiltrated the security institutions more deeply. A prison break on 19 January 2020, in which 75 inmates – including many PCC members – escaped from a jail in the city of Pedro Juan Caballero, was achieved through 'accomplices inside the prison', according to Paraguay's interior minister, Euclides Acevedo. He added that 'this is something that happens in all prisons' in Paraguay. Justice Minister Cecilia Pérez was even more categorical, saying corruption 'permeates' the prison system. Marcelo Pinheiro Veiga, a Brazilian man suspected of coordinating drug trafficking for years for the CV from his hideout in Paraguay, claimed before his extradition to Brazil that he had a vast network of collaborators in the police force who were paid to facilitate arms trafficking and to warn him of any operations to arrest him. One police officer reportedly received US$200,000.

The use of prisons as operational bases, from which PCC leaders continue to issue orders and recruit fellow inmates, is likely to spread beyond Brazil and Paraguay if this criminal expansion continues unchecked. The number of PCC members is still small in Colombia and Peru in comparison to Brazil and Paraguay. But their higher-quality coca base and cocaine make these countries more attractive for the PCC's drive to meet the higher demands of the European drug market. Furthermore, the success of the PCC's model in Bolivia and Paraguay makes it more likely that Colombia and Peru will in turn face the risk of serious corruption in their security institutions.

Public policies have lagged behind this expansion and internationalisation of organised crime, despite recent improvements in intelligence exchange. While Brazil's criminal groups possess the ability to establish corruption networks and even dominate parts of the prison systems in Brazil and Paraguay, the arrest of senior PCC leader Gilberto Aparecido dos Santos, alias 'Fuminho', in Mozambique in April 2020 demonstrated effective cooperation between Brazil's Federal Police, Mozambican authorities and the United States Drug Enforcement Administration, which assisted the operation. This type of robust intelligence work, however, remains limited, since the bulk of Brazil's police forces in states and municipalities have comparatively little investigative capacity. The Military Police, answering to state governors, is focused on repressive tactics against street criminals rather than intelligence-based operations. In some states, such as Rio de Janeiro, the Military Police has gained a reputation for shooting first and asking questions later – harming popular trust and therefore intelligence gathering. In 2019, police in Rio de Janeiro state were responsible for 1,814 killings. A Supreme Court judge issued an order in May 2020 banning police operations in Rio's slums during the COVID-19 pandemic after a 14-year-old boy was killed during an operation in the Salgueiro *favela* area.

Nevertheless, Fuminho's arrest was a major blow to the PCC's ambitious international plans. He was by far the most experienced PCC member in terms of his understanding of international drug trafficking, having coordinated the group's operations in Bolivia for some time. When at large, he was also one of the most trusted lieutenants of the PCC's top leader, Marcos Willians Herbas Camacho, alias 'Marcola'. Marcola has long been able to command his criminal empire from prison, but his communications with the outside world were severely weakened after his transfer to a federal prison in the capital, Brasilia, in March 2019, where he is under solitary-confinement measures for more than 20 hours per day. This is therefore an ironic moment for the PCC: its top leadership has suffered important blows just as it is consolidating international operations in Paraguay and Bolivia.

Brazilian President Jair Bolsonaro's track record in the fight against organised crime has been less impressive than his emphasis on public security during the electoral campaign had suggested. Following the transfer of high-profile criminal leaders to federal prisons in the early months of his presidency, Bolsonaro's main security initiatives were contained in an anti-crime bill that was the brainchild of former anti-corruption judge, and justice and public security minister, Sérgio Moro (who later resigned). The bill, signed into law in late 2019, introduced longer sentences, more restrictions for convicts and measures to speed up court proceedings. It did not tackle the local challenges faced by states and municipalities affected by high levels of violence, nor did it suggest any alternatives to states' overly militarised and often corrupt police forces.

Bolsonaro's image as a 'tough on crime' leader has come under strain since Moro resigned in April 2020, accusing Bolsonaro of trying to interfere in Federal Police investigations involving his sons, who are also politicians. The accusation later turned into a formal investigation by the Federal Police, authorised by the Supreme Court. Moro's resignation removes an obstacle to some of Bolsonaro's more radical views on security. As minister, Moro had publicly disagreed with some of the president's views about the use of force by police as a solution to crime and urban violence.

Crime fighting in Brazil's vast territory remains extremely fragmented, with little intelligence sharing between local police forces – although the Federal Police and General Prosecutor's Office have shown greater investigative capabilities of late. An ambitious security law introduced by Bolsonaro's predecessor, Michel Temer, created an integrated national-security system (Sistema Nacional de Segurança Pública, SNSP) designed to increase information sharing and standards between different regional bureaucracies. But projects later lost steam, with several federal-government initiatives currently delayed.

During the COVID-19 pandemic, organised-crime groups have exploited opportunities to augment their influence and presence. Most Brazilian states passed stay-at-home orders in mid-March 2020, giving rise to expectations that poor and marginalised communities might experience a respite from the persistent criminal violence that often plagues them. But many states saw homicide numbers increase during April and May, at the height of COVID-19 lockdowns. Meanwhile, weak state institutions permitted criminal groups in Rio de Janeiro to boast openly of enforcing curfews, diffusing public-health advice via mobile loudspeakers, rolling out door-to-door campaigns to enforce hygiene, distributing personal protective equipment (PPE), warning store owners against price gouging of critical items, and fashioning and distributing hand sanitisers (made from repurposed precursor chemicals required for processing cocaine). In other words, the pandemic has afforded some organised-crime groups an opportunity to showcase their ability to provide socio-economic goods and basic levels of governance where state institutions are weak – if present at all.

The weakness of state institutions along the cocaine trade route creates additional opportunities for corruption of law enforcement and the judicial system, as well as for recruitment among young, disenfranchised men in cities. These opportunities are not new, but they have been more systematically exploited due to the expansion of Brazilian organised criminal groups in recent years. The prison breaks and heavily armed attacks conducted by the PCC in Paraguay are examples of the group's increasing boldness in acting against both rivals and the state.

The competitive expansion of Brazilian criminal groups across the Amazon, Bolivia, Paraguay and potentially beyond is a significant new transnational dynamic with security implications. These groups' corrupting and coercive methods threaten both state capacity and social order in the region. The aggressive management of international drug flows poses a threat to countries along these paths, primarily in South America but potentially in Africa and beyond.

Venezuela's Political Stalemate
How long can it endure?

No progress was made in resolving Venezuela's protracted political conflict in the year to June 2020. Strategic errors by Juan Guaidó of the opposition Popular Will (VP) party eroded the traction under his campaign to force President Nicolás Maduro from office. Guaidó had declared himself 'interim president' of Venezuela in January 2019, and more than 50 countries recognised him as the legitimate president after Maduro was controversially sworn in for a second term the same month. The United States government, the Organization of American States (OAS) and the European Union had earlier rejected Maduro's re-election, having judged the conditions of the May 2018 contest to have been unfree and unfair.

The presidential election had been scheduled for December 2018 but was brought forward by the National Constituent Assembly (NCA), a body convened by Maduro in May 2017 to consider constitutional change. This followed ongoing confrontation with the opposition-dominated National Assembly, of which Guaidó assumed the presidency in December 2018. Opposition parties boycotted elections to the NCA in July 2017 and the body was not recognised as legitimate by the EU, the US government or the Lima Group of 11 Latin American countries and Canada.

Eighteen months after his constitutional challenge to Maduro, Guaidó had not translated the support of international actors into regime change. This was despite ongoing and catastrophic deterioration of the national economy, infrastructure and oil sector. Indicative of how little had been gained, in June 2020 the pro-government Supreme Court named a new board of the National Electoral Council (CNE) charged with overseeing National Assembly elections scheduled for December 2020. This was without consultation with Guaidó and went against the principle of a mutually agreed CNE board as the starting point of all recent dialogue efforts.

Such contradictory outcomes since the declaration of the interim presidency focused attention on the potentially counterproductive impacts of US engagement in Venezuela, which had been stepped up during the year. Robust US sanctions imposed on the oil sector in February 2019 exacerbated the humanitarian impact of gross domestic misgovernance and enabled Maduro to mobilise local anti-imperialist sentiment. Conditions of scarcity and hardship diminished popular enthusiasm for Guaidó's continual round of social protests. Rallies organised in July and November 2019, and January and May 2020, failed to draw sizeable crowds. US sanctions on officials in the Maduro government and the Venezuelan armed forces appeared to erode the appetite for rebellion from within the administration.

Faltering opposition challenge

After his constitutional challenge failed to force Maduro's immediate flight, Guaidó reverted to the traditional opposition strategy of inducing a military uprising. A civil–military rebellion launched in Caracas in April 2019, fronted by Guaidó, collapsed amid lacklustre popular support and repression by security forces. In May 2020, a farcical plot to capture Maduro – named *Operation Gideon* – was organised by Silvercorp USA, a US private-security firm, with a contingent of Venezuelan exiles. Guaidó's political strategist Juan José Rendón and National Assembly deputy Sergio Vergara resigned from their positions in the interim government as evidence of payments to the US company and recorded discussions with its director were revealed. Guaidó denied foreknowledge of the plot. Eight of the rebels involved in *Operation Gideon* were killed and two American participants were captured by the Venezuelan military. More private, mercenary incursions were expected after the US attorney general indicted Maduro and senior government figures in March 2020 for drug trafficking and set a US$15 million reward for Maduro's capture.

The legitimacy of Guaidó's interim administration has also been undermined by political scandals. In June 2019, his reputation was damaged by a corruption case that involved claims that members of his VP party distributed funds to Venezuelan military exiles in Colombia. In December 2019 VP parliamentarians were linked to allegations of payments in a lobbying campaign by a Colombian businessman. In a public-relations blunder in September 2019, photographs emerged of Guaidó with members of the Rastrojos criminal organisation in Colombia.

The failure to push Maduro from power highlighted the limited mechanisms at Guaidó's disposal to achieve regime change. The continued unity behind Maduro within and across the service branches of the Venezuelan security sector remained the Achilles heel of regime-change ambitions. Guaidó did maintain a lead in opinion polls, however. The polling firm Datanalisis reported in February 2020 that 78% of 800 respondents would vote for Guaidó in a presidential election, and just 22% for Maduro. However, Guaidó's approval rating collapsed from a high of over 80% in early 2019 to 40% in February 2020, as popular expectations of a quick change of government diminished. A Meganálisis survey conducted in November 2019 made for more ominous reading, with an approval rating of 10% for both Guaidó and Maduro. In a follow-up survey in May 2020, 85% of respondents felt that the opposition's strategy was 'very poor'.

Tactical misjudgements increased the pressure on Guaidó from within the fractious Venezuelan opposition alliance, the Democratic Unity Roundtable. Lacking sufficient political authority to enforce his leadership across the unwieldy anti-government movement, Guaidó attempted to accommodate antagonistic views. It proved impossible to bridge the divide between those resolutely opposed to any form of engagement or compromise with the Maduro government, and more pragmatic elements that supported a negotiated solution to the political crisis.

The more outspoken elements of the anti-dialogue lobby included María Corina Machado of the Come Venezuela (VV) party, former Caracas mayor Antonio Ledezma and individuals appointed by Guaidó to serve as his ambassadors overseas and at the OAS. These pro-intervention voices cohered within the diaspora community, were influential in Washington DC, and intensified the demands for foreign-led military action to remove Maduro as Guaidó's gaffes accumulated.

Guaidó moved firmly behind military action as an option for regime change as part of a 'new stage of struggle' unveiled in September 2019. In July 2019, the opposition-dominated National Assembly had voted to reincorporate Venezuela into the 1947 Inter-American Treaty of Reciprocal Assistance (TIAR). Venezuela had withdrawn from this 18-country mutual-defence agreement in 2012. In August 2019, Guaidó deposited an instrument of accession at the OAS, the depositary of TIAR. This was accepted by the OAS, which in line with the Lima Group of countries and OAS president Luis Almagro recognised Guaidó as president of Venezuela.

The legality of multilateral military action in Venezuela under TIAR was bitterly contested by OAS members at a September 2019 meeting. Although a more modest resolution to capture, extradite and sanction senior Venezuelan officials followed, Uruguay left TIAR in September 2019, protesting the sanctions as a violation of international law. The country returned to TIAR in March 2020. The pro-government Supreme Court in Venezuela rejected the re-accession process.

While encouraging the Venezuelan military to rebel against Maduro and courting intervention from external actors, Guaidó also sought to placate more pragmatic elements of the domestic opposition movement. Both Maduro and Guaidó sent representatives to renewed dialogue efforts led by the Norwegian government in summer 2019. These broke down early in August 2019, with Maduro suspending participation after the US imposed tough, comprehensive sanctions during the talks.

In a communiqué issued on 16 September 2019, Guaidó rejected the possibility of any future discussions with Maduro's officials. The significance of this move was undercut by agreement between the Maduro government and some opposition parties to continue dialogue efforts. The fracturing of the informal opposition alliance reflected the diversity of parties and ideologies within the anti-government movement. The dominant influences around Guaidó supported military intervention, had US educational backgrounds, were strongly aligned to hawkish elements in US government and were neoliberal in economic policy. This sat uncomfortably with Venezuela's more domestically rooted parties, local civil-society organisations, the significant centre-left popular sentiment in the country, and with new organisations created by former Chavistas such as May 2018 presidential candidate Henri Falcón.

In early January 2020, while engaged in an international tour, Guaidó was outflanked by the decision of some opposition parties not to support his re-election as president of the National Assembly. This would have deprived him of the right to exercise the interim presidency. After decamping to another building with loyal National Assembly members for the casting of a new vote, Guaidó received the endorsement of the legislature. The US sanctioned those Assembly members who had not backed Guaidó.

Dual political authority and the issue of international recognition

With neither Maduro nor Guaidó capable of vanquishing the other, Venezuela remained in the unprecedented situation of having two presidents with parallel cabinets, supreme courts, popular assemblies and overseas diplomatic missions. However, despite these rival claims to political authority, Guaidó exercised no control over the state, the military or the national territory.

This situation created international legal and diplomatic complexity. For example, in recognising Guaidó as the legitimate head of state, in August 2019 the US government froze Venezuelan assets in the US that the Maduro government had controlled. This included CITGO, the US-based operations of the Venezuelan national oil company PDVSA, to which Guaidó appointed a new board in August 2019. The US asset freeze – and handing of management to Guaidó – was justified on the basis that it would prevent theft and corruption by Maduro officials. But the move gave Guaidó responsibility for addressing a series of claims made against CITGO by creditors, unpaid bond holders and foreign companies that were owed compensation for expropriation processes. This included ConocoPhillips and the Canadian mining company Crystallex. The Russian state oil company Rosneft also had a claim on CITGO, having received a 49.9% stake in the company as collateral for a US$1.5 billion loan to PDVSA in 2016. In May 2020, a US judicial ruling allowed for the sale of CITGO to pay the slew of financial demands on the Venezuelan state. This was blocked by the US Treasury's Office of Foreign Assets Control (OFAC).

Maduro condemned the responsibility granted to Guaidó for management of Venezuelan assets in the US and the Bank of England as 'piracy'. It was also criticised by moderate opposition groups due to the absence of oversight of Guaidó's team. In June 2020, Guaidó's attorney general resigned when it was revealed he had been in secret negotiations to settle the claim of the US energy multinational ConocoPhillips.

In a further illustration of the wider complexities of the political struggle in Venezuela, in a May 2020 determination of authority over US$1.8bn of Venezuelan gold held in the Bank of England, the UK Supreme Court ruled in favour of Guaidó on the basis that he was 'unequivocally recognised' as interim president by the British government.

Disputes over US sanctions and national-asset ownership intensified in the context of the COVID-19 pandemic and the wider humanitarian crisis in Venezuela. The Maduro government maintained that onerous US sanctions were illegal under international law. The Pope and senior United Nations and EU officials called for the US sanctions to be lifted in April and May 2020 as COVID-19 exacerbated health-service pressures and the country faced ongoing shortages of food, water, gasoline and medical supplies, energy blackouts and the return of 75,000 Venezuelan migrants from Colombia and Ecuador by June 2020.

The escalating humanitarian crisis increased pressure on Maduro and Guaidó to accept an agreement that would enable Venezuelan oil to be exchanged for food and water supplies. Guaidó rejected working with Maduro to ease the impact of US sanctions. His team deflected pressure to monetise or repatriate Venezuelan assets overseas and rejected a proposal for gold reserves held in the Bank of England to be paid to the UN in exchange for humanitarian assistance. In February 2020, Guaidó's representatives overseas launched a high-profile media and diplomatic campaign protesting alleged corruption and criminality in the Maduro government. This focused on allegations of illegal gold-mining operations and drug trafficking involving the Venezuelan military and senior government figures.

US escalates pressure

The presidency of Donald Trump marked a dramatic tightening of mechanisms intended to force the process of political change in Venezuela. Selective sanctions on individual officials first introduced by then-president Barack Obama in 2015 were extended to the oil, gold and financial sectors in 2018 and 2019, and the personal sanctions were widened from seven individuals during the Obama presidency to more than 70 under Trump.

In August 2019, US sanctions became yet more severe amid speculation of a breakthrough in Norway-led negotiations, an initiative that was not popular with senior US officials including then-national security advisor John Bolton. Guaidó was under intense pressure not to make concessions in the negotiations, and US officials were emphatic that the US would reject any outcome that enabled Maduro to remain in office. Sponsors of the dialogue interpreted the extension of US sanctions during the negotiations as an intentional derailment of the talks by the US.

The collapse of dialogue efforts in August enabled hawkish US and Venezuelan elements to steer a more radical strategy of escalated pressure on Maduro. In the same month, the US announced the opening of an interim diplomatic office of the US government to Venezuela, the Venezuela Affairs Unit, in Bogotá, and rumours circulated of back-channel communications between the Trump administration and Diosdado Cabello, one of the most influential figures in the Maduro government. Amid intensifying speculation that Maduro would be forced to step down from the presidency due to the combination of sanctions and military threat, Bolton was removed from his post on 9 September 2019.

With Bolton's departure, the US temporarily pivoted back to a strategy of shoring up Guaidó's authority through bilateral development assistance. An agreement was signed in October 2019 between USAID and Carlos Vecchio, Guaidó's ambassador to the US. This provided an additional US$98m to the US$632m the US had already committed to Venezuela for humanitarian programmes, distributed through Guaidó's interim administration. The bulk of the new revenues, which also supported the administrative costs of Guaidó's team, were reallocated by

USAID from Central American countries, which had their funding cut for failing to prevent migration to the US.

The US returned to more aggressive posturing at the start of 2020 as Maduro stabilised his presidency through a series of pragmatic economic adjustments that were intended to contain a hyperinflationary spiral and increase the availability of goods. This included allowing US dollars sent back to Venezuela by the 4.5m members of the diaspora community to be used in domestic transactions.

In February 2020, Venezuelan state airline Conviasa and its fleet of 40 aircraft were added to the US sanctions list, with secondary sanctions precluding other countries from supplying parts to the ailing fleet. On 18 February, OFAC sanctioned the Russian oil company Rosneft, its Swiss-incorporated trading arm and the president of the company's board of directors for continuing to trade with Venezuela.

As OFAC squeezed the opportunities for sanctions busting and the US attorney general indicted Maduro for drug trafficking in March 2020, US Secretary of State Mike Pompeo unveiled a Framework for a Peaceful Democratic Transition in Venezuela. The proposed pathway acknowledged a role for the Venezuelan military in overseeing the transition process; marked a move away from the perennial US insistence on the exclusion of Maduro from future political arrangements; and looked to the 'establishment of broadly acceptable transitional government to administer free and fair presidential elections and a pathway to lifting Venezuela-related US sanctions'.

The Maduro government rejected the Framework as unacceptable US interference. In April, the US returned to the familiar theme of corruption and drug trafficking within the Maduro family, Maduro government and the Venezuelan armed forces. Two US Navy frigates were dispatched to the Caribbean to support drug-interdiction efforts.

The mobilisation by the US government and Guaidó's interim administration of a 'narco-state' narrative was not supported by data from the US inter-agency Consolidated Counterdrug Database, which indicated that 210 metric tonnes of cocaine passed through Venezuela in 2018. By contrast, US State Department figures showed that over six times that volume (1,400 metric tonnes) passed through Guatemala the same year. The deployment to Colombia in June 2020 of a contingent of the US Security Force Assistance Brigade (SFAB) escalated Venezuelan government concerns that Colombia would be used as a beachhead for a US-led or -supported invasion of Venezuela. Bilateral relations between Venezuela and Colombia remained tense and unpredictable amid insecurity in border areas, allegations of cross-border movements by left- and right-wing paramilitary forces, and the close relations between Guaidó and Colombian President Iván Duque.

Accounting for the limitations of US strategy

The US has provided extensive support to Guaidó in the 18 months since the declaration of the interim presidency, but this has not succeeded in accelerating the removal of Maduro. In accounting for the limitations of US strategy, the ongoing problems are incoherence and inconsistency within the Trump administration, and the failure of the US government to build a broad alliance around its Venezuela policy. Uptake of US sanctions by Latin American and European countries was limited, and little energy was expended by Guaidó or the US government to widen the coalition of support or imprint US action with democratic commitment. As a result, most countries in the Caribbean, Middle East, Africa and South Asia continued to recognise Maduro as the legitimate head of state.

The Trump administration underestimated both the unity of the Venezuelan armed forces behind Maduro, and the strategic, financial and diplomatic support that countries such as China, Cuba, India, Iran, Russia and Turkey would continue to provide to the Venezuelan government. After China suspended oil shipments from Venezuela in September 2019 in response to the August US sanctions, Russian companies stepped into the gap. When the US sanctioned Rosneft's trading arm in February 2020 for handling Venezuelan oil, Iran moved into the breach, shipping 1.5m barrels of petroleum to Venezuela in June 2020 and supplying parts for the recovery and reactivation of Venezuelan domestic refining capacity.

Both China and Russia had a significant level of exposure in Venezuela after an estimated US$100bn in loans and credit to the Chávez and Maduro governments. Stakes in Venezuela's energy and mining sectors were an important element of the geo-economic footprint of both countries in Latin America. Despite rumours of frustration with mismanagement by the Maduro government, China and Russia remained important allies, providing technical assistance to the Venezuelan military and energy infrastructure, dispatching humanitarian assistance during the COVID-19 pandemic, and blocking US efforts to sanction Venezuela in fora such as the UN Security Council. But despite their influence over the Maduro government, neither China nor Russia was party to dialogue efforts convened by European and Latin American countries.

In resisting Guaidó's challenge and surviving US sanctions, Maduro has sought to build a new layer of nationalist myth and anti-imperialist legitimacy for his administration. His calculation by mid-2020 was that this would carry the ruling United Socialist Party of Venezuela to victory in elections for a new National Assembly in December 2020.

Mexico's 'Fourth Transformation'
Can López Obrador deliver change while distrusting institutions?

In December 2018, Andrés Manuel López Obrador won a historic victory in Mexico's presidential election. He did so by promising to deliver what he dubbed the country's 'Fourth Transformation' – the previous three being independence from Spain in 1821, the liberal reforms carried out by statesman Benito Juárez in the mid-nineteenth century and the Mexican Revolution of 1910–20. This Fourth Transformation aimed to deliver Mexico from long-standing corruption, insecurity, inequality and poverty.

López Obrador is the most powerful president in a generation. He won 53% of the vote in a four-way electoral race. His 30-point margin of victory was the largest for a presidential candidate since the return of competitive elections in 2000. His National Regeneration Movement (MORENA) party, in coalition with smaller, ideologically diverse parties, commands a two-thirds majority in Mexico's Congress, which enables them to pass constitutional amendments. This coalition and its partners also control 19 of Mexico's 32 state legislatures (constitutional amendments must be ratified by 17 state legislatures to be valid).

The decimation of the opposition in the 2018 election left the major opposition parties, the Institutional Revolutionary Party (PRI) and the National Action Party (PAN), leaderless and powerless. Against a background of weak opposition and acute political polarisation, the president's populist and personalistic political style (he gives a press briefing every morning) has allowed him to dominate the national conversation. He has been able to control the narrative, champion his policies, attack detractors and place loyalists in key positions of political authority. However, despite enjoying the power to effect change, a year and a half into his presidency his success in delivering the promised transformation has been mixed.

Underlying this failure is López Obrador's top-down governing style, which does not engage in the institution-building necessary to achieve far-reaching change. Furthermore, his insistence on 'republican austerity' entails massive cuts to ministerial budgets and government salaries, depriving institutions of the funds necessary to operate and to retain public-sector expertise. López Obrador has also accumulated a worrying concentration of power in the executive branch. He has taken a highly personalistic approach to governing and has been wary of delegating. He has proven sensitive to criticism, publicly clashing with ministers who disagree with him and strongly criticising the media. Developments in López Obrador's priority policy areas – corruption, security and the economy, which voters consistently list as Mexico's greatest challenges – illustrate his presidency's mixed record so far. These limitations, and the government's inability to change course, became more evident amid the health and economic crises that resulted from the COVID-19 pandemic.

Powerful president, but little progress

In the year to June 2020, there was a notable lack of progress in combatting Mexico's perennial ills, and even backsliding in some areas, such as the economy (even before the global recession brought on by COVID-19).

Corruption

In 2019, a Transparency International report ranked Mexico 130th out of 180 countries for corruption, below such countries as Ukraine and Pakistan. It noted that 34% of Mexicans reported paying a bribe in 2019 – the highest rate in Latin America after Venezuela. Of a sample of Mexicans who had contact with the police in the last year, 52% reported having paid bribes.

López Obrador has highlighted corruption as the most serious challenge Mexico faces. He has drawn attention to it in a way his predecessors were loath to, treating it as a national issue and making it a centrepiece of his daily press briefings. There have been some notable successes in the fight, including the issue of an arrest warrant for Emilio Lozoya, the former head of Mexico's state-owned oil company, Pemex, on allegations of corruption in July 2019. But overall, the administration's efforts to address corruption have been characterised by the president's personalistic style. He has promised to 'lead by example', as corruption in Mexico starts 'from above, not below'. Officials in his government have, surprisingly, declared corruption in the federal government to be officially eradicated. He has not strengthened Mexico's anti-corruption infrastructure, nor appointed an official to head the National Anti-Corruption System, a body created by Congress in 2015 in response to a series of corruption scandals. As under the previous administration, it remains unable to work effectively.

The president's opposition to these – and most other – institutions is ideological: he believes them to be inherently corrupt as they were initiated under his 'neoliberal' predecessors. López Obrador's preferred method of eradicating supposed corruption in these bodies has been to reduce their funding. There has therefore been little investment in law enforcement or the judiciary. Indeed, under the government's austerity measures, many institutions have had their budgets cut, including US$80 million from the Attorney General's office. The courts remain slow and inefficient, with little new emphasis on training or investment.

López Obrador has also proven suspicious of independent watchdog groups as they are financed by the private sector. He has clashed with and worked to discredit the media when criticised. However, the media proved invaluable in unearthing corruption under the previous administration led by Enrique Peña Nieto (2012–18).

Despite his campaign against corruption, the president has been forgiving of corruption within his inner circle. He has overlooked the failure of some officials to declare assets or explain their wealth. Most recently, this has involved the president of MORENA, Yeidckol Polevnsky, who faces charges of tax evasion and the misuse of US$17.5m in party funds (she denies wrongdoing). Meanwhile, López Obrador has been harsh on opponents for similar misdeeds, most notably former supreme court judge Eduardo Medina Mora, who resigned under accusations of tax evasion. If the president fails to address corruption in his ranks and to build anti-corruption institutions, the situation is unlikely to improve.

Security

The president's hostility to institution-building and insistence on budget cuts have also undermined Mexico's fight against crime. 2018 and 2019 broke records for the number of homicides in the country. The rate of other crimes also remained high: according to the Mexican Institute for Competitiveness (IMCO), there were some 857,000 reported crimes in Mexico in 2019, with polls showing that almost 80% of Mexicans lived in fear of crime. Organised-crime groups have increasingly moved beyond drug trafficking into other activities, including extortion, kidnapping and fuel theft, increasing the average citizen's chances of becoming a victim of crime.

Since former president Felipe Calderón declared a fight against the cartels in 2006, efforts to combat drug trafficking have become increasingly militarised. The approach has mostly relied on a 'kingpin strategy' of capturing leaders of drug-trafficking groups. This strategy had the effect of increasing, rather than reducing, violence as groups splintered and fought among themselves for territory. The increased use of the armed forces also led to an increase in human-rights abuses. A November 2018 report by the Mexican Commission for the Defense and Promotion of Human Rights noted 148 severe human-rights violations by the armed forces between 2007 and 2017, as well as dozens of disappearances and nearly 400 instances of torture.

On the campaign trail, López Obrador had argued against this strategy, promising instead to focus on addressing the poverty and corruption that both drive people to criminal activity and lead to the high rate of impunity for crimes in Mexico. A civil-society group, Impunidad Cero (Zero Impunity), used National Institute of Statistics and Geography (INEGI) data on the performance of Mexico's state-level attorney-general offices to highlight that in 2018 just 6.8% of crimes were reported, and of these, just 19.4% were solved. It therefore concluded that the probability that any crime will be reported, investigated and solved is just 1.3%. According to a 2018 Latinobarómetro poll, just 19% of Mexicans trust the police, compared to an average of 35% across Latin America, while just 23% trust the judicial system. The president's record on security suffers from the same lack of institution-building and investment seen in his anti-corruption strategy. Mexico's police forces remain poorly paid, poorly trained and under-equipped. A 2018 INEGI report found that on average Mexican police earn just three times the national minimum wage, and that 86% had acquired material necessary for the performance of their duties from their own salaries. It found that in 2017, only 80% of police officers had received at least one training course upon joining the police, and only 57.8% had received refresher courses.

Local research has emphasised the need to invest in local police forces to combat crime. This approach would help increase citizen trust and make the forces more effective. Despite this advice, López Obrador's most notable move to combat public insecurity has been the creation in early 2019 of a 60,000-member National Guard to battle drug cartels. National Guard members are primarily drawn from the armed forces but are responsible to the Ministry of Public Security rather than the Secretariat of National Defense.

This approach has been controversial. It seemingly contradicts López Obrador's promises to change the country's approach to crime and risks repeating his predecessors' mistakes. Moreover, the force has so far produced little success. Its most significant action so far was its deployment to Mexico's southern border in June 2019 to control migration from Central America, following an agreement with the Trump administration that Mexico made to avert the threat of tariffs.

A visible failure of the new security strategy came in October 2019, when government forces battled members of the Sinaloa Cartel in the town of Culiacán following the arrest of Ovidio Guzmán López, the son of infamous drug kingpin Joaquín 'El Chapo' Guzmán. As the fighting spun out of control, the government was embarrassingly forced to release Guzmán López in order to halt the violence. Despite expectations that COVID-19 and social-distancing measures would lead to a decline in crime and homicides, local reports indicated that criminal groups had moved into extortion and related activities, and that homicide rates remained high. The Jalisco New Generation Cartel undertook a brazen assassination attempt, though unsuccessful, on the Mexico City chief of police in June, showing the extent of their power.

Economy

The third leg of the Fourth Transformation was López Obrador's promise to deliver Mexico from its chronic economic underperformance and create an economy that delivered real GDP growth of 4% a year to eradicate poverty and inequality.

The president passed a variety of reforms aimed at improving the situation of poorer Mexicans, including a cumulative rise in the minimum wage of 39% (while few Mexicans are paid the minimum wage, which remains low at just US$5.45 per day, many workers' wages are benchmarked against it). He extended income support for students, disabled people and the elderly. He also expanded labour rights, including extending protection to housekeepers and other domestic workers.

But the larger growth story has been mixed. In the first year of López Obrador's administration – even before the onset of the COVID-19 pandemic – real GDP contracted by 0.3%. The pandemic will worsen the situation: the IMF forecasts real GDP will contract by 10.5% in 2020, with a slow recovery predicted.

The weak growth pre-COVID-19 can be linked to unease among private-sector investors and López Obrador's strategy of republican austerity. This unease has been fuelled by the administration's continuous changing of the rules of the game. Even before taking office, López Obrador held a questionable referendum (with the participation of less than 1% of registered voters) to cancel a partially completed new airport project – estimated to cost US$13 billion – outside Mexico City. He has used similar referendums since to cancel other multibillion-dollar projects amid local opposition.

The president has also proven hostile to private investment more broadly, particularly in the energy sector. Although he has not rolled back the 2013 energy reform that opened the energy sector to foreign investment for the first time since the 1930s, he has cancelled further auctions of oilfields. He has injected billions of dollars into Pemex to keep the company afloat amid plans to use it as an engine for economic development. Changes to the tax code that allow for business owners to be imprisoned or have assets frozen simply on suspicion of tax evasion (and which makes tax evasion a crime similar in severity to drug trafficking) have also alarmed investors.

Republican austerity has also made investors wary. Although some social expenditure has been increased (at the expense of spending elsewhere), much new spending has been directed not at health and education, which could strengthen the long-term foundations to improve the Mexican economy, but on big-ticket infrastructure projects of questionable value. These include the construction of a tourist train costing US$7bn in the Yucatán Peninsula as well as a refinery costing US$8bn in López Obrador's home state of Tabasco. Both have been criticised for their environmental impact and frivolity (the latter project in particular, given Mexico's already underutilised refining infrastructure), and for sapping funds needed elsewhere.

The government doubled down on its austerity programme during the COVID-19 pandemic. While the pandemic should present a golden opportunity for a leftist president to expand spending and the role of the state, López Obrador has instead insisted that the government should not contract public debt to bail out the private sector. He has not expanded support for displaced workers and has only provided minimal support to small businesses. The economic fallout of these decisions will be significant and will harm Mexico's poorer citizens, especially as some 50% of Mexicans work in the informal sector. A 2019 report by Mexico's National Council for the Evaluation of Social Development Policy (CONEVAL) showed that in 2018, 41.9% of Mexicans lived in poverty and another 36.2% were at risk of falling into poverty. A telephone survey by INEGI showed that some 12m Mexican workers became economically inactive in April 2020 after the onset of the COVID-19 pandemic. The government's austerity programme is likely to exacerbate inequality and poverty in the post-crisis era. Scant government support and lack of fiscal stimulus mean that Mexico is likely to recover more slowly than other Latin American countries after the pandemic eases as businesses struggle, with many unlikely to reopen at all.

A different sort of transformation

The government has also floundered in its efforts to mitigate the health impact of COVID-19. López Obrador denied the seriousness of the virus at first and only belatedly accepted the need for lockdowns and social distancing, which have been weakly enforced. By late June, Mexico was reporting 212,802 confirmed cases and 26,381 deaths; it also reported the highest daily death toll of any country in the world in the final week of June.

Despite these developments, voters still have high expectations for López Obrador's ability to deliver the Fourth Transformation. His approval rating has remained above 60% for most of his presidency, suggesting that so far voters are willing to give him time to deliver results. Still, as Mexico's deep-seated social ills remain unresolved, voter patience is likely to run out. His MORENA party has already seen a precipitous drop in support.

The 2021 midterm elections are likely to be more competitive than previously expected. The race will see elections to every seat in lower house the Chamber of Deputies, as well as 13 state governorships and 27 state legislatures. However, although MORENA may lose its majority, it is unclear who would stand to benefit, given the continued weakness of PAN and PRI, and López Obrador's domination of the national conversation.

The opposition's inability to capitalise on the government's failures suggests that the López Obrador presidency could change Mexico in ways different from those intended by the Fourth Transformation. Prior to the 2018 election, Mexico was dominated for decades by a three-way politics between the centre-right PAN, the centre-left Democratic Revolutionary Party (PRD) and the ideologically flexible PRI. MORENA's landslide win in 2018 disrupted the three-party system. MORENA, although nominally left wing, lacks ideology and consists of many opportunistic hangers-on riding López Obrador's coattails. Its coalition partners included the far-left Labour Party (PT) as well as the right-wing evangelical Social Encounter Party (PES, since dissolved and incorporated into MORENA). The party is therefore largely a vehicle for López Obrador and likely lacks staying power. As evidence of the splintering of Mexico's politics, some ten political parties have registered to contest the 2021 elections.

This breakdown is likely to deliver more fractured congresses, raising the risks of political stability and legislative gridlock. It is also likely to further exacerbate Mexicans' already low trust in the government and its ability to deliver change.

This raises the question of who should follow López Obrador. The president's project has been populist and personalistic, and he will struggle to find a successor. Mexican presidents are limited to a single six-year term, and the ban on re-election is a popular legacy of the Mexican Revolution. MORENA is unlikely to provide another convincing candidate who could escape López Obrador's shadow, and any attempt to lift term limits would likely prove unpopular. High levels of political polarisation, and lack of progress in instituting the promised Fourth Transformation, are likely to erode further Mexicans' trust in democracy.

This will set the stage for more significant political turmoil, and perhaps greater appetite for the strongman leadership evident elsewhere in the region, particularly in Brazil under President Jair Bolsonaro. The López Obrador administration may well prove to be transformational for Mexico, although perhaps not in the way he, or voters, intended.

Index

C

D

E

F

G

H

U